S0-CLR-786

Effect of Carbohydrates on Lipid Metabolism

Progress in Biochemical Pharmacology

Vol. 8

Series Editor: R. Paoletti, Milan

S. Karger · Basel · München · Paris · London · New York · Sydney

Effect of Carbohydrates on Lipid Metabolism

Edited by I. MACDONALD, Department of Physiology, Guy's Hospital Medical School, London

List of Contributors: MARGARET J. ALBRINK, Morgantown, W. Va.; P. FÁBRY, Prague; F. HEINZ, Hannover-Kleefeld; I. R. KUPKE, Hannover-Kleefeld; I. MACDONALD, London; P. J. NESTEL, Canberra; E. A. NIKKILÄ, Helsinki; A. VRÁNA, Prague; D. ZAKIM, San Francisco, Calif.

24 figures and 38 tables

S. Karger · Basel · München · Paris · London · New York · Sydney 1973

Progress in Biochemical Pharmacology

Vol. 2: Second International Symposium on Drugs Affecting Lipid Metabolism, Milan, September 1965. Part I: Cholesterol and Atherosclerosis – Plasma Triglycerides and Lipoproteins – Drugs Affecting Plasma Lipids. XII + 516 p., 2 cpl., 168 fig., 133 tab., 1967. ISBN 3-8055-0381-4

Vol. 3: Second International Symposium on Drugs Affecting Lipid Metabolism, Milan, September 1965. Part II: Fatty Acid – Prostaglandins – FFA Mobilization – FFA and Triglyceride Transport – Liposoluble Vitamins. XII + 528 p., 183 fig., 160 tab., 1967. ISBN 3-8055-0382-2

Vol. 4: International Symposium on Atherosclerosis, Athens, May/June 1966. Epidemiology – Intact Organisms – Lipoproteins – Plasma Lipids – Whole Artery – Tissue Cultures – Hormones – Primates – Non Primates – Platelets – Clinical. VIII + 636 p., 2 cpl., 236 fig., 151 tab., 1968. ISBN 3-8055-0383-0

Vol. 5: Synthesis and Use of Labelled Lipids and Sterols. Symposium, Milan 1968. Editor: GROSSI-PAOLETTI, E. (Milan). VIII + 176 p., 93 fig., 40 tab., 1969. ISBN 3-8055-0384-9

Vol. 6: Biochemistry and Pharmacology of Free Fatty Acids. XII + 395 p., 55 fig., 41 tab., 1971. ISBN 3-8055-1211-2

Vol. 7: Drugs Affecting Kidney Function and Metabolism. Editor: EDWARDS, K. D. G. (Sydney). XIV + 538 p., 73 fig., 20 tab., 1972. ISBN 3-8055-1386-0

S. Karger · Basel · München · Paris · London · New York · Sydney
Arnold-Böcklin-Strasse 25, CH-4011 Basel (Switzerland)

Printed in Switzerland by Buchdruckerei Reinhardt, 4000 Basel
ISBN 3-8055-1600-2

Contents

Dietary Carbohydrates and Adipose Tissue Metabolism

Effects of Dietary Carbohydrates on Serum Lipids

Dietary Carbohydrates in Lipid Disorders in Man

Progr. biochem. Pharmacol., vol. 8, pp. 1–56 (Karger, Basel 1973)

The Enzymes of Carbohydrate Degradation

F. Heinz

Medizinische Hochschule Hannover, Institut für Klinische Biochemie und Physiologische Chemie, Hannover-Kleefeld

Contents

Introduction

It is the purpose of this part of the review to describe the possible pathways of the degradation of the carbon skeleton of the following hexoses: glucose, galactose, fructose, xylitol and sorbitol (fig. 1). The main interest will be stressed on the enzymes involved in these degradation processes, especially on the regulative enzymes. The subject will be the biosynthesis of L-glycerol-3-phosphate and acetyl-CoA. These intermediates serve as direct precursors for lipid biosynthesis.

Glucose Metabolism

It is not the object of this review to describe the digestion of high molecular carbohydrates; numerous facts are described in textbooks of biochemistry and physiological chemistry. In experiments concerning the levels of glycolytic enzymes, as related to the substrate contents it was demonstrated that the enzyme levels are sufficient for an optimal glycolytic rate except for the following enzymes: glucokinase-hexokinase, phosphofructokinase and pyruvate kinase. The glycolytic rate and, indirectly, the synthesis of lipids are controlled at these three points. Accordingly, the properties of these three enzymes will be discussed in detail. As far as the transformation of carbohydrates to lipids is concerned, two other enzymes have to be discussed: L-glycerol-3-phosphate dehydrogenase yielding the glycerol moiety of the fat, and pyruvate dehydrogenase forming acetyl-CoA, and therefore influencing fatty acid synthesis.

Hexokinase

(EC 2. 7. 1. 1. Hexokinase: ATP: D-glucose-6-phosphotransferase; EC 2. 7. 1. 2. Glucokinase: ATP: D-glucose-6-phosphotransferase.)

All the animal tissues utilizing free sugars contain enzymes belonging

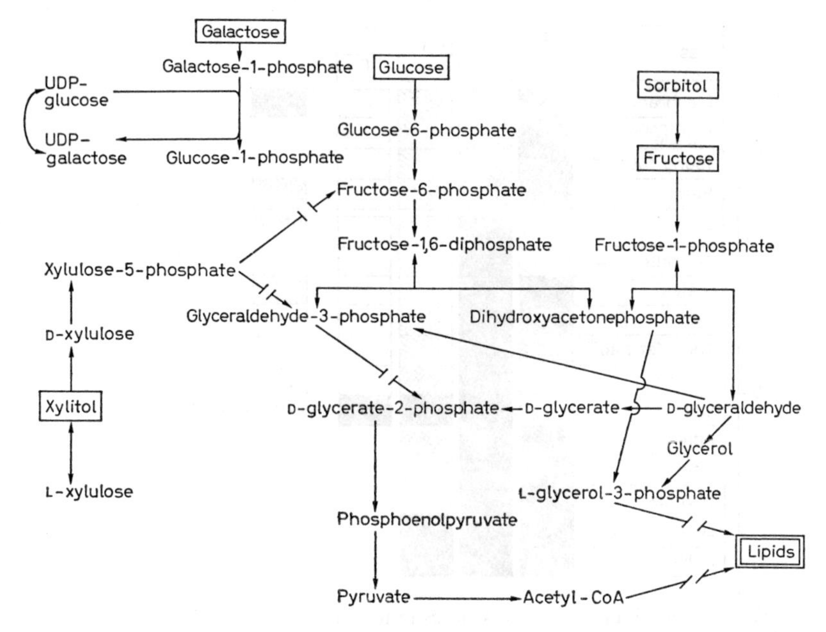

Fig. 1. Metabolic pathways of galactose, glucose, sorbitol, xylitol and fructose.

to a group called hexokinase. They catalyze the transfer of phosphate from ATP to a monosaccharide acceptor, especially glucose, thereby forming glucose-6-phosphate and ADP. This catalytic process requires magnesium ions.

CH_2OH — H, O, OH, H, OH, H, HO, H, H, OH
ATP ADP →
CH_2OⓅ — H, O, OH, H, OH, H, HO, H, H, OH

In glucose metabolism, the phosphorylating step is rate-limiting. In mammalian tissues, especially in tissues which are freely permeable to glucose [111], two groups of hexokinase were detected: the classical low K_M hexokinase, and another high K_M type present in liver, which is termed glucokinase [239]. In liver, the hexokinase is localized in the epithelial cells of the bile ducts, and the glucokinase in the hepatocytes [174, 217]. In the liver of all animals and of man, the activity of the hexokinases is approximately 0.4 U/g fresh weight. High glucokinase activities

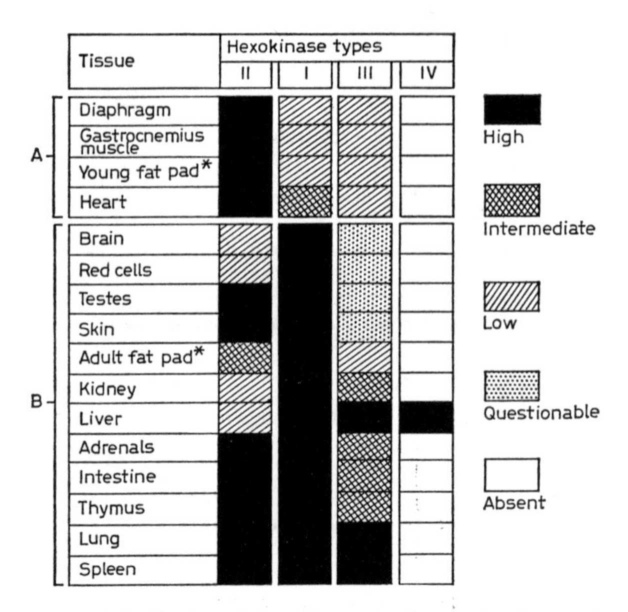

Fig. 2. Distribution of hexokinases in the rat. A refers to tissues which are highly insulin-sensitive and B to those which are relatively insensitive to insulin [according to reference 123].

(approximately 2 U/g fresh weight) were measured in the liver of omnivorous and non-ruminant herbivorous mammals, whilst in carnivorous and herbivorous mammals the enzyme levels were low [10]. In human liver only low glucokinase activity was found (0.2–0.4 U/g fresh weight) [81, 253]. Recent investigations revealed that the so-called hexokinase is a set of three isoenzymes which can be separated by gel electrophoresis. According to their increasing mobilities at pH 8.4, they were designated as hexokinases I, II and III.

Type IV, which moves with the highest velocity, is identical to glucokinase. Figure 2 shows the isoenzyme patterns of different rat tissues [122, 123]. In rat liver all four isoenzymes were detected [66, 123]. This was also verified for human liver [188]. In fat pads of young rats, three types of hexokinases are present [173], in older animals, however, only two types were seen [248]. Type III is lost during the ageing process [173]. The four enzymes were purified from different tissues [122, 202, 206]. Hexokinases I–III of mammalian tissues are different from the hexokinase of yeast [71].

Table I. Hexokinases

	Glucose (K_M) M	ATP (K_M) M
Hexokinase I	4.4×10^{-5}	4.2×10^{-4}
Hexokinase II	1.3×10^{-4}	7.0×10^{-4}
Hexokinase III	2.1×10^{-2}	12.9×10^{-4}
Glucokinase	1.8×10^{-2}	4.9×10^{-4}

Glucokinase: only in liver. Hexokinases: in all tissues, particulate – soluble. *Inhibitors:* Hexokinases: glucose-6-phosphate, ADP. Glucokinase: acetyl-CoA and phosphoenolpyruvate.

Kinetical data are given in table I. The three hexokinases also differ in their stabilities. Type I is heat-stable, type III is stabilized by glucose against heat denaturation, whilst type II is very sensitive. The molecular weight of hexokinases I–III was 96,000 daltons. No cleavage of the enzymes into subunits was possible [49], in contrast to glucokinase which seems to be composed of two subunits having each a molecular weight in the region of 50,000 [18, 188]. Hexokinases are inhibited by their products, glucose-6-phosphate and ADP [51, 159] and it is evident that the inhibitor effect of glucose-6-phosphate has a regulative function in glucose metabolism. The inhibition by this product could be reversed by inorganic phosphate [51]. Recently, an allosteric activation of brain hexokinase by magnesium ions and magnesium adenosine triphosphate was described [8].

Glucokinase is not inhibited by its products, but acetyl-CoA and phosphoenolpyruvate cause a feedback inhibition [243]. In brain, heart, adipose tissue and intestinal mucosa, hexokinases were found in particulate and soluble cell fractions. Types I and II were found in the soluble and particulate bound fractions [124], whereas type III was only detected in the cytosol. In brain, heart and intestinal mucosa the insoluble matrix are the mitochondria from which the enzymes were solubilized by treatment with detergents, substrates and salt solutions [87, 134, 172, 206, 228, 229, 254]. The ratio particulate to soluble is species-dependent [236]. In small intestinal mucosa hexokinase is localized in the epithelial cells of the brush border zone [16]. In perfusion experiments with the small intestine of fasted rats which has a higher portion of particulate hexokinase than that of normal rats, glucose raised the soluble part of the enzyme [16, 169]. In adipose tissue, where hexokinases are localized in the microsomes, mitochondria and soluble fractions, they were solubilized by glucose-6-phosphate, whilst inorganic phosphate antagonizes this procedure [219]. Recently, it was demonstrated that glucokinase is localized in the microsomal fraction of the liver [17, 19]. In addition to the metabolic regulation of particulate-soluble hexokinases, hexokinase II and glucokinase are under dietary and hormonal control. In diabetic animals and in man the glucokinase activity is reduced [3, 178, 189, 200, 217, 222, 223, 241, 253] and can be restored by application of insulin. A low-carbohydrate diet also diminishes the glucokinase activity. The fall in activity is reversed by feeding glucose [178]. The hexokinase activity in mucosa cells is also reduced by fasting and low-carbohydrate diet. This process could be normalized by feeding glucose [3, 210, 221]. In adipose tissue, the activity of hexokinase II is reduced in diabetic animals and after fasting, it is restored by insulin treatment and by feeding glucose [143, 189]. In the rat, it was shown that the response to feeding or to insulin treatment is age-dependent, the hormonal sensitivity is faster in the fat pads and in the liver of young animals as compared to the older ones [2, 173]. It is also known that liver glucokinase is absent in fetal liver [240].

Phosphofructokinase

(EC 2.7.1.11. ATP: D-fructose-6-phosphate-1-phosphotransferase.)

Phosphofructokinase catalyzes the transfer of the terminal group of ATP to fructose-6-phosphate in the presence of Mg^{++}. The products are fructose-1,6-diphosphate and ADP:

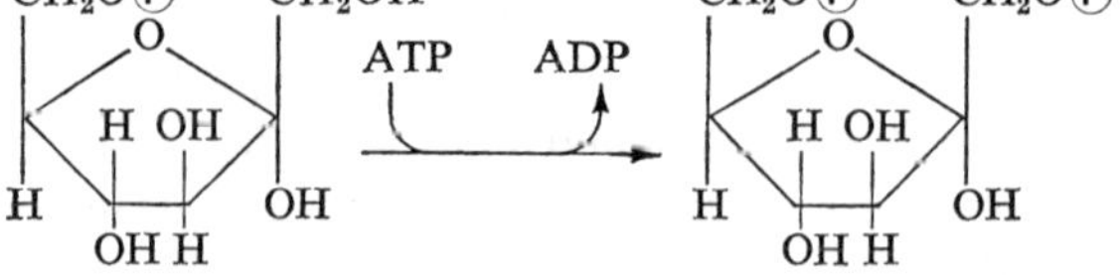

CH_2OⓅ CH_2OH O H OH H OH OH H —ATP → ADP→ CH_2OⓅ CH_2OⓅ O H OH H OH OH H

Under physiological conditions, the phosphorylation step is essentially irreversible with a $K_{eq} \approx 10^3$. This enzyme is present at different activities in all tissues. In muscle, its activity is 100 times as high as in the liver. Phosphofructokinase can be purified from muscle [61, 98, 150, 151], brain [156, 176], kidney [235], heart [161, 162, 163] and liver [25, 128,

194, 234]. A series of feedback mechanisms clearly indicates that the regulation of the phosphofructokinase activity is fundamental in the control of glycolysis. As was proved for other so-called regulative enzymes at least two different species exist: the muscle (M) type is suited to high glycolytic rates and is localized in skeletal muscle, heart and brain; the liver (L) type is present in tissues with high anabolic rates as in liver, kidney and adipose tissue. In the liver 90–95 %, in the kidney 50 % and in adipose tissue 58 % of the enzyme activity is of the L-type [227]. These two isoenzymes show a different behaviour in immunological and electrophoretical studies and with respect to kinetic parameters [25, 128]. The localization of the enzyme in the particulate or soluble part of the cell is not clear. It was demonstrated that the particulate material of heart muscle can be solubilized and activated by ATP and other nucleotides [163, 164]. Apparently, ATP stabilizes the soluble and active aggregation of the enzyme and the rôle of sulphide-disulphide interaction in the solubilizing process is also discussed [186]. A possible conversion of the subunits to the oligomers probably has regulative functions. The equilibrium is controlled by the protein concentration, the pH, the presence of substrates and by modifier molecules [48, 165]. The smallest molecular weight for the active enzyme is 360,000. MAIER [160] found that in the undiluted cell sap 90 % of the muscle phosphofructokinase activity is associated with a molecular weight greater than 1,000,000. In addition, HOFER and PETTE [97] reported that one part of the skeletal muscle enzyme contains ribonucleic acid. It is postulated that this substance makes the enzyme more sensitive towards positive and negative effectors.

At pH values higher than 7.5, phosphofructokinase gives a normal Michaelis-Menten kinetic, but at smaller pH values the kinetic properties of the enzyme are sigmoid [99, 165] (table II). Small pH changes produce considerable effects on the allosteric behaviour of this enzyme [232].

According to Monod's theory on allosteric enzymes, the smallest active species, having a molecular weight of 360,000, consists of 4 protomers each with 90,000 daltons [183]. Each protomer has 4 peptide chains, one containing the catalytic centre [184], which binds the substrates fructose-6-phosphate, MgATP [125] and three regulator units which bind the allosteric effectors. There is some evidence that in agreement with the theory mentioned above, the enzyme exists in two interconvertible allosteric conformations, with different affinities for its substrates and its allosteric effectors, PFK_R and PFK_T. These two types are separated by gel electrophoresis [180]. ATP concentrations, higher than necessary for the

Table II. Phosphofructokinase

	K_M M
Fructose-6-phosphate	5.6×10^{-5}
ATP	5×10^{-5}

In all tissues. Particulate? – soluble. Two types: M-type in muscle heart, brain and L-type in liver, kidney, adipose tissue.
Kinetic properties:
pH > 7.5. Michaelis-Menten kinetic.
pH < 7.5. Allosteric control.
(PFK_R: high affinity to fructose-6-phosphate)
(PFK_T: low affinity to fructose-6-phosphate)

$$PFK_R \underset{3',5'\text{ AMP, AMP}}{\overset{\text{citrate, ATP}}{\rightleftharpoons}} PFK_T$$

saturation of the catalytic side, are potent inhibitors. Thus, 2 moles ATP are bound by the allosteric sides of each promoter thus reducing the affinity to fructose-6-phosphate [125–127, 155]. (The number of bound ATP molecules contrasts to the number of allosteric polypeptide chains mentioned above.) By the binding of ATP, the reactivity of one sulph-hydryl group is reduced [33, 126], which may have consequences in the sulphide-disulphide interaction discussed above. Citric acid favours the dissociation of all ATP-binding sites and replaces the ATP at the allosteric sites and lowers the affinity to fructose-6-phosphate. Furthermore, it was demonstrated that AMP, 3′, 5′-AMP and inorganic phosphate increase the affinity to fructose-6-phosphate. In the activation process, the cyclic nucleotide is three times as effective as AMP. The activators are bound at the allosteric regulator site and can also replace ATP or citrate [165]. A pH-independent NAD and NADH inhibition [25] was also shown. However, NEWSHOLME and SUYDEN [177] proved that this effect is due to the inhibition of aldolase, which was used in the test system for the phosphofructokinase assay. An inhibitor effect, demonstrated for fatty acids [25], was unspecific [213] and should not be taken to interpret metabolic regulation. The data discussed apply to skeletal muscle and heart enzyme. The liver enzyme shows a higher sensitivity to the ATP in-hibition. It is relatively insensitive against other inhibitors and activators such as ADP, AMP, etc.

From the regulation point of view, it is evident that during muscle contraction [40] as well as during anoxia [157], when the AMP and ADP levels are elevated the enzyme is fully activated. In the resting muscle, however, the activity is reduced by high ATP levels. The lower sensitivity of the liver enzymes against the nucleotide activators is also understandable. In tissues with high anabolic rates the AMP and ADP levels do not play an important rôle.

In rat liver, the activity is increased by a high-fat diet [214] whilst in human adipose tissue the enzyme activity is reduced in hyperlipaemic patients [133].

Pyruvate Kinase

(EC 2. 7. 1. 40. ATP: pyruvate phosphotransferase.)

The enzyme pyruvate kinase catalyzes the transfer of the phosphate group from phosphoenolpyruvate to ADP. The products are pyruvate and ADP:

$$\begin{array}{c} COOH \\ | \\ C{=}O \\ | \\ CH_3 \end{array} \quad \xrightarrow{\;ATP \;\curvearrowright\; ADP\;} \quad \begin{array}{l} COOH \\ | \\ C{-}O{\sim}\textcircled{P} \\ \| \\ CH_2 \end{array}$$

Pyruvate kinase is localized in the soluble part of the cell. From thermodynamic considerations, this step is irreversible under physiological conditions. Like phosphofructokinase it is especially suited to the control of glycolysis.

Two immunologically distinct forms designated as M and L can be detected. Electrofocusing [43] and electrophoresis [218] indicate that there are at least 6 isoenzymes; three of them correspond to the L- and three to the M-type. It was pointed out that the liver enzyme is allosterically modulated by metabolites as well as pyruvate kinase in kidney and adipose tissue.

M-Type

This type was first isolated by NEGELEIN in crystalline form as mentioned by BÜCHER and PFLEIDERER [28]. In the meantime crystalline preparations were obtained from human muscle, rabbit muscle [15] and pig heart [137]. Table III presents kinetic data.

Table III. Pyruvate kinase

	M-type(K_M) M	L-type(K_M) M
Phosphoenolpyruvate	0.7×10^{-4}	0.8×10^{-3}
ADP	0.3×10^{-3}	0.3×10^{-3}

In the soluble part of the cell. At least two different isoenzymes: muscle type (M) in skeletal muscle, brain, heart; liver type (L) in liver, kidney cortex, adipose tissue, erythrocytes.
Kinetic properties:
M-type (PK_M) Michaelis-Menten kinetic.
L-type (PK_L) pH < 7.0 Michaelis-Menten kinetic.
L-type: pH > 7.0 allosteric control, two forms.
($PK_{L,R}$: high affinity to phosphoenolpyruvate, K^+ and Mg-ADP. $PK_{L,T}$: low affinity to phosphoenolpyruvate, high affinity to ATP)

$$PK_{L,R} \underset{\text{fructose-1,6-diphosphate}}{\overset{\text{alanine}}{\rightleftharpoons}} PK_{L,T}$$

L-Type

Seasonal and nutritional changes in the pyruvate kinase level of rat liver encouraged TANAKA *et al.* [224] to purify pyruvate kinase from liver. This group succeeded in separating two isoenzymes in crystalline homogeneous form. One enzyme corresponds to the M-type and the other was called the L-type.

Similar to the M-type, liver pyruvate kinase is a tetrameric enzyme. According to Monod's theory on allosteric enzymes, pyruvate kinase L exists in two interconvertible conformations L_R and L_T showing different affinities to substrates and allosteric modifiers [14, 45, 199]. The R-state has a high affinity to phosphoenolpyruvate, potassium ions, and ADP magnesium, in contrast to the T-state with a low affinity to the binding of phosphoenolpyruvate and no affinity to potassium ions and ADP magnesium. The product ATP also has a high affinity to the T-state. By the action of the allosteric modifier fructose-1,6-diphosphate [14, 32, 114, 115, 129, 166, 225, 226, 245], the equilibrium between the two forms is shifted to the active R-state, whilst the binding of alanine [32, 115, 154, 167, 225, 245, 246] favours the inactive T-state. At pH values lower than

7, the allosteric behaviour of pyruvate kinase L is transferred to a normal Michaelis-Menten type. In addition to the metabolic regulation discussed above, acetyl-CoA is an inhibitor of pyruvate kinase L [245]. The mechanism is not clear. Presumably, this inhibition is due to a chemical modification of the pyruvate kinase. Fatty acids are also inhibitors of the L-type [245, 246]. In addition to liver, the allosteric L-type is also present in kidney cortex [46], in adipose tissue [190] and in erythrocytes. In normal rat-liver, the ratio L:M types is approximately 3. A high-protein diet [224] reduces especially the concentration of the L-type [253] and the ratio L:M is lowered to 1. In diabetic animals, the L:M ratio of 0.6 can be restored by insulin treatment of the animals thereby raising the concentration of the L-type [242, 243]. In contrast to glucose, a diet rich in fructose causes an increase of the L-levels in the liver [132, 247]. A high-fat diet reduces the enzyme level of the L-type in adipose tissue and liver [67]. In the rat jejunum pyruvate kinase activity is induced by fructose [222].

L-Glycerol-3-Phosphate Dehydrogenase

(EC 1. 1. 1. 8. L-glycerol-3-phosphate NAD oxidoreductase.)

L-Glycerol-3-phosphate dehydrogenase catalyzes the NAD-dependent dehydrogenation of L-glycerol-3-phosphate to dihydroxyacetone phosphate:

$$\begin{array}{c} H_2COH \\ | \\ C{=}O \\ | \\ H_2CO\text{Ⓟ} \end{array} \xrightarrow{NADH_2 \quad NAD} \begin{array}{c} H_2COH \\ | \\ HOCH \\ | \\ H_2CO\text{Ⓟ} \end{array}$$

At physiological pH values the equilibrium is far towards the L-glycerol-3-phosphate site and favours the production of the acyl acceptor for triglycerides and phospholipid synthesis. The enzyme is present in the soluble part of the cell in all animal and human tissues. The highest enzyme level is found in muscle. As in muscle, the activity found in other tissues is sufficient to prevent, under normal physiological conditions, this reaction from being rate-limiting in lipid synthesis.

L-Glycerol-3-phosphate dehydrogenase has been crystallized from skeletal muscle of the rabbit [15, 259] and the rat [55]. Homogeneous preparations were also obtained from the liver of the rabbit [144, 181] and man [145].

Table IV. L-Glycerol-3-phosphate dehydrogenase

Substrate	K_M M
L-Glycerol-3-phosphate	0.68×10^{-3}
Dihydroxyacetone phosphate	0.018×10^{-3}
NAD	0.19×10^{-3}
NADH	0.0042×10^{-3}

Data for the rabbit-liver enzyme at pH 7.5 [144]. In the soluble part of all mammalian tissues.

Kinetic data are given in table IV. It was found that the muscle enzyme consists of two identical polypeptide chains. Both together have a molecular weight of 68,000 daltons [56, 144, 255]. Muscle glycerol-3-phosphate dehydrogenase contains 2–3 acetyl groups and binds 1 mole of adenosine diphosphoribose [4, 52]. Kinetic experiments demonstrate that the substrate L-glycerol-3-phosphate can only be bound by the enzyme if NAD is added. Consequently, dihydroxyacetone phosphate is bound by the addition of NADH [20]. The muscle enzyme is inhibited competitively by L-glycerol-3-phosphate [22]. It is likely that the chemical composition of the muscle enzymes is species-dependent. It also has been found that differences exist between the isoenzymes of muscle and liver. However, in electrofocusing experiments identical values for the isoelectric point (IP) [57] were measured for the two enzymes. In addition, different enzyme patterns were found in the liver [198].

Recent experiments proved that the liver enzyme is subjected to allosteric control [144]. In addition to the catalytic sites, which bind L-glycerol-3-phosphate, dihydroxyacetone phosphate, NAD and NADH, there exists at least one regulator site for L-glycerol-3-phosphate, which lowers the K_M for this substrate and increases the K_M for dihydroxyacetone phosphate. Similar results were obtained for NAD. When added to the allosteric NAD binding site, the K_M for NAD is lowered but the K_M for NADH is increased.

In addition, it was found that the enzyme is inhibited by free fatty acids [205, 238]. But there are some doubts regarding the specificity of the inactivation mechanism. It cannot be excluded that it is an unspecific detergent effect on the surface of the enzyme. L-Glycerol-3-phosphate dehydrogenase is also subjected to dietary control. Its activity is lowered by feeding high-fat diets.

Pyruvate Dehydrogenase Complex

The pyruvate dehydrogenase complex catalyzes the following overall reaction:

Pyruvate + CoA + NAD → acetyl-CoA + CO_2 + NADH.

In lipid biosynthesis, this reaction yields acetyl-CoA, the starting material for fatty acids.

In animal tissues the pyruvate dehydrogenase multi-enzyme complex is localized in mitochondria. The molecular weight for the heart muscle enzyme is 9,500,000 dalton. The functional unit consists of 3 different enzymes: the pyruvate dehydrogenase (E_1), dihydrolipoyl transacetylase (E_2), and the FAD-containing dihydrolipoyl dehydrogenase (E_3) which catalyze the following individual reactions:

1. $CH_3C(=O)-COOH + TPP-E_1 \longrightarrow CH_3CH(OH)-TPP-E_1 + CO_2$

2. $CH_3CH(OH)-TPP-E_1 + CH_2-CH_2-CH(-S-S-)-R-E_2$ (cyclic disulfide, S bridging the first CH_2 and CH) $\searrow$ $CH_2(S-C(=O)-CH_3)-CH_2-CH(SH)-R-E_2 + TPP-E_1$.

3. $CH_2(S-C(=O)-CH_3)-CH_2-CH(SH)-R-E_2 + HS-CoA$ $\searrow$ $CH_2(SH)-CH_2-CH(SH)-R-E_2 + CH_3-C(=O)-S\,CoA$.

4. $CH_2(SH)-CH_2-CH(SH)-R-E_2 + FAD-E_3$ $\searrow$ $CH_2-CH_2-CH(-S-S-)-R-E_2 + FADH_3-E_3$ (cyclic disulfide).

5. $FADH_2-E_3 + NAD^+ \longrightarrow FAD-E_3 + NADH$.

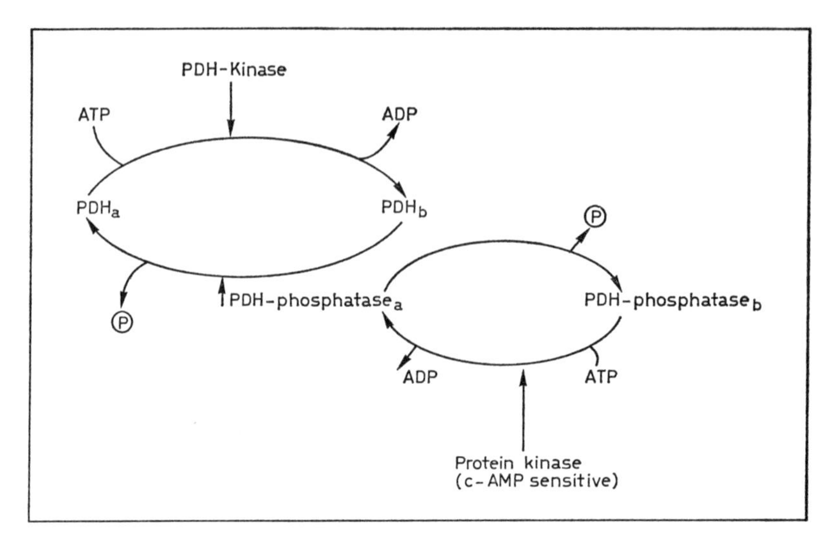

Fig. 3. Interconversions of pyruvate dehydrogenase.

The multi-enzyme complex has been purified from skeletal muscle [117], heart [72], kidney [113] and liver. It is also present in adipose tissue. It is likely that the architecture of the complex isolated from animal tissues is similar to the *Escherichia coli* enzyme, which has been studied in great detail. However, in contrast to *E. coli,* the animal complex is under metabolic and hormonal regulation. It exists in two interconvertible forms (fig. 3): the active a-form and the inactive b-form. The active a-form is inactivated and transferred to the b-form by ATP-dependent phosphorylation of the pyruvate dehydrogenase unit [72, 152, 153, 251]. This phosphorylation is dependent on magnesium ions and is catalyzed by pyruvate dehydrogenase kinase. The substrate pyruvate is a powerful protector against this inactivation. The inactive b-type is activated via pyruvate dehydrogenase phosphatase action thereby removing the phosphate residue. The phosphatase has been purified by SIESS and WIELAND [211] and needs magnesium for optimal action. This enzyme is present in two interconvertible forms: the a-type, which is active and represents a phosphoenzyme, and the inactive dephospho b-type. The phosphatase b is converted to the a-form by an ATP-dependent phosphorylation catalyzed by a 3′,5′-AMP-sensitive protein kinase. The last reaction is responsible for the hormonal sensitivity of the pyruvate dehydrogenase complex. Because of the action of some hormones on the AMP-cyclase system at the cell wall, the 3′,5′-AMP level in the cell can be influenced. Insulin, for example, reduces the 3′,5′-AMP level in fat cells and increases the conversion of pyruvate to acetyl-CoA and CO_2 [118, 119]. Epinephrine elevates the 3′,5′-AMP levels and reduces the pyruvate oxidation.

In addition to the chemical modification, acetyl-CoA is a competitive inhibitor of the multi-enzyme complex [60]. Citrate, as a metabolite of the Krebs cycle, also reduces its activity [167, 212]. As was shown by WIELAND *et al.*, the a-type in rat heart and kidney is decreased by fasting [250].

Galactose Metabolism

Reviews on this subject are given by FISCHER and WEINLAND [54] and KALCKAR [120].

In human and animal nutrition, galactose is normally present as a part of the lactose molecule. The milk of mammals has a high lactose concentration and, therefore, a maximum in galactose metabolism can be observed in the neonatal period. Following oral administration, the lactose molecule is hydrolyzed to glucose and galactose in the brush border zone of the small intestinal mucosa. Afterwards, the galactose moiety is converted into a glucose intermediate, mainly in the liver. The mobilization of insulin by galactose is only possible after its conversion to glucose [63].

The subject of this part of the review will be the conversion of galactose into intermediates of the metabolic pathway of glucose.

In the first step, galactose is phosphorylated to α-galactose-1-phosphate. In a second reaction, the uridyl part of the uridine diphosphoglucose is transferred to the galactose-1-phosphate and uridine diphosphogalactose and glucose-1-phosphate are formed. The latter is an intermediate of glucose metabolism. The enzyme epimerase catalyzes the reconversion of uridine diphosphogalactose into uridine diphosphoglucose (fig. 1).

Galactokinase

(EC 2. 7. 1. 6. ATP: D-galactose-1-posphotransferase.)

Galactokinase catalyzes the formation of α-galactose-1-phosphate from adenosine triphosphate and D-galactose.

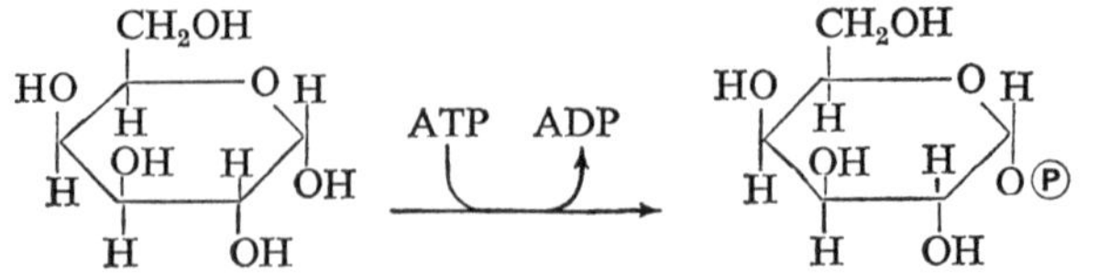

Galactokinase is present in liver, kidney, small intestinal mucosa, heart, skeletal muscle and brain of mammals [54]. The highest activity is found in the liver [44]. The enzyme level is dependent on age. Young animals and humans have higher activities than older ones. Galactokinase is localized in the soluble part of the cell and is purified from the liver [12, 44]. Kinetic data are shown in table V. The products galactose-1-phosphate and ADP are non-competitive inhibitors of the enzyme [13]. In the rat jejunum it was demonstrated that the activity of galactose could be raised by feeding galactose [220].

Hexose-1-Phosphate Uridylyl Transferase

(Galactose-1-phosphate uridylyl transferase, EC 2. 7. 7. 12. UDP glucose: α-D-galactose-1-phosphate uridylyl transferase.)

The enzyme catalyzes a reaction which can be described as a uridyl transfer between the two nucleophilic agents galactose-1-phosphate and glucose-1-phosphate.

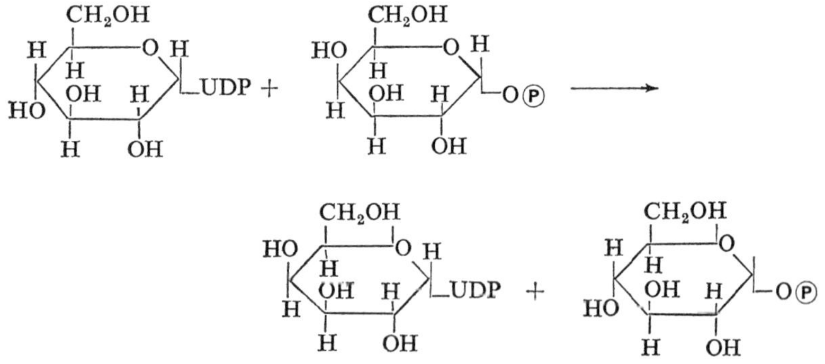

The equilibrium constant approximates 1. The uridyl transferase is detected in liver, kidney, small intestinal mucosa, skeletal muscle and heart of animals and men. The enzyme can be purified from the liver of animals [170] and from normal and galactosaemic subjects [207]. Kinetic

Table V. Galactokinase

Substrate	K_M M
Galactose	$1\text{–}3 \times 10^{-4}$
ATP-Mg^{++}	2×10^{-4}
In liver, kidney small intestinal mucosa, etc.	

data are shown in table VI. Uridine tri-, di- and monophosphates are inhibitors of this enzyme [208]. In the jejunum of the rat, the enzyme level is elevated by the application of galactose [220].

Uridine Diphosphoglucose Epimerase
(EC 5. 1. 3. 2. UDP-glucose-4-epimerase.)

The enzyme catalyzes the following step:

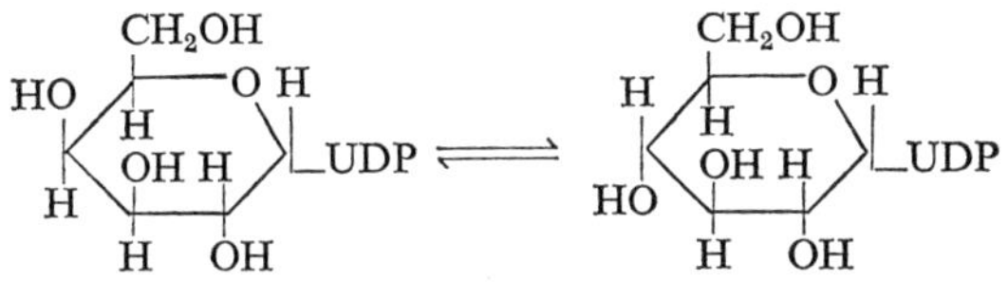

The reaction mechanism is not clear. However, it was demonstrated that the mammalian enzyme requires NAD. The epimerization is probably an oxidation-reduction mechanism at the catalytic centre of the en-

Table VI. Uridine diphosphoglucose epimerase

Substrate	K_M M
UDP glucose	9.0×10^{-5}
UDP galactose	5.0×10^{-5}
In liver, and small intestinal mucosa.	

Table VII. Hexose-1-phosphate uridylyl transferase

Substrate	K_M M
Galactose-1-phosphate	0.3×10^{-3}
UDP glucose	0.2×10^{-3}
UDP galactose	0.3×10^{-3}

In liver, kidney, small intestinal mucosa, skeletal muscle and heart ($K_{eq} \backsim 1$).

zyme. The epimerase can be detected in liver and intestinal mucosa [35, 209]. The enzyme was purified from liver [168]. Kinetic data are given in table VII.

NADH is an inhibitor of this enzyme. Its activity is possibly regulated by the NAD : NADH ratio in the cell [36]. Similar to the other 'galactose' enzymes, its activity can be increased by galactose feeding in the small intestinal mucosa.

Appendix

A uridine diphosphate galactose pyrophosphatase which catalyzes the formation of uridine diphosphogalactose from galactose-1-phosphate and uridine triphosphate is present in liver. Its activity, however, is too low to have any considerable function in the galactose metabolism.

Fructose Metabolism

The following reviews dealing with this subject have been published: 'Le métabolisme du fructose' by HERS [89] and 'Biochemie, physiologische und klinische Probleme des Fructosestoffwechsels' by LEUTHARDT and STUHLFAUTH [148], and a series of papers by HERMAN and ZAKIM [85, 86, 260].

Since KÜLZ [136] reported in 1874 that fructose is tolerated by human diabetics, efforts have been undertaken to distinguish between the metabolic pathways of fructose and glucose. In medicine, fructose is used to substitute starch in the nutrition of diabetic persons and, in addition, fructose is one-half of the sucrose molecule. It is assumed that in middle Europe, 25 % of the nutritional carbohydrate consists of fructose.

Experiments demonstrated that fructose is metabolized in liver, kidney and small intestinal mucosa. In these tissues, a special metabolic pathway was postulated by Cori *et al.* [37–39, 41], Hers [89], Leuthardt and Testa [146a, b], Leuthardt *et al.* [147] and our group. This fructolytic pathway is shown in figure 1 and includes the following steps: ATP-dependent phosphorylation to fructose-1-phosphate by ketohexokinase (formerly called fructokinase). This substance is then split by a special aldolase into D-glyceraldehyde and dihydroxyacetone phosphate. The metabolism of D-glyceraldehyde, however, can be catalyzed by 4 different enzymes:

1. Triokinase phosphorylates ATP-dependent D-glyceraldehyde to D-glyceraldehyde-3-phosphate.

2. Aldehyde dehydrogenase is able to oxidize D-glyceraldehyde to glyceric acid. The latter compound is then phosphorylated, ATP-dependent, by glycerate kinase to glycerate-2-phosphate or it is oxidized by D-glycerate dehydrogenase to hydroxypyruvate, an intermediate of serine metabolism.

3. The NAD-dependent alcohol dehydrogenase is capable of reducing D-glyceraldehyde to glycerol, which is then phosphorylated by glycerol kinase to L-glycerol-3-phosphate.

4. The NADP-dependent alcohol dehydrogenase, which is different from the enzyme described above, catalyzes the same step.

The reduction followed by the phosphorylation is in variance to isotope studies with fructose-6-^{14}C, or fructose-1-^{14}C [91, 195].

Ketohexokinase

(Fructokinase, ketokinase; EC 2.7.1.12. ATP: D-fructose-1-phosphotransferase.)

Ketohexokinase catalyzes the ATP-dependent phosphorylation of fructose to fructose-1-phosphate.

HOCH$_2$ O H$_2$COH — H OH — H, OH — OH H →(ATP → ADP)→ HOCH$_2$ O H$_2$CO(P) — H OH — H, OH — OH H

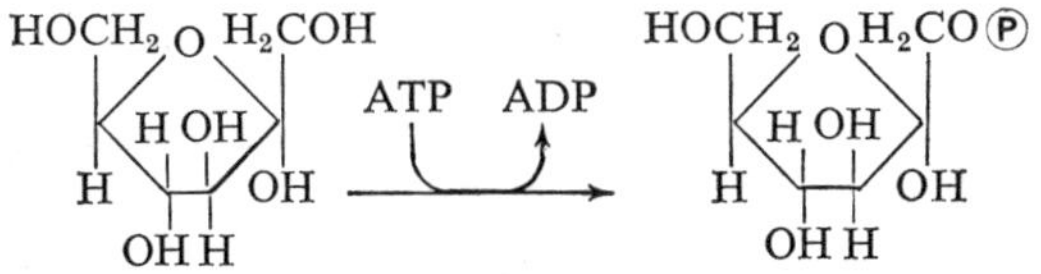

The enzyme is found in the liver of warm-blooded animals and in man [81, 82], and it is absent in reptilia, amphibia and fishes. In the cell, ketohexokinase is localized in the cytoplasm [78]. Ketohexokinase is also

Table VIII. Enzyme activities involved in fructose metabolism in liver, kidney and small intestinal mucosa of man. The enzyme activities are expressed as international units (U). Determinations are performed at 25 °C. (In kidney and small intestinal mucosa glycerate kinase was not detectable.)

Enzyme	Liver		Kidney			
			cortex		medulla	
	wet weight U/g	U/100 mg protein	wet weight U/g	U/100 mg protein	wet weight U/g	U/100 mg protein
Ketohexokinase	1.23	1.2	1.31	1.53	0.67	0.90
Aldolase, S: F-1,6-P	3.46	4.1	2.82	3.37	1.00	1.35
Aldolase, S: F-1-P	2.08	2.5	1.24	1.47	0.20	0.29
Alcohol dehydrogenase (NAD), S: D-glyceraldehyde	3.10	4.0	0.18	0.25	0.19	0.26
Alcohol dehydrogenase (NADP), S: D-glyceraldehyde	3.60	4.6	1.39	1.62	0.49	0.66
Aldehyde dehydrogenase, S: D-glyceraldehyde	1.04	1.3	0.37	0.44	0.05	0.07
D-Glycerate kinase, S: D-glyceride acid	0.13	0.07	–	–	–	–
Glycerol kinase	0.62	0.72	–	–	–	–
Triokinase	2.07	2.8	0.48	0.57	0.20	0.29

S = substrate, F-1,6-P = fructose-1,6-diphosphate, F-1-P = fructose-1-phosphate.

present in kidney [79, 204], and in small intestinal mucosa [74, 80, 204]. The localization in these three tissues is in agreement with *in vivo* experiments on fructose metabolism.

Table VIII demonstrates the enzyme levels in liver [81], kidney and small intestinal mucosa [204] in man. In table IX, the enzyme activities in these tissues are also given for some laboratory animals. Comparing the ketohexokinase activities in rat liver and kidney as related to the wet

Table VIII (continued)

Small intestinal mucosa									
duodenum		jejunum I		jejunum II		ileum I		ileum II	
wet weight U/g	U/100 mg protein	wet weight U/g	U/100 mg protein	wet weight U/g	U/100 mg protein	wet weight U/g	U/100 mg protein	wet weight U/g	U/100 mg protein
0.82	1.46	0.66	1.36	0.76	1.52	0.41	0.94	0.65	1.86
1.54	2.74	2.18	4.49	2.65	4.67	0.91	1.92	1.21	3.02
0.61	1.20	0.60	1.24	0.64	1.14	0.25	0.54	0.32	0.80
0.30	0.53	0.24	0.48	0.21	0.43	0.08	0.19	0.13	0.39
0.75	1.33	0.90	1.75	0.71	1.34	0.44	0.96	0.44	1.07
0.18	0.32	0.10	0.23	0.13	0.24	0.08	0.16	0.08	0.21
–	–	–	–	–	–	–	–	–	–
–	–	–	–	–	–	–	–	–	–
0.64	1.14	0.58	1.16	0.80	1.38	–	–	–	–

weight of the tissue, similar results are obtained. In human kidney we have separated the cortex from the medulla; in the medulla, only one-half of the activity was found, as compared to the cortex. Other authors failed to detect ketohexokinase in the human medulla [131]; but BURCH *et al.* [30] were able to estimate fructose-1-phosphate in this part of the kidney after fructose was administered intraperitoneally to rats. In human small intestinal mucosa, the activity was identical in all areas [204].

Table IX. Enzyme activities involved in fructose metabolism in liver, kidney and small intestinal mucosa of different laboratory animals. The enzyme activities are expressed as international units (U). Determinations are carried out at 25 °C. In rabbit liver, glycerate kinase and glycerol kinase were not estimated

Enzyme	Liver				
	rabbit		rat		guinea-pig
	wet weight U/g	U/100 mg protein	wet weight U/g	U/100 mg protein	wet weight U/g
Ketohexokinase	0.98	1.6	2.20	1.80	1.07
Aldolase, S: F-1,6-P	2.78	2.63	4.34	3.50	2.58
Aldolase, S: F-1-P	1.41	1.42	1.63	1.20	1.12
Alcohol dehydrogenase (NAD), S: D-glyceraldehyde	1.88	2.47	1.64	1.33	5.37
Alcohol dehydrogenase (NADP), S: D-glyceraldehyde	1.79	1.70	0.38	0.31	3.48
Aldehyde dehydrogenase, S: D-glyceraldehyde	0.60	0.53	1.40	1.14	1.35
D-Glycerate kinase, S: D-glyceride acid	–	–	3.14	3.60	0.05
Glycerol kinase			2.08	1.63	ND
Triokinase	–	–	1.65	1.34	0.43

ND = no data. Other abbreviations as for table VIII.

Comparisons of the activity of the enzyme, as related to the protein content in liver, in the cortex of the kidney and in the small intestinal mucosa (mucosa contains more water than the other two tissues), leads to similar results in the three organs.

The enzyme activities given in tables VIII and IX were measured at 20 °C; when the experiments were performed at 37 °C, 2.2 U/g were found in the liver of the rat, and 1.7 U/g in human liver. In the rat, the

Table IX (continued)

Rat kidney		Small intestinal mucosa							
		rabbit		rat		guinea-pig		golden hamster	
wet weight U/g	U/100mg protein	wet weight U/g	U/100mg protein	wet weight U/g	U/100mg protein	wet weight U/g	U/100mg protein	wet weight U/g	U/100mg protein
1.62	1.69	0.74	1.32	0.25	0.44	1.19	4.06	0.89	3.59
4.35	4.53	1.85	3.38	0.31	0.53	4.41	15.0	1.37	5.37
0.94	1.00	0.76	1.39	0.081	0.13	1.81	6.2	0.65	2.49
0.29	0.3	0.16	0.31	–	–	0.11	0.39	–	–
2.94	3.07	0.73	1.30	0.40	0.68	1.40	4.8	0.59	2.13
1.16	1.21	0.14	0.25	0.20	0.33	0.23	0.74	0.13	0.42
2.32	2.50	0.19	0.33	0.44	0.74	0.52	1.38	0.19	0.33
2.12	2.25	–	–	0.27	0.45	0.52	1.49	–	–
1.31	1.34	0.74	1.36	0.37	0.62	1.00	3.42	0.32	1.33

extraction rate of fructose during the perfusion of the liver agrees well with the enzyme activity measured at 37 °C. In human liver, the total capacity for fructose metabolism was calculated to be 0.45 g/min at 37 °C. from *in vivo* experiments reported by TYSTRUP *et al.* [233] and CRAIG *et al.* [42], an extraction rate of 0.41–0.55 g/min fructose can be calculated. These comparisons give excellent agreements between *in vitro* and *in vivo* conditions. If the total capacity for fructose intermediate

Table X. Ketohexokinase

Substrate	K_M M
Liver enzyme	
Fructose	0.46×10^{-3} (0.4 M K^+)
	0.8×10^{-3} (0.1 M K^+)
ATP-Mg	1.56×10^{-3} (0.4 M K^+)
	1.33×10^{-3} (0.1 M K^+)
Mucosa	
Fructose	0.7×10^{-3}

In liver, kidney, and small intestinal mucosa. Localized in the cytoplasm.

metabolism is calculated from the data given in table VI for the particulate tissues, 77 % of the total human capacity is present in liver, 14 % in kidney and 9 % in small intestinal mucosa. These data also correspond to *in vivo* experiments.

If taken *per os,* fructose is resorbed by the intestine more slowly than glucose. If calculated from the ketohexokinase activity for the whole human small intestine, 2.5 g fructose can be phosphorylated and metabolized per hour. In laboratory animals, the enzyme levels of ketohexokinase in small intestinal mucosa is species-dependent. In the rat [73], only one-third to one-quarter of the activity found in guina-pigs, golden hamster and rabbit could be measured [80]. Feeding rats a diet rich in fructose for 3 weeks caused an adaption of the ketohexokinase [74] in the liver and kidney.

Ketohexokinase was purified by LEUTHARDT and TESTA [146a, b], VESTLING *et al.* [237], HERS [90], KUYPER [138], PARKS *et al.* [185], and ADLEMAN *et al.* [1] from liver. Recently, a highly purified enzyme preparation was described by SÁNCHEZ *et al.* [203]. From intestinal mucosa, ketohexokinase was purified by CADENAS and SOLS [31]. The two enzymes seem to be identical. In addition to magnesium ions, which form an active complex with ATP, potassium is required for optimal activity. There is some evidence that potassium is an allosteric effector in the enzymatic mechanism. ADP, as a product, inhibits the enzyme. Kinetic data are given in table X.

L-Sorbose, D-tagatose, D-xylulose and L-galactoheptulose are also phosphorylated at the carbon-1 position by ketohexokinase.

Aldolase

(Ketose-1-phosphate aldolase, phosphofructo-aldolase; EC 4. 1. 2. 7; Ketose-1-phosphate aldehyde-lyase, fructose-1,6-diphosphate aldolase, EC 4. 1. 2. 13; fructose-1,6-diphosphate: D-glyceraldehyde-3-phosphate lyase.)

The name ketose-1-phosphate aldolase was attached especially to the isoenzyme in liver, which splits fructose-1-phosphate at nearly the same rate as fructose-1,6-diphosphate. The term fructose-1,6-diphosphate aldolase describes the enzymes which are present in skeletal muscle and yeast, and which split fructose-1,6-diphosphate more than 100-fold faster than fructose-1-phosphate [149]. Much has been done in isolating and characterizing the aldolases from different tissues, and this resulted in the discovery of several isoenzymes with great variations in the affinity to the two substrates [5, 34, 187]. It seems to be more meaningful to use the name aldolase for all isoenzymes and to differentiate between them according to Rutter's A, B, C nomenclature [201]:

Aldolase 1-A: localized especially in skeletal muscle.

Aldolase 1-B: localized especially in liver.

Aldolase 1-C: localized especially in brain.

Fructose-1-phosphate is converted by aldolase 1-B to dihydroxyacetone phosphate and D-glyceraldehyde [95, 147].

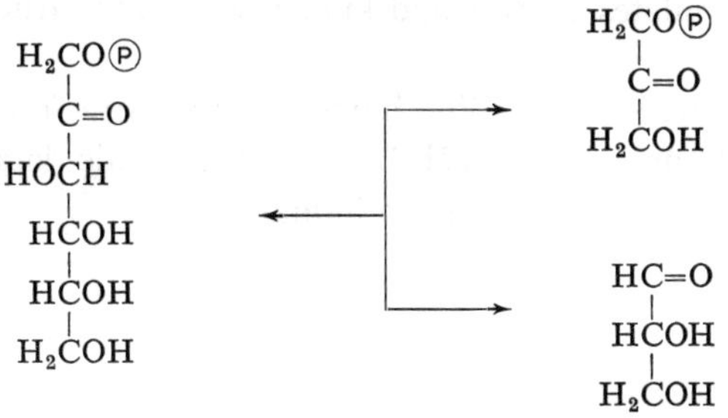

H_2COⓅ – $C{=}O$ – $HOCH$ – $HCOH$ – $HCOH$ – H_2COH ⇄ H_2COⓅ – $C{=}O$ – H_2COH + $HC{=}O$ – $HCOH$ – H_2COH

The same enzyme is also responsible for splitting fructose-1,6-diphosphate to dihydroxyacetone phosphate and D-glyceraldehyde-3-phosphate.

Aldolase is present in all tissues; the special type which splits fructose-1-phosphate as fast as necessary in order to participate in fructose metabolism, however, only exists in the liver, the kidney and in the small intestinal mucosa (table VIII, IX). In these tissues the isoenzyme patterns were different due to naturally occurring hybrids between polypeptide chains from the A and B types.

Table XI. Aldolase

Substrate	Aldolase (K_M)		
	A M	B M	C M
Fructose-1,6-diphosphate	5×10^{-5}	1×10^{-5}	5×10^{-5}
Fructose-1-phosphate	4.5×10^{-2}	7.5×10^{-3}	2.5×10^{-2}

In all mammalian tissues, with different isoenzyme patterns. Three types, aldolase A, B and C.

From a kinetic point of view, the activity of aldolase in liver, kidney and small intestinal mucosa seems to be high enough for fructose metabolism, but the equilibrium is more on the fructose-1-phosphate site [149]. In addition, products of ATP degradation, like allantoin and uric acid, whose concentrations were elevated during fructose loading, inhibit the aldolase [258]. For these reasons, the splitting of fructose-1-phosphate by aldolase is the rate-limiting step in fructose metabolism and, accordingly, fructose-1-phosphate accumulates in liver and kidney when high fructose amounts are offered [30, 76, 258].

Aldolase 1-B was isolated in crystalline homogeneous form from ox [182], rabbit [64, 193] and human liver [47]. In table XI, kinetic data are known, especially for the enzyme isolated from human tissues.

D-Triokinase

(Triosekinase, EC 2. 7. 1. 28. ATP: D-glyceraldehyde-3-phosphotransferase.)

Triokinase phosphorylates D-glyceraldehyde using ATP to glyceraldehyde-3-phosphate.

$$\begin{array}{c} HC{=}O \\ | \\ HCOH \\ | \\ H_2COH \end{array} \xrightarrow{ATP \;\to\; ADP} \begin{array}{c} HC{=}O \\ | \\ HCOH \\ | \\ H_2CO\text{Ⓟ} \end{array}$$

Table XII. D-Triokinase

Substrate	K_M M
D-Glyceraldehyde	$8\text{–}35 \times 10^{-6}$ (dependent on ATP-Mg^{++} concentration)
Dihydroxyacetone	6×10^{-6} (2 mmole ATP)
ATP-Mg^{++}	7.7×10^{-4}

In the cytoplasm of liver, kidney and small intestinal mucosa.

The enzyme is present in the soluble fraction of vertebrate liver (table VIII, IX) with the exception of Amphibia [78, 81, 82]. Kidney [79, 204] and the small intestine mucosa [80, 204] are also able to phosphorylate D-glyceraldehyde (table VI, VII). In human liver, kidney and intestinal mucosa, the phosphorylation of D-glyceraldehyde seems to be the only way to metabolize [81, 204] D-glyceraldehyde derived from fructose. This is supported by the observation that there is no D-glycerate kinase and that the reduction step forming glycerol is at variance with isotope studies using labelled fructose [91, 195]. For human kidney and liver, the enzyme levels are high enough to guarantee an optimal phosphorylation rate for D-glyceraldehyde derived from fructose; but in the small intestine, only the jejunum contains triokinase. If fructose is phosphorylated in the ileum forming D-glyceraldehyde, the latter would have to enter the blood stream for further metabolism. Feeding a diet rich in fructose to rats, the activity of the triokinase is increased in the liver more than 3 times and doubled in the kidney, as compared to the normal enzyme level [74].

Purified enzyme preparations were obtained by HERS [89, 93] from the liver of rabbit and guinea-pig, and by our group from ox liver [77], and recently by FRANDSEN and GRUNNERT [58] from the rat. The enzyme preparations from the different animals seem to be similar. In addition to ATP, magnesium ions are necessary for optimal activity. Similar to glycerol kinase, triokinase can phosphorylate dihydroxyacetone. Kinetic data are given in table XII.

Table XIII. Aldehyde dehydrogenase

Substrate	K_M M
Acetaldehyde	1.7×10^{-6}
D-Glyceraldehyde	4×10^{-4}

In liver kidney and small intestinal mucosa. Localized in cytoplasm and mitochondria.

Aldehyde Dehydrogenase

(Glyceraldehyde dehydrogenase, EC 1.2.1.3. Aldehyde: NAD oxidoreductase.)

In fructose metabolism, aldehyde dehydrogenase oxidizes D-glyceraldehyde to D-glyceric acid [121, 139] using NAD.

$$\begin{array}{c} HC{=}O \\ | \\ HCOH \\ | \\ H_2COH \end{array} \xrightarrow{NAD \;\to\; NADH_2} \begin{array}{c} COOH \\ | \\ HCOH \\ | \\ H_2COH \end{array}$$

The enzyme is localized in numerous tissues. From the pharmacological point of view, it is involved in the detoxication of aldehydes, such as acetaldehyde and formaldehyde. The activities in liver, kidney and small intestine are of particular interest, especially as far as fructose metabolism is concerned. In the liver, 60 % of the aldehyde dehydrogenase is localized in the cytoplasm, and 40 % in the mitochondria [78]. In all tissues (table VI, VII) which were examined the activity is sufficient to guarantee an optimal oxidation rate for D-glyceraldehyde. The physiological rôle of this step in fructose metabolism in man remains unclear, because glycerate kinase could not be detected.

Aldehyde dehydrogenase was purified from liver of animals [100, 139, 191] and man [21, 130]. FELDMANN and WERNER [53] succeeded in isolating an enzyme in a homogeneous form from horse liver. Two different isoenzymes were found by BLAIR and BODLEY [21] in human liver. In addition to glyceraldehyde, other aliphatic aldehydes also serve as excellent substrates; benzaldehyde is oxidized to a lesser degree. Kinetic data are given in table XIII.

Table XIV. D-Glycerate kinase

Substrate	K_M M
D-Glycerate	3×10^{-4}
ATP	9.7×10^{-3}

In liver, kidney and small intestinal mucosa of rat; localized in mitochondria and cytoplasm. Needs Mg^{++} for optimal activity.

In liver mitochondria, an aldehyde oxidase is present which catalyzes the oxygen-dependent oxidation of various aldehydes to the corresponding acids: however, no detailed information is available whether this enzyme will participate in D-glyceraldehyde oxidation or not.

D-Glycerate Kinase

(EC 2. 7. 1. ATP: D-glycerate-2-phosphotransferase.)

D-Glycerate kinase phosphorylates D-glycerate, ATP-dependent, to D-glycerate-2-phosphate.

$$\begin{array}{c} COOH \\ | \\ HCOH \\ | \\ H_2COH \end{array} \xrightarrow{ATP \quad ADP} \begin{array}{c} COOH \\ | \\ HCO\textcircled{P} \\ | \\ H_2COH \end{array}$$

This enzyme is found in the liver, kidney and small intestinal mucosa [74, 78, 79] of the rat. In rat liver, it is distributed to equal parts between the cytoplasm and mitochondria [78]; only traces can be detected in human and guinea-pig livers [81], and in human kidney and in small intestinal mucosa [204] (table VI). For this reason, the participation of the enzyme in human fructose metabolism seems to be impossible.

Purified enzyme preparations isolated from liver by ICHIHARA and GREENBERG [112], HOLZER and HOLLDORF [106] and LAMPRECHT *et al.* [140]. The identity of the reaction product with D-glycerate-2-phosphate was demonstrated by LAMPRECHT *et al.* [141].

The soluble and particulate enzymes appear to be similar. For optimal activity, in addition to D-glycerate and ATP, magnesium ions are needed.

Kinetic data are given in table XIV.

D-Glycerate Dehydrogenase

(EC 1. 1. 1. 29. D-glycerate: NAD oxidoreductase.)

In liver, an enzyme could be detected which oxidizes D-glyceric acid to hydroxypyruvate using NAD or NADP as a coenzyme.

$$\begin{matrix} COOH \\ | \\ HCOH \\ | \\ H_2COH \end{matrix} \xrightarrow{NAD(P) \rightarrow NAD(P)H_2} \begin{matrix} COOH \\ | \\ C{=}O \\ | \\ H_2COH \end{matrix}$$

The enzyme was purified by WILLIS and SALLACH [252] and HEINZ *et al.* [75]. It introduces D-glyceric acid into the pathway of serine biosynthesis. The localization at the crossover point to serine makes possible the metabolic regulation of this enzyme [197].

Alcohol Dehydrogenase (NAD-Dependent)

(Glycerol dehydrogenase. EC 1. 1. 1. 1. Alcohol: NAD oxidoreductase.)

In addition to the NAD-dependent oxidation of alcohol to acetaldehyde, the NADH-dependent reduction of D-glyceraldehyde can be catalyzed by liver alcohol dehydrogenase [108, 256].

$$\begin{matrix} HC{=}O \\ | \\ HCOH \\ | \\ H_2COH \end{matrix} \xrightarrow{NADH_2 \rightarrow NAD} \begin{matrix} H_2COH \\ | \\ HCOH \\ | \\ H_2COH \end{matrix}$$

In liver, kidney and small intestine (table VI, VII) high activities could be detected with D-glyceraldehyde as substrate. As far as carbohydrate metabolism is concerned, it appears to be obvious from isotope studies with fructose-6-^{14}C and fructose-1-^{14}C that glycerol is an intermediate of fructose metabolism. When glycogen was isolated after the administration of fructose-6-^{14}C or fructose-1-^{14}C, the radioactivity was only present in carbon atoms 1 and 6 of the glucose moiety [91, 195]. The reduction of D-glyceraldehyde to glycerol, together with the sterospecific attack of glycerokinase, would shift the labelling to the positions 3 and 4 of the glucose molecule of glycogen, but this was not observed.

The K_M for D-glyceraldehyde is extremely high [108a, b] and, for this

Table XV. Alcohol dehydrogenase (NAD)

Substrate	K_M M
D,L-Glyceraldehyde	3×10^{-2}
Ethanol	5.4×10^{-4}
In liver, kidney and small intestinal mucosa.	

reason, the reduction seems to be doubtful, too. Alcohol dehydrogenase can be isolated, in crystalline form, from the liver of various animals and man [e.g., 175]. Kinetic data are shown in table XV.

Alcohol Dehydrogenase (NADP-Dependent)

(Glycerol dehydrogenase, aldehyde reductase. EC 1. 1. 1. 2. Alcohol: NADP oxidoreductase.)

Glycerol dehydrogenase is an NADP-dependent enzyme, which seems to be highly specific for the reduction of D-glyceraldehyde to glycerol.

$$\begin{array}{c} HC{=}O \\ | \\ HCOH \\ | \\ H_2COH \end{array} \xrightarrow{NADPH_2 \;\to\; NADP} \begin{array}{c} H_2COH \\ | \\ HCOH \\ | \\ H_2COH \end{array}$$

Glycerol dehydrogenase is found in the liver [92] kidney, small intestinal mucosa (table VIII, XI), skeletal muscle, adipose tissue, lung, brain and heart [92, 230]. Its physiological function is not clear. From a kinetic point of view, D-glyceraldehyde is a good substrate for the enzyme, because the K_M is low, and the enzyme levels are sufficient for a high reduction rate. It is thought that the glycerol dehydrogenase reaction can be an initiating step for glycerol metabolism in skeletal muscle and heart, but in view of the thermodynamic equilibrium constant, together with the high K_M for glycerol, the reaction strongly favours the reduction step.

The enzyme was first purified by MOORE [171] from liver. In the meantime, enzyme preparations were obtained from skeletal muscle by TOEWS [230] and KORMANN *et al.* [135]. Kinetic data are shown in table XVI.

Table XVI. Alcohol dehydrogenase (NADP)

Substrate	K_M M
Glycerol	0.63
D-Glyceraldehyde	1.5×10^{-4}
NADPH	1.6×10^{-5}

In skeletal muscle, liver, kidney, small intestinal mucosa, adipose tissue, lung, brain and heart.

Glycerol Kinase

(L-triokinase. EC 2. 7. 1. 30. ATP: glycerol phosphotransferase. Glycerol kinase catalyzes the ATP-dependent phosphorylation of glycerol to L-glycerol 3-phosphate.)

$$\begin{matrix} H_2COH \\ | \\ HCOH \\ | \\ H_2COH \end{matrix} \xrightarrow{ATP \quad ADP} \begin{matrix} H_2COH \\ | \\ HCOH \\ | \\ H_2CO(P) \end{matrix}$$

The enzyme levels in liver, kidney and in small intestinal mucosa are shown in tables VI and VII. In addition to its possible function in fructose metabolism, glycerol kinase rephosphorylates glycerol, which is formed in lipolysis. Recently glycerol kinase was also detected in adipose tissue

Table XVII. Glycerol kinase

Substrate	K_M M
Glycerol	4×10^{-4}
Dihydroxyacetone	6×10^{-5}
ATP	2.6×10^{-5}

In liver, kidney, small intestinal mucosa and adipose tissue.

[196], where its activity seems to be sufficient for the rephosphorylation of glycerol [88] derived from fat.

Glycerol kinase was purified by BUBLITZ and KENNEDY [26] from rat liver. WIELAND and SUYTER [249] obtained a crystalline preparation from pigeon liver. Glycerol, dihydroxyacetone and L-glyceraldehyde are phosphorylated by this enzyme. Kinetic data are shown in table XVII. ADP and L-glycerol-3-phosphate are inhibitors of glycerol kinase [69, 70]; AMP acts as a negative modifier.

Further Metabolism of Triosephosphates Derived from Fructose

From the reactions mentioned above, fructose enters the glycolytic pathway at the glyceraldehyde-3-phosphate, D-glycerate-2-phosphate or L-glycerol-3-phosphate level. Due to the high metabolic capacity of liver, kidney and small intestinal mucosa for fructose and the impossibility to store fructose in animal tissues, fructose loading causes the elevation of the substrate levels of intermediates of the glucose, fructose and gluconeogenic pathways [30, 76, 258].

Metabolic Regulation of Fructose Metabolism at the Carbohydrate Level

When fructose is given to rats intravenously, intraperitoneally or orally, the fructose-1-phosphate concentrations in liver and kidney [30, 76, 258] are elevated to levels as high as 10 μM/g wet weight. This phosphorylation is coupled with a diminution of the ATP concentration [23, 27, 65, 158, 192]. By adenylate kinase, the ATP system is coupled with ADP and AMP in the cell. Consequently, the AMP concentration is raised after fructose loading. An increased AMP level causes higher degradation rates of AMP, the sum of adenine nucleotides is reduced and uric acid and allantoin concentrations [23, 192] are augmented. These changes interfere with the metabolic enzyme reaction. AMP and its degradation products inhibit the aldolase [258]. Together with the equilibrium which is at the fructose-1-phosphate side, this metabolic block will result in the formation of high amounts of fructose-1-phosphate. In addition to reduced fructose-1-phosphate metabolism, the metabolism of fructose-1,6-diphosphate will also be reduced.

The high fructose-1-phosphate content inhibits the conversion of fructose-6-phosphate to glucose-6-phosphate by the glucose phosphate isomerase [261]. These two inhibitions, the condensation of glyceraldehyde-3-phosphate and dihydroxyacetone phosphate by the aldolase reaction, and the isomerization of fructose-6-phosphate to glucose-6-phosphate, are responsible for a reduced gluconeogenic rate at high fructose levels present in the first minutes after fructose loading. Especially in liver and kidney, the activity of pyruvate kinase L [50], an allosteric enzyme, will be increased by fructose-1-phosphate and this will stimulate the metabolic rate for pyruvate and lactate production. The diminished ATP concentration changes the energy state of the cell. ATP-dependent enzyme systems are switched on or off. Pyruvate dehydrogenase, for instance, is pushed from the inactive b-form (phospho-form) into the active a-form (dephospho-form) [216]. The formation of acetyl-CoA, together with ketogenesis and fatty acid synthesis, is enhanced [84, 216]. The influence on the glycogen enzyme system remains unclear. Like pyruvate dehydrogenase, glycogen synthetase can be shifted to the active dephospho-form [107] and therefore stimulates glycogen synthesis and diminish the uridine phosphate glucose concentration [29]. The low ATP concentration probably has the same effect on uridine phosphate glucose.

The influence of an ATP decrease on the phosphorylase system is the inhibition of the glycogenolysis, because the phosphorylated a-form is the active enzyme, and the dephospho b-form the inactive one. This effect is fortified by a low phosphate concentration [158, 192, 257], the cosubstrate of phosphorylase, after fructose application.

In the hereditary fructose intolerance [e.g., 59], which is due to a muscle-like aldolase in kidney and liver [59, 179], the blood sugar is reduced to very low levels after fructose loading. It is likely that this effect due to an inhibited glycogenolysis rather than to a reduced glucogenogenesis induced by fructose-1-phosphate.

Xylitol Metabolism

The following reviews dealing with this particular subject have been published: 'Polyols' by TOUSTER and SHAW [231] and 'Xylit, Stoffwechsel und klinische Verwendung' by LANG [142]. The abstracts of the symposia on pentites and pentitol [109] also give a good survey, especially on the use of these substances in medicine.

The organism is capable of metabolizing large amounts of xylitol. Due to the fact that its metabolism is not dependent on insulin, it is suitable for the substitution of sucrose and starch in the diet of diabetics. In addition, xylitol is an intermediate of the glucuronate cycle [110]; the blood of healthy humans contains 0.03–0.06 mg%. Following injection, xylitol is eliminated from blood with the same velocity as glucose. When taken *per os,* 40 g is resorbed slowly but completely by a healthy person.

The first step in xylitol metabolism is the dehydrogenation to L- or D-xylulose by an enzyme system which has its highest activity in liver. This oxidation step shifts the redox state, mainly of the NAD-NADH system, to the reduced form and elevates the L-glycerol-3-phosphate content to a high level [83, 116]; this change influences fat metabolism. The well-known antiketogenic effect of xylitol is also due to this shift of the NAD-NADH system.

D-Xylulose enters the pentosephosphate cycle via phosphorylation by xylulokinase. The metabolism of L-xylulose via carboxylation and the reversibility of the glucuronate cycle seems to be impossible. In the case of L-xylulose formation during a first rapid dehydrogenation process, D-xylulose could be produced from this substance with xylitol as an intermediate [215] (fig. 1).

Dehydrogenation of Xylitol

There are two possibilities for the dehydrogenation of xylitol by two groups of enzymes.

1. L-Xylulose reductase or NADP-dependent xylitol dehydrogenase (xylitol: NADP oxidoreductase; L-xylulose forming EC. 1. 1. 10).

2. D-Xylulose reductase or NAD-dependent xylitol dehydrogenase (xylitol: NAD oxidoreductase; D-xylulose forming EC 1. 1. 10) and L-iditol dehydrogenase or sorbitol dehydrogenase (L-iditol: NAD oxidoreductase, EC 1. 1. 1. 14).

L-Xylulose Reductase

In 1957 HOLLMANN and TOUSTER [102] reported the solubilization of a NADP-dependent dehydrogenase from guinea-pig liver mitochondria which catalyzes the reduction of L-xylulose to xylitol.

Table XVIII. Activities of enzymes involved in xylitol and sorbitol metabolism. The enzyme activities are expressed as international milliunits (mU)

	L-Xylulose reductase, guinea pig [105]	L-Iditol dehydrogenase, guinea pig [104]	D-Xylulose reductase rat [9]			Human [62]	D-Xylulokinase, guinea pig [105]	Transaldolase, rabbit [25]		Transketolase, rabbit [25]	
	mU/g wet tissue	mU/mg protein	mU/g wet tissue	mU/mg protein	mU/whole tissue	mU/g wet tissue	mU/mg protein	mU/g wet tissue	mU/mg protein	mU/g wet tissue	mU/mg protein
Liver	806 (male) 703 (female)	64	2.77	0.025	27	5.73	53	705	6	355	2
Kidney	563	75	0.86	0.007	2.1	1.24	20	645	11	430	7
Lung	177	–	0.01	0.0001	0.02			465	8		
Spleen	152	6				0.38	13				
Adrenal gland	199							255	12	1,160	13
Testis	39	3	0.8	0.002	0.24	0.17	13	595	11	435	48
Ovary	49	–	–								
Brain	41	4	0.003	0.0001	0.01		30	300	10	230	8
Cardiac muscle	12	6		0.0004	0.024	0.09 (left) 0.21 (right)	10				
Skeletal muscle	–	2				0.103	17				
Fat	–	2	0.007	0.0006	0.07		86				
Small intestinal mucosa	–	–	0.007	0.0003	0.076						

Table XIX. L-Xylulose reductase

Substrate	K_M M
L-Xylulose	2.9×10^{-5}
Xylitol	2.5×10^{-2}

In all tissues except muscle. Highest activity in liver and kidney; in mitochondria and cytoplasm (kinetic data are given for the liver enzyme).

$$\begin{array}{c} H_2COH \\ | \\ HCOH \\ | \\ HOCH \\ | \\ HCOH \\ | \\ H_2COH \end{array} \xrightleftharpoons[]{NADP \quad NADPH_2} \begin{array}{c} H_2COH \\ | \\ HCOH \\ | \\ HOCH \\ | \\ C{=}O \\ | \\ H_2COH \end{array}$$

The enzyme was further purified by HOLLMANN [103] and was found to be highly specific for xylitol. The localization in the mitochondrium matrix either confined to the inner membrane or bound to the inner membrane, was described by ARSENIS *et al.* [6]. ARSENIS and TOUSTER [7] were able to separate a cytoplasmic enzyme from the bound particle. Both enzymes have similar properties. The distribution of the NADP-dependent enzyme in different tissues of guinea-pigs is shown in table XVIII. The highest activities were measured in liver and kidney. Kinetic data are shown in table XIX.

D-Xylulose Reductase and L-Iditol Dehydrogenase

There is some confusion in the nomenclature of these enzymes. The name D-xylulose reductase was given to the NAD-specific dehydrogenase weakly bound to the outer membrane or present between the inner and outer membranes of mitochondria [6] and the name L-iditol dehydrogenase, which is identical with sorbitol-, polyoldehydrogenase and ketoreductase, was given to the NAD-specific dehydrogenase localized in the cytoplasm of the cells of different tissues. Apart from localizing these enzymes within the cell, no further efforts have been made to separate and

Table XX. D-Xylulose reductase

Substrate	K_M M
L-Iditol	4.18×10^{-3}
Sorbitol	3.38×10^{-3}
Xylitol	5.96×10^{-3}
Ribitol	1.78×10^{-2}

In all tissues. Highest activity in liver and kidney; in mitochondria and cytoplasm (also known as sorbitol, polyol and L-iditol dehydrogenase).

compare them. The two enzymes seem to have similar properties and possess a broad specificity against a number of a cyclic polyols like xylitol, ribitol, sorbitol and iditol which are oxidized to corresponding ketosugars; D-xylulose, D-ribulose, D-fructose and L-sorbose.

$$\begin{array}{c} H_2COH \\ | \\ HCOH \\ | \\ HOCH \\ | \\ HCOH \\ | \\ H_2COH \end{array} \quad \underset{}{\overset{NAD \quad\quad NADH_2}{\longleftrightarrow}} \quad \begin{array}{c} H_2COH \\ | \\ C{=}O \\ | \\ HOCH \\ | \\ HCOH \\ | \\ H_2COH \end{array}$$

The particulate enzyme was purified from guinea-pig liver by HOLLMANN [103], and the non-particulate enzyme was cristallized by SMITH [215]. With sorbitol as a substrate, 65 % of the activity was found in the cytoplasm and 35 % in the mitochondria [78]. The kinetic data are given in table XX.

Table XVIII shows the enzyme levels of L-iditol dehydrogenase or D-xylulose reductase in different tissues of the guinea-pig [105], rat [9], and man [62].

For all tissues which were examined, the highest enzyme activity as related to its wet weight or its protein content was found in the liver, followed by the kidney. If the activities are calculated for the individual tissues as a whole, 90 % of the body capacity for xylitol oxidation resides in the liver.

D-Xylulokinase

(EC 2. 7. 1. 17. ATP: D-xylulose-5-phosphotransferase.)

$$\begin{array}{c} H_2COH \\ | \\ C{=}O \\ | \\ HOCH \\ | \\ HCOH \\ | \\ H_2COH \end{array} \xrightarrow{\text{ATP} \quad \text{ADP}} \begin{array}{c} H_2COH \\ | \\ C{=}O \\ | \\ HOCH \\ | \\ HCOH \\ | \\ H_2CO\text{Ⓟ} \end{array}$$

This is an enzyme which catalyzes the ATP-dependent phosphorylation of D-xylulose to D-xylulose-5-phosphate. It was isolated by HICKMANN and ASHWELL [96]. The kinase is specific for D-xylulose and requires magnesium ions for optimal activity. Kinetic data are given in table XXI.

Table XXI. D-Xylulokinase

K_M D-xylulose: 4×10^{-3} M
In all tissues. Needs Mg^{++} for optimal activity.

Enzyme levels in various tissues of guinea-pigs are shown in table XVIII. According to BÄSSLER *et al.* [9], the phosphorylation of D-xylulose is the rate-limiting step of xylitol metabolism in mammalian liver and kidney. D-Xylulose-5-phosphate is the first intermediate of xylitol metabolism in the pentosephosphate cycle.

Further Metabolism of Xylitol Intermediates in the Pentosephosphate Cycle

D-Xylulose-5-phosphate is converted to ribulose-5-phosphate by the the action of ribulose-5-phosphate-3-epimerase (D-ribulose-phosphate-5-epimerase, EC 5. 1. 3. 1).

This compound is converted to ribose-5-phosphate by ribose-5-phosphate isomerase (D-ribose phosphate ketolisomerase, EC 5. 3. 1. 6). The activities of these enzymes appear to be as high as is necessary in order to guarantee an equilibrium between these three substances. Due to the action of transketolase (EC 2.2.2.1, sedoheptulose-7-phosphate: D-glyceraldehyde-3-phosphate glycolaldehyde transferase), 1 mole xylulose-5-phosphate and 1 mole ribulose-5-phosphate are converted to glyceraldehyde-3-phosphate, a compound of the glycolytic pathway, and sedoheptulose-7-phosphate. The further metabolism of these two products, catalyzed by the transaldolase (EC 2. 2. 1. 2, sedoheptulose-7-phosphate: D-glyceraldehyde-3-phosphate dihydroxyacetone transferase) leads to fructose-6-phosphate and tetrose-4-phosphate. The latter compound can be added to another molecule of xylulose-5-phosphate by further action of transketolase and, finally, fructose-6-phosphate and glyceraldehyde-3-phosphate, both intermediates of the glycolytic pathway, are formed.

The enzymes which oxidize xylitol and phosphorylate D-xylulose are present at different concentrations in different tissues (e.g., high concentration in liver), whilst the enzymes of the pentosephosphate cycle such as transketolase and transaldolase [24] show similar concentrations in different tissues (table XVIII). The well-known localization of xylitol metabolism to the liver could also be well understood from an enzymological point of view. Summarizing the results described above it can be said that from 3 moles of xylitol 3 moles of ATP and 3 moles of NADP, 2 moles of fructose-6-phosphate and 1 mole of glyceraldehyde-3-phosphate are formed. In addition, 3 moles of NADH and 3 moles of ADP are produced.

---

3 xylitol + 3 NAD → 3 D-xylulose + 3 NADH
3 D-xylulose + 3 ATP → 3 D-xylulose-5-phosphate + 3 ADP
D-Xylulose-5-phosphate → D-ribulose-5-phosphate
D-Ribulose-5-phosphate → D-ribose-5-phosphate
D-Xylulose-5-phosphate + D-ribose-5-phosphate →
sedoheptulose-7-phosphate + D-glyceraldehyde-3-phosphate
Sedoheptulose-7-phosphate + D-glyceraldehyde-3-phosphate →
fructose-6-phosphate + tetrose-4-phosphate
D-Xylulose-5-phosphate + tetrose-4-phosphate →
D-fructose-6-phosphate + D-glyceraldehyde-3-phosphate

---

3 Xylitol + 3 NAD + ATP → 2 fructose-6-phosphate + D-glyceraldehyde-3 phosphate + 3 NADH + 3 ADP

Sorbitol Metabolism

Like xylitol, large amounts of sorbitol are rapidly metabolized by the human organism. This metabolism is not dependent on insulin. Like xylitol, sorbitol is an antiketogenic substance, and this effect is due to the first step in sorbitol metabolism, which is the oxidation to fructose (fig. 1). This causes an alteration of the redox state of the NAD-NADH system which is shifted to the reduced site. Consequently, higher amounts of L-glycerol-3-phosphate can be produced via carbohydrate degradation, therefore influencing fat metabolism. In addition, the levels of other intermediates of the glycolytic pathway and of fructose metabolism are augmented [83]. Sorbitol is also a physiological intermediate in the synthesis of fructose from glucose in the seminal vesicles and in the lens of the eye.

Dehydrogenation of Sorbitol

Sorbitol is oxidized by the same enzymes which catalyze the NAD-dependent dehydrogenation of xylitol; the products are NADH and fructose.

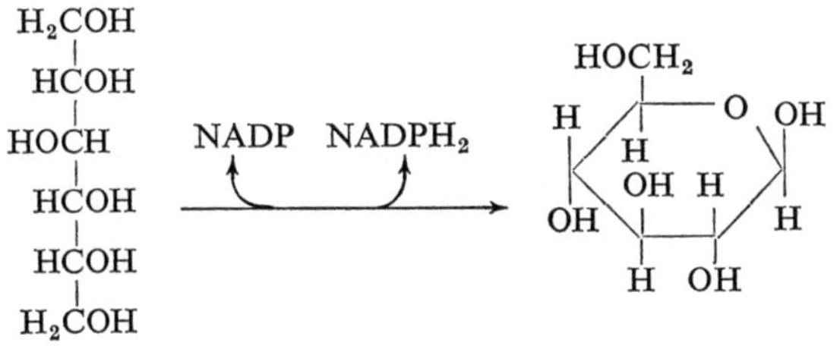

H_2COH
HCOH
HOCH
HCOH
HCOH
H_2COH

NADP → $NADPH_2$

$HOCH_2$
H O OH
H
OH H
OH H
H OH

For kinetic data consult the section on xylitol, and for aspects concerning further sorbitol metabolism the section on fructose metabolism.

References

1 Adleman, R. C.; Ballard, F. J., and Weinhouse, S.: Purification and properties of rat liver fructokinase. J. biol. Chem. *242:* 3360–3365 (1967).

2 Adleman, R. C.: An age dependent modification of enzyme regulation. J. biol. Chem. *245:* 1032–1035 (1970).

3 Anderson, J. W. and Zakim, D.: The influence of alloxan diabetes and fasting

on glycolytic and glucogenic enzyme activities of out intestinal mucosa and liver. Biochem. biophys. Acta *201:* 236–240 (1970).

4 ANKEL, H.; BÜCHER, T. und CZOK, R.: Ueber das Baranowski Enzym. Biochem. Z. *332:* 315–327 (1960).

5 ANSTALL, H. B.; LAPP, C., and TRUJILLO, J. M.: Isoenzymes of aldolase. Science *154:* 657–658 (1966).

6 ARSENIS, C.; MANIATIS, T., and TOUSTER, O.: Intramitochondrial localization of the nicotinamide adenine dinucleotide and nicotine amide adenine dinucleotide phosphate-dependent xylitol dehydrogenases. J. biol. Chem. *243:* 4396–4409 (1968).

7 ARSENIS, C. and TOUSTER, O.: Nicotinamide adenine dinucleotide phosphate-linked xylitol dehydrogenases in guinea pig liver cytosol. J. biol. Chem. *244:* 3895–3899 (1969).

8 BACHELARD, H. S.: Allosteric activation of brain hexokinase by magnesium ions and magnesium ion-adenosine triphosphate complex. Biochem. J. *125:* 249–254 (1971).

9 BÄSSLER, K. H.; STEIN, G. und BELZER, W.: Xylitstoffwechsel und Xylitresorption. Stoffwechseladaption als Ursache für Resorptionsbeschleunigung. Biochem. Z. 346: 171–185 (1966).

10 BALLARD, F. J.: Glucose utilization in mammalian liver. Comp. Biochem. Physiol. *14:* 437–443 (1965).

12 BALLARD, F. J.: Purification and properties of galactokinase from pig liver. Biochem. J. *98:* 347–352 (1966).

13 BALLARD, F. J.: Kinetic studies with liver galactokinase. Biochem. J. *101:* 70–75 (1966).

14 BAILEY, E.; STIRPE, F., and TAYLOR, C. B.: Regulation of rat liver pyruvate kinase. The effect of preincubation, pH cooperations, fructose-1,6-diphosphate and dietary changes on enzyme activity. Biochem. J. *108:* 427–436 (1968).

15 BEISENHERZ, G.; BOLTZE, H. J.; BÜCHER, T.; CZOK, R.; GARBADE, K. H.; MEYER-ARENDT, E. und PFLEIDERER, G.: Diphosphofructosealdolase, Phosphoglyceraldehyd-Dehydrogenase, Milchsäure-Dehydrogenase, Glycerophosphat-Dehydrogenase und Pyruvat-Kinase aus Kaninchenleber in einem Arbeitsgang. Z. Naturforsch. *8:* 555–577 (1953).

16 BERG, J. W. O. VAN DEN and HÜLSMANN, W. C.: Insoluble hexokinase in the brush border region of rat intestinal epithelial cells. FEBS Letters *12:* 173–175 (1971).

17 BERTHILLIER, G. and COLOBERT, L.: Intracellular localization of glucokinase in rat liver. C. R. Acad. Sci. *267:* 794–796 (1968).

18 BERTHILLIER, G.; COLOBERT, L.; RICHARD, M. et GOT, R.: Glucokinases du foie de rat purification et propertiétés des formes particulées. Biochim. biophys. Acta *206:* 1–16 (1970).

19 BERTHILLER, G. et GOT, R.: Glucokinase microsamique du foie de rat. Un enzyme régulateur de la glycolyse? FEBS Letters *8:* 122–124 (1970).

20 BLACK, W. J.: Kinetic studies on the mechanism of cytoplasmic L-α-glycerol-phosphate dehydrogenase of rabbit skeletal muscle. Canad. J. Biochem. *44:* 1301–1317 (1966).

21 BLAIR, A. H. and BODLEY, F. H.: Human liver aldehyde dehydrogenase. Partial purification and properties. Canad. J. Biochem. *47:* 265–272 (1969).

22 BLANCHAER, M. C.: The competitive inhibition of the cytoplasmic L-α-glycerol-phosphate dehydrogenase of skeletal muscle by L-α-glycerol-phosphate. Canad. J. Biochem. *43:* 17–24 (1965).

23 BODE, C.: Fructose induced depletion of liver adenine nucleotides in man. Horm. Metab. Res. *3:* 285–290 (1971).

24 BRAND, K.: Transaldolase; in BERGMEIER Methoden der enzymatischen Analyse, pp. 674–678 (Verlag für Chemie, Weinheim 1970).

25 BROCK, D. J. H.: Purification and properties of sheep liver phosphofructokinase. Biochem. J. *113:* 235–242 (1969).

26 BUBLITZ, C. and KENNEDY, E. P. J.: Synthesis of phosphatides in isolated mitochondria. III. The enzymatic phosphorylation of glycerol. J. biol. Chem. *211:* 951–961 (1954).

27 BUDSZINSKI, W.-D.: Metabolitmuster in Rattenniere nach Fructosebelastung; Diss. Hannover (1971).

28 BÜCHER, T. and PFLEIDERER, G.: Pyruvate kinase from muscle. Meth. Enzymol. *1:* 435–440 (1950).

29 BURCH, H. B.; MAX, P.; CHU, K., and LOWRY, O. H.: Metabolic intermediates in liver of rats given large amounts of fructose or dihydroxyacetone. Biochem. biophys. Res. Commun. *34:* 619–616 (1969).

30 BURCH, H. B.; LOWRY, O.; MEINHARDT, L.; MAX, P., and CHUY, K. J.: Effects of fructose, dihydroxyacetone, glycerol, and glucose on metabolites and related compounds in liver and kidney. J. biol. Chem. *245:* 2092–2105 (1970).

31 CADENAS, E. and SOLS, A.: Ketokinase activity of the small intestine mucosa. Biochim. biophys. Acta *112:* 490–498 (1960).

32 CARMINATTI, H.; DE ASÚA, L. J.; RECONDO, E.; PASSERON, S., and ROZENGURT, E.: Some kinetic properties of liver pyruvate kinase (type L). J. biol. Chem. *243:* 3051–3056 (1968).

33 CHAPMAN, A.; SCANNER, T., and PHIL, A.: Modification of regulatory properties of phosphofructokinase by acetylation. Europ. J. Biochem. *7:* 588–598 (1969).

34 CHRISTEN, P.; RENSING, U.; SCHMID, A. und LEUTHARDT, F.: Multiple Formen der Aldolase in Organextrakten der Ratte. Ueber Aldolasen. 8. Mitteilung. Helv. chim. Acta *49:* 1872–1875 (1966).

35 COHN, R. and SEGAL, S.: Some characteristic and developmental aspects of rat uridine diphosphogalactose-4-epimerase. Biochim. biophys. Acta *171:* 333–341 (1969).

36 COHN, R. and SEGAL, S.: Regulation of mammalian liver uridine diphosphogalactose-4-epimerase by pyrimidine nucleotides. Biochim. biophys. Acta *222.* 533–536 (1970).

37 CORI, C. F. and CORI, G. T.: The rate of glycogen formation in the liver of normal and insulinized rats during absorption of glucose, fructose and galactose. J. biol. Chem. *70:* 577–585 (1926).

38 CORI, C. F. and CORI, G. T.: The influence of insulin on the utilization of glucose, fructose and dihydroxyacetone. J. biol. Chem. *76:* 755–795 (1928).

39 CORI, G. T.; CLOSE, J. O., and CORI, C. F.: Fermentable sugar in heart and skeletal muscle. J. biol. Chem. *103:* 13–24 (1933).

40 CORI, C. F.: Phosphorylation of carbohydrates; in Symposium on Respiratory Enzymes, pp. 175–189 (University of Wisconsin Press, Madison 1942).

41 CORI, G. T.; OCHOA, S.; SLEIN, M. W., and CORI, C. F.: The metabolism of fructose in liver. Isolation of fructose-1-phosphate and inorganic pyrophosphate. Biochim. biophys. Acta *7:* 304–317 (1951).

42 CRAIG, J. W.; DRUCKER, W. R.; MILLER, M.; OWENS, J. E.; WOODWARD, H., jr.; BORFMANN, B., and PRITCHARD, W. H.: Metabolism of fructose by liver of diabetic and non-diabetic subjects. Proc. Soc. exp. Biol. Med. *78:* 698–702 (1951).

43 CRISS, W. E.: A new pyruvate kinase isoenzyme in hepatomas. Biochem. biophys. Res. Commun. *35:* 901–905 (1969).

44 CUATRECASAS, P. and SEGAL, S.: Mammalian galactokinase. Developmental and adaptive characteristics in the rat liver. J. biol. Chem. *240:* 2382–2388 (1965).

45 DE ASÚA, L. J.; ROZENGURT, E., and CARMINATTI, H.: Some kinetic properties of liver pyruvate kinase (type L). III. Effect of monovalent cations on its allosteric behavior. J. biol. Chem. *245:* 3901–3905 (1970).

46 DE ASÚA, L. J.; ROZENGURT, E., and CARMINATTI, H.: Two different forms of pyruvate kinase in rat kidney cortex. FEBS Letters *14:* 22–24 (1971).

47 DIKOW, A. U.; JECKEL, D. und PFLEIDERER, G.: Isolierung und Charakterisierung von Aldolase A, B und C aus menschlichen Organen. Z. Physiol. Chem. *352:* 1151–1156 (1971).

48 DÖNNICKE, M.; HOFER, H. W., and PETTE, D.: Influence of the enzyme concentration on the reaction of rabbit muscle phosphofructokinase with antibodies. FEBS Letters *20:* 187–190 (1972).

49 EASTERBY, J. S.: The polypeptide chain molecular weight of mammalian hexokinase. FEBS Letters *18:* 23–26 (1971).

50 EGGLESTON, L. V. and WOODS, H. F.: Activation of liver pyruvate kinase by fructose-1-phosphate. FEBS Letters *6:* 43–45 (1970).

51 ENGLAND, P. J. and RANDLE, P. J.: Effectors of rat heart hexokinase and the control of rats of glucose phosphorylation in the perfused rat heart. Biochem. J. *105:* 907–920 (1967).

52 EYS, J. VAN; JUDD, J.; FORD, J., and WOMACK, W. B.: On the chemistry of rabbit muscle. Glycerolphosphate dehydrogenase. Biochemistry *3:* 1755–1763 (1964).

53 FELDMANN, R. J. and WERNER, H.: Horse liver aldehyde dehydrogenase. I. Purification and characterization. J. biol. Chem. *247:* 260–266 (1972).

54 FISCHER, W. und WEINLAND, H.: Stoffwechsel der Galaktose und ihrer Derivate (Thieme, Stuttgart 1965).

55 FONDY, T. P.; LEVIN, L.; SOLLOHUTS, S. J., and ROSS, R. C.: Structural studies on nicotinamide adenine dinucleotide L-glycerol-3-phosphate dehydrogenase crystallized from rat skeletal muscle. J. biol. Chem. *243:* 3148–3610 (1968).

56 FONDY, T. F.; ROSS, C. R., and SOLLOHUTS, S. J.: Structural studies on rabbit muscle glycerol-3-phosphate dehydrogenase and a comparison of chemical and physical determinations of its molecular weight. J. biol. Chem. *244:* 1631–1644 (1969).

57 FONDY, T. F.; SOLOMON, J., and ROSS, C. R.: Comparism of glycerol-3-phosphate dehydrogenase from rat liver and muscle. Arch. Biochem. *145:* 604–611 (1971).

58 FRANDSEN, E. K. and GRUNNERT, N.: Kinetic properties of triokinase from rat liver. Europ. J. Biochem. *23:* 588–592 (1971).

59 FROESCH, E. R.; PRADER, A.; WOLF, H. P. und LABHART, A.: Die hereditäre Fructoseintoleranz. Helv. paediat. Acta *14:* 99–112 (1959).

60 GARLAND, P. B. and RANDLE, P. J.: Control of pyruvate dehydrogenase in the perfused rat heart by intracellular concentration of acetyl-CoA. Biochem. J. *91:* 6C–7C (1964).

61 GATT, S. and RACKER, E.: Regulatory mechanisms in carbohydrate metabolism. I. Crabtree effect in reconstructed systems. J. biol. Chem. *234:* 1015–1023 (1959).

62 GERLACH, U.: Zur klinischen Bedeutung der Aktivitätsmessung von Sorbitdehydrogenase im menschlichen Blutserum. Klin. Wschr. *7:* 93–98 (1959).

63 GITZELMAN, G. and ILLIG, R.: Inability of galactose to mobilize insulin in galactokinase-deficient individuals. Diabetes, N.Y. *5:* 143–145 (1969).

64 GÖSCHKE, H. und LEUTHARDT, F.: Kristallisation einer Aldolase aus Kaninchenleber. Helv. chim. Acta *46:* 1791–1792 (1963).

65 GOLDBLATT, P. J.; WITSCHI, H.; FRIEDMANN, M. A.; SCILLI, R. J. VON, and SULL, K. H.: Some structural and functional consequences of hepatic adenosine triphosphate deficiency induced by interperitoneal D-fructose administration. Lab. Invest. *23:* 378–385 (1970).

66 GONZALES, C.; URETA, T.; BABUE, J.; RABAJILLE, E., and NIEMEYER, H.: Characterization of isoenzymes of adenosine triphosphate. D-hexose-6-phosphotransferase from rat liver. Biochemistry *6:* 460–468 (1967).

67 GRIGLIO, S.; GASQUET, P.; LAVAU, M. et LOWY, R.: Activité de la kinase du fructose 6-phosphate et de la kinase pyruvate du foie et du tissu adipeux epididymaire chez le rat rendu obèse par un oégine hyperlipidique. Arch. int. Physiol. *79:* 287–300 (1971).

68 GROSSBARD, L. and SCHIMKE, R. T.: Multiple hexokinases of rat tissues. Purification and comparison of soluble forms. J. biol. Chem. *241:* 3546–3560 (1966).

69 GRUNNET, N. and LUNDQUIST, F.: Kinetics of glycerol kinase from mammalian liver and *Candida mycoderma*. Europ. J. Biochem. *3:* 78–84 (1967).

70 GRUNNET, N.: Inhibition of glycerol kinase by α-glycerol phosphate. Biochem. J. *119:* 927–928 (1970).

71 HANSEN, R. J.; CREIGHTON, S. R., and MCGRATH, C. J.: Hexose-ATP phosphotransferases. Comparative aspects. II. Cross-reactivity of hexokinase from a variety of species with antisera to yeast hexokinase. Comp. Biochem. Biophys. *39:* 607–615 (1971).

72 HAYAKAWA, T.; HIRSAHIMA, M.; IDE, S.; HAMADA, M.; OKABE, K., and KOIKE, M.: Mammalian α-keto acid dehydrogenase complexes. I. Isolation, purification and properties of pyruvate dehydrogenase complex of pig heart muscle. J. biol. Chem. *241:* 4694–4699 (1966).

73 HEINZ, F.: Enzyme des Fructosestoffwechsels. Messung der Enzymaktivitäten in der Dünndarmmukosa der Ratte. Z. Physiol. Chem. *349:* 339–344 (1968).

74 HEINZ, F.: Enzyme des Fructosestoffwechsels. Änderung von Enzymaktivitäten

in Leber und Niere der Ratte bei fructose- und glucosereicher Ernährung. Z. Physiol. Chem. *349:* 399–404 (1968).

75 HEINZ, F.; BARTELSEN, K. und LAMPRECHT, W.: D-Glycerat-Dehydrogenase aus Leber. Beitrag zum Serinstoffwechsel. Z. Physiol. Chem. *329:* 222–240 (1962).

76 HEINZ, F. und JUNGHÄNEL, J.: Metabolitmuster in Rattenleber nach Fructose-applikation. Z. Physiol. Chem. *350:* 859–866 (1969).

77 HEINZ, F. und LAMPRECHT, W.: Anreicherung und Charakterisierung einer Triosekinase aus Leber. III. Zur Biochemie des Fructosestoffwechsels. Z. Physiol. Chem. *324:* 88–100 (1961).

78 HEINZ, F. und LAMPRECHT, W.: Enzyme des Fructosestoffwechsels: Aktivität und Verteilung in der Leber der Ratte. Z. Physiol. Chem. *348:* 855–863 (1967).

79 HEINZ, F. und LAMPRECHT, W.: Enzyme des Fructosestoffwechsels. Messung der Enzymaktivität in der Rattenniere. Z. Physiol. Chem. *348:* 1179–1182 (1967).

80 HEINZ, F. und LAMPRECHT, W.: Enzyme des Fructosestoffwechsels. Enzymaktivitäten in der Dünndarmmukosa verschiedener Laboratoriumstiere. Comp. Biochem. Physiol. *27:* 319–327 (1968).

81 HEINZ, F.; LAMPRECHT, W., and KIRSCH, J.: Enzymes of fructose metabolism in human liver. J. clin. Invest. *47:* 1826–1882 (1968).

82 HEINZ, F. and WEINER, F.: Enzymes of fructose metabolism in the liver of some vertebrates. Comp. Biochem. Physiol. *31:* 283–296 (1969).

83 HEINZ, F. und WITTNEBEN, E.: Metabolitmuster in der Rattenleber nach Sorbitapplikation. Z. Physiol. Chem. *351:* 1215–1220 (1970).

84 HELMREICH, G.; GOLDSCHMIDT, S.; LAMPRECHT, W. und RITZE, F.: Der Einfluss von Kohlehydraten, insbesondere Fructose, auf den Umfang und den zeitlichen Ablauf der Bildung von aktivierter Essigsäure und Brenztraubensäure in der Rattenleber. II. Vergleichende Untersuchungen über die Dissimilation der Fructose und der Glucose. Z. physiol. Chem. *292:* 184–206 (1953).

85 HERMAN, R. H. and ZAKIM, D.: 1. The fructose metabolic pathway. J. amer. clin. Nutr. *21:* 245–249 (1968).

86 HERMAN, R. H. and ZAKIM, D.: Fructose metabolism. 3. Enzyme deficiences. J. amer. clin. Nutr. *21:* 516–519 (1968).

87 HERNANDEZ, A. and CRANE, R. K.: Association of heart hexokinase with subcellular structure. Arch. Biochem. *113:* 223–229 (1966).

88 HERRERA, E. and LAMA, L.: Utilization of glycerol by rat and adipose tissue *in vitro.* Biochem. J. *120:* 433–434 (1970).

89 HERS, H. G.: Le métabolisme du fructose (Arscia, Bruxelles 1948).

90 HERS, H. G.: La fructokinase du foie. Biochim. biophys. Acta *8:* 416–423 (1952).

91 HERS, H. G.: The conversion of fructose-1-^{14}C and sorbitol-1-^{14}C to liver and muscle glycogen in the rat. J. biol. Chem. *214:* 373–381 (1955).

92 HERS, H. G.: L'aldose-réductase. Biochim. biophys. Acta *37:* 120–126 (1960).

93 HERS, H. G.: Triokinase; in BOYER, LARDY and MYRBÄCK The enzymes, vol. 5, pp. 75–83, 2nd ed. (Academic Press, New York 1962).

94 HERS, H. G.: Triokinase; in COLLOWICK and KAPLAN Methods in enzymology, vol. V, pp. 362–364 (Academic Press, New York 1962).

95 HERS, H. G. et KUSAKA, T.: Le métabolisme du fructose-1-phosphate dans le foie. Biochim. biophys. Acta *11:* 427–437 (1953).

96 HICKMAN, J. and ASHWELL, G.: Purification and properties of D-xylulokinase in liver. J. biol. Chem. *232:* 737–748 (1958).

97 HOFER, H. W. and PETTE, D.: The role of nucleic acid in phosphofructokinase as an enzyme modulator. Life Sci. *5:* 199–204 (1966).

98 HOFER, H. W. und PETTE, D.: Verfahren einer standartisierten Extraktion und Reinigung der Phosphofructokinase aus Kaninchen-Muskel. Z. Physiol. Chem. *349:* 995–1012 (1968).

99 HOFER, H. W. und PETTE, D.: Wirkung und Wechselwirkung von Substraten und Effektoren an der Phosphofructokinase des Kaninchen-Skelettmuskels. Z. Physiol. Chem. *349:* 1378–1392 (1968).

100 HOLLDORF, A.; HOLLDORF, C.; SCHNEIDER, G. und HOLZER, U.: Aldehyd-Dehydrogenase aus Leber. Ein Enzym des Fructosestoffwechsels. Z. Naturforsch. *14:* 229–234 (1959).

101 HOLLMANN, S. and TOUSTER, O.: An enzymatic pathway from L-xylulose to D-xylulose. J. amer. chem. Soc. *78:* 3544–3545 (1956).

102 HOLLMANN, S. and TOUSTER, O.: The L-xylulose-xylitol enzyme and other polyol dehydrogenases of guinea pig liver mitochondria. J. biol. Chem. *225:* 87–102 (1957).

103 HOLLMANN, S.: Trennung, Reinigung und Eigenschaften der mitochondrialen Xylit-Dehydrogenasen der Meerschweinchenleber. Z. Physiol. Chem. *317:* 193–216 (1959).

104 HOLLMANN, S. und LAUMANN, G.: Ueber Vorkommen und Aktivität der Xylit: NADP-Oxydoreduktase (L-Xylulose-bildend) in Wirbeltieren. Z. Physiol. Chem. *348:* 1073 (1967).

105 HOLLMANN, S. und REINAUER, H.: Stoffwechsel der Pentosen und Pentitole. Z. Ernährungsw. suppl. 11, pp. 1–7 (1971).

106 HOLZER, H. und HOLLDORF, A.: Anreicherung, Charakterisierung und biologische Bedeutung einer D-Glyceratkinase aus Rattenleber. Biochem. Z. *329:* 283–291 (1957).

107 HOLZER, H.: Regulation of enzymes by enzyme-catalyzed chemical modification. Adv. Enzymol. *32:* 297–326 (1969).

108a HOLZER, H. und SCHNEIDER, S.: Funktion der Leber-Alkoholdehydrase. Angew. Chemie *67:* 276–277 (1955).

108b HOLZER, H. und SCHNEIDER, S.: Zum Mechanismus der Beeinflussung der Alkoholoxydation in der Leber durch Fructose. Klin. Wschr. *33:* 1006–1009 (1955).

109 HORECKER, B. L.; LANG, K., and TAKAYI, Y. (ed.): Metabolism, physiology and clinical use of pentoses and pentitols (Springer, Berlin 1969).

110 HORECKER, B. L.: Stoffwechselwege der Kohlenhydrate, ihre Regulierung und physiologische Bedeutung, Teil III. Med. Ernähr. *10:* 127–129 (1969).

111 HORNICHTER, R. D. and BROWN, J.: Relationship of glucose tolerance to hepatic glucokinase activity. Diabetes, N.Y. *18:* 257–261 (1969).

112 ICHIHARA, A. and GREENBERG, D. M.: Studies on the purification and properties of D-glyceric acid kinase. J. biol. Chem. *225:* 949–958 (1957).

113 ISHIKAWA, E.; OLIVER, R. M., and REED, L. J.: α-keto acid dehydrogenase complexes. V. Macromolecular organization of pyruvate and α-keto-glutarate

dehydrogenase complexes isolated from beef kidney mitochondria. Proc. nat. Acad. Sci., Wash. *56:* 534–541 (1966).

114 IRVING, M. G.; WILLIAMS, J. F.; LAYDE, M. M., and UTTER, M. F.: Studies on the interaction between rabbit liver pyruvate kinase and its allosteric effector fructose-1,6-diphosphate. Proc. austr. Biochem. Soc. *3:* 2 (1970).

115 JACOBSON, K. W. and BLACK, J. A.: Conformational differences in the active sites of muscle and erythrocyte pyruvate kinase. J. biol. Chem. *246:* 5504–5509 (1971).

116 JAKOBS, A.; WILLIAMSON, J. R., and ASAKURA, T.: Xylitol metabolism in perfused rat liver. Interactions with gluconeogenesis and ketogenesis. J. biol. Chem. *246:* 7623–7633 (1971).

117 JAYANNATHAN, V. and SCHWERT, R. S.: Pyruvic oxidase of pigeon breast muscle. I. Purification and properties of the enzyme. J. biol. Chem. *196:* 551–562 (1952).

118 JUNGAS, R. L.: Effect of insulin on fatty acid synthesis from pyruvate, lactate or endogenous sources in adipose tissue. Evidence for the hormonal regulation of pyruvate dehydrogenase. Endocrinology *86:* 1368–1375 (1970).

119 JUNGAS, R. L.: Hormonal regulation of pyruvate dehydrogenase. Metabolism *20:* 43–53 (1971).

120 KALCKAR, H. M.: Uridine diphosphogalactose, metabolism, enzymology and biology. Adv. Enzymol. *20:* 111–134 (1958).

121 KATTERMANN, R.; DOLD, U. und HOLZER, H.: D-Glycerat beim Fructosestoffwechsel in der Leber. Biochem. Z. *334:* 218–226 (1961).

122 KATZEN, H. M. and SCHIMKE, R. T.: Multiple forms of hexokinase in the rat. Tissue distribution, age dependency and properties. Proc. nat. Acad. Sci., Wash. *54:* 1218–1225 (1965).

123 KATZEN, H. M.: The multiple forms of mammalian hexokinase and their significance to the action of insulin. Adv. Enzyme Reg. *5:* 335–356 (1968).

124 KATZEN, H. M.; SODERMAN, D. D., and WILEG, C. E.: Multiple forms of hexokinase. Activities associated with subcellular particulate and soluble fractions of normal and streptozotocin diabetes rat tissues. J. biol. Chem. *245:* 4081–4096 (1970).

125 KEMP, R. G. and KREBS, E. G.: Binding of metabolites by phosphofructokinase. Biochemistry *6:* 423–434 (1967).

126 KEMP, R. G.: Allosteric properties of muscle phosphofructokinase. I. Binding of magnesium adenosine-tri-phosphate to the inhibitory site. Biochemistry *8:* 3162–3167 (1969).

127 KEMP, R. G.: Allosteric properties of muscle phosphofructokinase. II. Kinetics of native and thiol-modified enzyme. Biochemistry *8:* 4490–4496 (1969).

128 KEMP, R. G.: Rabbit liver phosphofructokinase. Comparisons of some properties with those of muscle phosphofructokinase. J. biol. Chem. *246:* 245–252 (1971).

129 KOLER, R. D. and VANBELLINGHEN, P.: The mechanism of precurse or modulation of human pyruvate kinase I by fructose diphosphate. Adv. Enzyme Reg. *6:* 127–142 (1968).

130 KRAEMER, R. J. and DIETRICH, R. A.: Isolation and characterization of human aldehyde dehydrogenase. J. biol. Chem. *243:* 6402–6408 (1968).

131 KRANHOLD, J. F.; LOH, D., and MORRIS, R. C., jr.: Renal fructose-metabolizing enzymes. Significance in hereditary fructose intolerance. Science *165:* 402–403 (1969).

132 KREBS, H. A. and EGGLESTON, L. V.: The role of pyruvate kinase in the regulation of gluconeogenesis. Biochem. J. *94:* 3C–4C (1965).

133 KRONE, W.; DOERR, H. W., and SCHWANDT, P.: Enzymaktivitäten im Fettgewebe bei Hyperlipidämie. Klin. Wschr. *48:* 503–504 (1970).

134 KROPP, E. S. and WILSON, J. E.: Hexokinase binding sites on mitochondrial membranes. Biochem. biophys. Res. Commun. *38:* 74–79 (1970).

135 KORMANN, A. W.; HURST, R. O., and FLYNN, T. G.: Purification and properties of an $NADP^+$ dependent glycerol dehydrogenase from rabbit skeletal muscle. Biochim. biophys. Acta *258:* 40–55 (1972).

136 KÜLZ, E.: Beiträge zur Pathologie und Therapie des Diabetes mellitus (Marburg 1874).

137 KUPIECKI, F. P. and COON, M. J.: Hydroxylamine kinase and pyruvic kinase. J. biol. Chem. *235:* 1944–1947 (1960).

138 KUYPER, C. M. A.: Studies on fructokinase. 1. Substrate specifity. Koninkl. ned. Akad. Wetensch. B *62:* 137–144 (1959).

139 LAMPRECHT, W. und HEINZ, F.: Isolierung von Glycerinaldehyd-Dehydrogenase aus Rattenleber. Zur Biochemie des Fructosestoffwechsels. Z. Naturforsch. *13:* 464–465 (1958).

140 LAMPRECHT, W.; DIAMANTSTEIN, T.; HEINZ, F. und BALDE, P.: Phosphorylierung von D-Glycerinsäure zu 2-Phospho-D-Glycerinsäure mit Glyceratkinase in Leber. I. Zur Biochemie des Fructosestoffwechsels. Z. Physiol. Chem. *316:* 97–112 (1959).

141 LAMPRECHT, W.; HEINZ, F. und DIAMANTSTEIN, T.: Phosphorylierung von D-Glycerinsäure zu 2-Phospho-D-Glycerinsäure mit Glyceratkinase in Leber. II. Identifizierung des Reaktionsproduktes durch Papierchromatographie. Z. Physiol. Chem. *328:* 204–206 (1962).

142 LANG, K.: Xylit, Stoffwechsel und klinische Verwendung, Klin. Wschr. *49:* 233–245 (1971).

143 MCLEAN, P.; BROWN, J.; WALTERS, E., and GREENSLANDE, K.: Effect of alloxan-diabetes on multiple forms of hexokinase in adipose tissue and lung. Biochem. J. *105:* 1301–1305 (1967).

144 LEE, G.-P. and CRANE, J. E.: L-Glycerol 3-phosphate dehydrogenase. I. Effects of the substrates on the catalytic properties of the hepatic nicotinamide adenine dinucleotide-linked enzyme from the rabbit. J. biol. Chem. *246:* 7616–7622 (1971).

145 LEHMANN, F. G. und PFLEIDERER, G.: Kristallisierung und biochemische Eigenschaften von Glycerin-3-Phosphat-Dehydrogenase aus menschlicher Leber. Z. Physiol. Chem. *349:* 1777–1785 (1968).

146a LEUTHARDT, F. und TESTA, E.: Ueber die Phosphorylierung der Ketosen. Helv. physiol. pharmacol. Acta *8:* 67 (1950).

146b LEUTHARDT, F. und TESTA, E.: Die Phosphorylierung der Fructose in der Leber. Helv. chim. Acta *34:* 931–938 (1951).

147 LEUTHARDT, F.; TESTA, E. und WOLF, H. P.: Der enzymatische Abbau des Fructose-1-Phosphats in der Leber. Helv. chim. Acta *36:* 227–251 (1953).

148 LEUTHARDT, F. und STUHLFAUTH, K.: Biochemie, physiologische und klinische Probleme des Fructosestoffwechsel; in BAUER Medizinische Grundlagenforschung, vol. 3, pp. 418–488 (Thieme, Stuttgart 1960).

149 LEUTHARDT, F.: Aldolasen; in HOPPE-SEYLER und THIERFELDER Handbuch der physiologisch- und pathologisch-chemischen Analyse, pp. 607–623, 10th ed. (Springer, Berlin 1966).

150 LING, K. H.; MARCUS, F., and LARDY, H. A.: Purification and some properties of rabbit skeletal muscle phosphofructokinase. J. biol. Chem. *240:* 1893–1899 (1965).

151 LING, K. H.; PAETKAN, V.; MARCUS, F., and LARDY, H. A.: Phosphofructokinase. I. Skeletal muscle. Meth. Enzymol. *9:* 425–429 (1966).

152 LINN, T. C.; PETTIT, F. H.; HUCHO, F., and REED, L. J.: α-Keto acid dehydrogenase complexes. XI. Comparative studies of regulatory properties of pyruvate dehydrogenase complexes from kidney, heart and liver mitochondria. Proc. nat. Acad. Sci., Wash. *64:* 227–234 (1969).

153 LINN, T. C.; PETTIT, F. H., and REED, L. J.: α-Keto acid dehydrogenase complexes. X. Regulation of the activity of the pyruvate dehydrogenase complex from beef kidney mitochondria by phosphorylation and dephosphorylation. Proc. nat. Acad. Sci., Wash. *64:* 234–241 (1969).

154 LLORENTE, P.; MARCO, R., and SOLS, A.: Regulation of liver pyruvate kinase and the phosphoenol pyruvate crossroads. Europ. J. Biochem. *13:* 45–54 (1970).

155 LORENSON, M. Y. and MANSOUR, T. E.: Studies on heart phosphofructokinase. Binding properties of native enzyme and of enzyme desensitized to allosteric control. J. biol. Chem. *244:* 6420–6431 (1969).

156 LOWRY, O. H. and PASSONNEAU, J. V.: Kinetic evidence for multiple binding sites on phosphofructokinase. J. biol. Chem. *241:* 2268–2279 (1966).

157 LYNEN, F.; HARTMANN, G.; NETTER, K. F., and SCHNEGRAF, A.: Ciba Foundation Symp. on Regulation of Cell Metabolism, pp. 256–273 (Little, Brown, Boston 1959).

158 MÄENPÄÄ, P. H.; RAIVIO, K. O., and KEKOMÄKI, M. P.: Liver adenine nucleotides. Fructose induced depletion and its effect on protein synthesis. Science *161:* 1253–1254 (1968).

159 MAGER, S. E.; MAYFIELD, A. C., and HAAS, J. A.: Heart muscle hexokinase. Subcellular distribution and inhibition by glycose-6-phosphate. Molec. Pharmacol. *2:* 393–405 (1966).

160 MAIER, K. P.: Electrophoretic differentiation of multiple forms of phosphofructokinase in the tissue fluid of rat skeletal muscle. Biochim. biophys. Acta *250:* 75–85 (1971).

161 MANSOUR, T. E.: Phosphofructokinase. II. Heart muscle. Meth. Enzymol. *9:* 430–436 (1966).

162 MANSOUR, T. E.: Studies on heart phosphofructokinase. Purification, inhibition and inactivation. J. biol. Chem. *238:* 2285–2292 (1963).

163 MANSOUR, T. E.; WAKID, N. B., and SPROUSE, H. M.: Purification, crystallization and properties of activated sheep heart phosphofructokinase. Biochem. biophys. Res. Commun. *19:* 721–727 (1965).

164 MANSOUR, T. E.; WAKID, N., and SPROUSE, H. M.: Studies on heart phospho-

fructokinase. Purification, crystallization and properties of sheep heart phosphofructokinase. J. biol. Chem. *241:* 1512–1521 (1966).

165 MANSOUR, T. E. and AHLFORS, C. E.: Studies on heart phosphofructokinase. Some kinetic and physical properties of the crystalline enzyme. J. biol. Chem. *243:* 2523–2533 (1968).

166 MARCO, L.; CARBONELL, J., and LLORENTE, P.: Allosteric properties of adipose tissue pyruvate kinase. Biochem. biophys. Res. Commun. *43:* 126–132 (1971).

167 MARTIN, D. B. and SILBERT, C. K.: Control of pyruvate dehydrogenase in rat liver mitochondria by citrate. 6th Diabetes Drug Congr. (1967).

168 MAXWELL, E. S.: The enzymic interconversion of uridine diphosphogalactose and uridine diphosphoglucose. J. biol. Chem. *229:* 139–151 (1957).

169 MAYER, R. J.; SHAKESPEARE, P., and HÜBSCHER, G.: Glucose metabolism in the mucosa of small intestine. Changes of hexokinase activity during perfusion of the proximal half of the small intestine. Biochem. J. *116:* 43–48 (1970).

170 MAYES, J. S. and HANSEN, R. G.: Galactose-1-phosphate uridyl transferase. Meth. Enzymol. *9:* 708–713 (1966).

171 MOORE, B. W.: ATPN$^+$ specific glycerol dehydrogenase from liver. J. amer. chem. Soc. *81:* 5837–5838 (1959).

172 MOORE, C. L.: Purification and properties of two hexokinases from beef brain. Arch. Biochem. *128:* 734–744 (1968).

173 MOORE, R. O.: Effect of age of rats on the response of adipose tissue to insulin and the multiple forms of hexokinase. J. Geront. *23:* 45–49 (1968).

174 MORRISON, G. P.: Hexokinase and glucokinase activities in bile duct epithelial cells and hepatic cells from normal rat and human livers. Arch. Biochem. *122:* 569–573 (1968).

175 MOURAD, N. and WORONICK, C. L.: Crystallization of human liver alcohol dehydrogenase. Arch. Biochem. *121:* 431–439 (1967).

176 MUNZ, J. A.: Purification and some properties of brain phosphofructokinase. Arch. Biochem. *42:* 435–455 (1953).

177 NEWSHOLME, E. A. and SUYDEN, P. H.: The apparent inhibition of phosphofructokinase by reduced nicotinamide-adenine dinucleotide. A problem of coupled-enzyme assays. Biochem. *J. 119:* 787–789 (1970).

178 NIEMEYER, H. A.; CLARK-TURRI, L.; PÉREZ, N., and RABAJELLE, E.: Studies on factors affecting the induction of ATP D-hexose-6-phosphotransferase in rat liver. Arch. Biochem. *109:* 634–645 (1965).

179 NORDMAN, Y. and SCHAPIRA, F.: Liver aldolase in fructose intolerance. Immunological and kinetic studies. Committee abstracts, p. 261. FEBS Meeting (1969).

180 ODEIDE, R.; GUILLOTON, M.; DUPUIS, B.; RAVON, D., and ROSENBERG, A. J.: Study of an allosteric enzyme with 2 substrates. Rat muscle phosphofructokinase. I. Enzyme preparation and crystallization. Bull. Soc. Chim. biol. *50:* 2023–2033 (1968).

181 OTTO, J.: Reinigung der Kaninchenleber-Glycerophosphat-Dehydrogenase und Vergleich mit dem Kaninchenmuskel. Z. Physiol. Chem. *348:* 1240 (1967).

182 PEANASKY, R. J. and LARDY, H. A.: Bovine liver aldolase. I. Isolation, crystallization and some general properties. J. biol. Chem. *233:* 365–373 (1958).

183 PAETKAU, V. and LARDY, H. A.: Phosphofructokinase. Correlation of physical and enzymatic properties. J. biol. Chem. *242:* 2035–2042 (1966).

184 PAETKAU, V. H.; YOUNATHAN, E. S., and LARDY, H. A.: Phosphofructokinase. Studies on the subunit structure. J. molec. Biol. *33:* 721–736 (1968).

185 PARKS, R. E.; BEN-GERSHOM, E., and LARDY, H.: Liver fructokinase. J. biol. Chem. *227:* 231–242 (1957).

186 PARMEGGIANI, A.; LUFT, J. H.; LOVE, D. S., and KREBS, E. G.: Crystallization and properties of rabbit skeletal muscle phosphofructokinase. J. biol. Chem. *241:* 4625–4637 (1966).

187 PENHOET, E.; RAJKUMAR, T., and RUTTER, W. J.: Multiple forms of fructose diphosphate aldolase in mammalian tissues. Proc. nat. Acad. Sci., Wash. *56:* 1275–1282 (1966).

188 PILKIS, S. J.: Identification of human hepatic glucokinase and some properties of the enzyme. Proc. Soc. exp. Biol. Med. *129:* 681–684 (1968).

189 PILKIS, S. J.: Hormonal control of hexokinase activity in animal tissues. Biochim. biophys. Acta *215:* 461–476 (1970).

190 POGSON, E. J.: Adipose-tissue, pyruvate kinase. Properties and interconversion of two active forms. Biochem. J. *110:* 67–77 (1968).

191 RACKER, E.: Aldehyde dehydrogenase. A diphosphopyridine nucleotide linked enzyme. J. biol. Chem. *177:* 883–892 (1943).

192 RAIVIO, K. O.; KEKOMÄKI, M. P., and MÄENPÄÄ, P. H.: Depletion of liver adenine nucleotides induced by D-fructose. Dose-dependence and specificity of the fructose effect. Biochem. Pharmacol. *18:* 2615–2624 (1969).

193 RAJKUMAR, T.; WOODFIN, B. M., and RUTTER, W. J.: Aldolase B from (adult) rabbit liver. Meth. Enzymol. *9:* 491–498 (1966).

194 RAMAIAH, A. and TEJWANI, G. A.: Interconvertible forms of phosphofructokinase of rabbit liver. Biochim. biophys. Res. Commun. *39:* 1149–1156 (1970).

195 RAUSCHENBACH, P. und LAMPRECHT, W.: Einbau von ^{14}C-markierter Glucose und Fructose in Leberglykogen. Zum Fructosestoffwechsel in der Leber. Z. Physiol. Chem. *339:* 277–292 (1964).

196 ROBINSON, J. and NEWSHOLME, E. A.: Glycerol kinase activities in rat heart and adipose tissue. Biochem. J. *104:* 2c–4c (1967).

197 ROSENBLUM, J. Y.; ANTIKOWIAK, D. H.; SALLACH, H. J.; FLANDERS, L. E., and FAHIEN, L. A.: Purification and regulatory properties of beef D-glycerate dehydrogenase. Arch. Biochem. *144:* 375–383 (1967).

198 ROSS, C. R.; CURRY, S.; SCHWARTZ, A. W., and FONDY, T. P.: Multiple molecular forms of cytoplasmic glycerol-3-phosphate dehydrogenase in rat liver. Arch. Biochem. *145:* 591–603 (1971).

199 ROZENGURT, E.; DE ASÚA, L. J., and CARMINATTI, H.: Some kinetic properties of liver pyruvate kinase (type L). II. Effect of pH on its allosteric behaviour. J. biol. Chem. *244:* 3142–3147 (1969).

200 RUDERMAN, N. B. and LAURIS, V.: Effect of acute insulin deprivation on rat liver glucokinase. Diabetes *17:* 611–616 (1968).

201 RUTTER, W. J.: Evolution of aldolases. Fed. Proc. *23:* 1248–1257 (1964).

202 SALAS, J.; SALAS, M.; VIÑUELA, E., and SOLS, A.: Glucokinase of rabbit liver. Purification and properties. J. biol. Chem. *240:* 1014–1018 (1965).

203 Sánchez, J. J.; González, N. S., and Pontis, H.: Fructokinase from rat liver. I. Purification and properties. II. The role of K^+ on the enzyme activity. Biochem. biophys. Acta *227:* 67–78, 79–85 (1971).

204 Schlegel, F.: Enzyme des Fructosestoffwechsels. Aktivitäten in der Dünndarmmukosa und der Niere des Menschen. Diss. Hannover (1970).

205 Schwark, W. S.; Singhal, R. L., and Ling, G. M.: Fat acid inhibition of α-glycerophosphate dehydrogenase activity in rat brain. J. Pharm. Pharmacol. *22:* 458–460 (1970).

206 Schwarz, G. P. and Basford, R. E.: The isolation and purification of solubilized hexokinase from brain. Biochemistry *6:* 1070–1078 (1967).

207 Segal, S.; Rogers, S., and Holtzapple, P.: Liver galactose-1-phosphate uridyl transferase activity in normal and galactosemic subjects. J. clin. Invest. *50:* 500–506 (1970).

208 Segal, S. and Rogers, S.: Nucleotide inhibition of mammalian liver galactose-1-phosphate uridyl transferase. Biochim. biophys. Acta *250:* 351–360 (1971).

209 Segal, S.; Weinstein, A., and Koldovsky, O.: Developmental aspects of uridine diphosphate galactose 4-epimerase in rat intestine. Biochem. J. *125:* 935–941 (1971).

210 Shakespeare, P.; Srivastova, L. M., and Huebscher, G.: Glucose metabolism in the mucosa of the small intestine. Effect of glucose on hexokinase activity. Biochem. J. *111:* 63–67 (1969).

211 Siess, E. A. and Wieland, O. H.: Purification and characterization of pyruvate dehydrogenase phosphatase from pig heart muscle. Europ. J. Biochem. *26:* 96–105 (1972).

212 Silbert, C. K. and Martin, D. B.: Inhibition by citrate of pyruvate dehydrogenase in rat liver mitochondria. Biochem. biophys. Res. Commun. *31:* 818–824 (1968).

213 Simard-Duquesne, N.: Effect of actanoate *in vitro* on rat liver supernatant phosphofructokinase activity. Biochem. biophys. Res. Commun. *33:* 874–876 (1968).

214 Simard-Duquesne, N.: Phosphofructokinase and glycolysis in the liver of rats fed a thrombogenic died. Canad. J. Biochem. *48:* 222–224 (1970).

215 Smith, M. G.: Polyol dehydrogenases. 4. Crystallization of the L-iditol dehydrogenase of sheep liver. Biochem. J. *83:* 135–144 (1962).

216 Sölling, H. D. and Bernhard, G.: Interconversion of inactive to active pyruvate dehydrogenase in rat liver after fructose application *in vivo*. FEBS Letters *13:* 201–203 (1971).

217 Spay-Hagar, M.; Marco, R., and Sols, A.: Distribution of hexokinase and glucokinase between parenchymal and nonparenchymal cells of rat liver. FEBS Letters *3:* 68–71 (1969).

218 Susor, W. A. and Rutter, W. J.: Some distinctive properties of pyruvate kinase purified from rat liver. Biochem. biophys. Res. Commun. *30:* 14–20 (1968).

219 Spydeoold, Ø. and Borrebæk, B.: Association of epididymal tissue hexokinase with subcellular structure. Biochim. biophys. Acta *166:* 291–301 (1968).

220 Stifel, F. B.; Herman, R. H., and Rosensweig, N. S.: Dietary regulation of galactose metabolizing enzymes: adaptive changes in rat jejunum. Biochim. biophys. Acta *170:* 221–227 (1968).

221 STIFEL, F. B.; ROSENSWEIG, N. S.; ZAKIM, D., and HERMAN, R. H.: Dietary regulation of glycolytic enzymes. I. Adaptive changes in rat jejunum. Biochim. biophys. Acta *170:* 221–227 (1968).

222 STIFEL, F. B.; HERMAN, R. H., and ROSENSWEIG, N. S.: Dietary regulation of glycolytic enzymes. III. Adaptive changes in rat jejunal pyruvate kinase, phosphofructokinase, fructose diphosphatase and glycerol-3-phosphate dehydrogenase. Biochim. biophys. Acta *184:* 29–34 (1969).

223 SZEPESI, B. and FREEDLAND, R.: Time course of enzyme adaption. I. Effects of substituting glucose and fructose at constant concentrations of dietary protein. Canad. J. Biochem. *46:* 1459–1470 (1968).

224 TANAKA, T.; HARANO, Y.; SUE, F., and MORIMURA, H.: Crystallization, characterization and metabolic regulation of two types of pyruvate kinase isolated from rat tissues. J. Biochem. *62:* 71–91 (1967).

225 TANAKA, T.; SUE, F., and MORIMURA, H.: Feed-forward activation and feed-back inhibition of pyruvate kinase L of rat liver. Biochem. biophys. Res. Commun. *29:* 444–449 (1967).

226 TAYLOR, C. B. and BAILEY, E.: Activation of liver pyruvate kinase by fructose-1,6-diphosphate. Biochem. J. *102:* 32C–33C (1966).

227 TAYLOR, C. B. and BEW, M.: The distribution of two chromatographically distinguishable forms of phosphofructokinase in the tissue of the rat. Biochem. J. *119:* 797–799 (1970).

228 THOMPSON, M. F. and BACHELARD, H. S.: Solubilization of mitochondrial hexokinase from ox cerebral cortex with Triton X-100. Biochem. J. *111:* 18–19 (1969).

229 THOMPSON, M. F. and BACHELARD, H. S.: Cerebral-cortex hexokinase. Comparison of properties of solubilized mitochondrial and cytoplasmic activities. Biochem. J. *118:* 25–34 (1970).

230 TOEWS, C. J.: The kinetics and reaction mechanism of the nicotinamide adenine dinucleotide phosphate-specific glycerol dehydrogenase of rat skeletal muscle. Biochem. J. *105:* 1067–1073 (1967).

231 TOUSTER, O. and SHAW, D.: Biochemistry of a cyclic polyols. Physiol. Rev. *42:* 181–225 (1962).

232 TRIVEDI, B. and DANFORTH, W. U.: Effect of pH on the kinetics of frog muscle phosphofructokinase. J. biol. Chem. *241:* 4110–4114 (1966).

233 TYSTRUP, N.; WINKLER, K., and LUNDQUIST, F.: The mechanism of fructose effect on the ethanol metabolism of the human liver. J. clin. Invest. *44:* 817–830 (1965).

234 UNDERWOOD, A. H. and NEWSHOLME, E. A.: Properties of phosphofructokinase from rat liver and their relation to the control of glycolysis and gluconeogenesis. Biochem. J. *95:* 868–875 (1965).

235 UNDERWOOD, A. H. and NEWSHOLME, E. A.: Some properties of phosphofructokinase from kidney cortex and their relation to glucose metabolism. Biochem. J. *104:* 296–299 (1967).

236 VALLEJO, C. G.; MARCO, R., and SEBASTIAN, J.: The association of brain hexokinase with mitochondrial membranes and its functional implications. Europ. J. Biochem. *14:* 478–485 (1970).

237 VESTLING, C. S.; MYLORIE, A. K.; IRISH, U., and GRANT, N. H.: Rat liver fructokinase. J. biol. Chem. *185:* 789–801 (1950).

238 VIJAYVARGIYA, R. and SINGHAL, R. L.: L-glycerophosphate dehydrogenase inhibition in rat heart and adipose tissue. Proc. Soc. exp. Biol. Med. *133:* 670–673 (1970).

239 VIÑUELA, E.; SOLS, M., and SOLS, A.: Glucokinase and hexokinase in liver related to glucogen synthesis. J. biol. Chem. *238:* 1175–1177 (1963).

240 WALKER, D. G.: On the presence of two soluble glucose-phosphorylating enzymes in adult liver and the development of one of these after birth. Biochim. biophys. Acta *77:* 209–226 (1963).

241 WALKER, D. G. and RAO, S.: The role of glucokinase in the phosphorylation of glucose by rat liver. Biochem. J. *90:* 360–368 (1964).

242 WEBER, G.; STAMM, N. B., and FISHER, E. A.: Insulin. Inducer of pyruvate kinase. Science *149:* 65–67 (1965).

243 WEBER, G.; SINGHAL, R. L.; STAMM, N. B.; LEA, M. A., and FISHER, E. A.: Synchronous behavior pattern of key glycolytic enzymes. Glucokinase, phosphofructokinase and pyruvate kinase. Adv. Enzyme Reg. *4:* 59–81 (1966).

244 WEBER, G.; LEA, M. A., and STAMM, N. B.: Inhibition of pyruvate kinase and glucokinase by acetyl-CoA and inhibition of glucokinase by phosphoenol pyruvate. Life Sci. *6:* 2441–2452 (1967).

245 WEBER, G.; LEA, M. A.; CONVERY, H. J. U., and STAMM, N. B.: Regulation of gluconeogenesis and glycolysis. Studies of mechanisms controlling enzyme activity. Adv. Enzyme Reg. *5:* 257–298 (1967).

246 WEBER, G.; LEA, M. A., and STAMM, N. B.: Sequential feedback inhibition and regulation of liver carbohydrate metabolism through control of enzyme activity. Adv. Enzyme Reg. *6:* 101–124 (1968).

247 WEBER, G.: Regulation of pyruvate kinase. Adv. Enzyme Reg. *7:* 15–40 (1969).

248 WELLMANN, E.; LOEFFLER, G., and TRAUTSCHOLD, I.: Electrophoretic separation of the ATP/D-hexose-6-phosphotransferase in adipose tissue. Fresenius Z. analyt. Chem. *243:* 568–572 (1968).

249 WIELAND, O. und SUYTER, M.: Glycerokinase. Isolierung und Eigenschaften des Enzyms. Biochem. Z. *329:* 320–331 (1957).

250 WIELAND, O.; SIESS, E.; SCHULZE-WETHMAR, H. G.; FUNCKE, H. G. VON, and WINTON, B.: Active and inactive forms of pyruvate dehydrogenase in rat heart and kidney. Effect of diabetes, fasting and refeeding on pyruvate dehydrogenase interconversion. Arch. Biochem. *143:* 593–601 (1971).

251 WIELAND, O. and SIESS, E.: Interconversion of phospho- and dephospho-forms of pig heart pyruvate dehydrogenase. Proc. nat. Acad. Sci., Wash. *65:* 947–954 (1970).

252 WILLIS, J. E. and SALLACH, H. J.: Evidence for a mammalian D-glyceric dehydrogenase. J. biol. Chem. *237:* 910–915 (1962).

253 WILLMS, B.; BEN-AMI, P., and SÖLING, H. D.: Hepatic enzymes activities of glycolysis and gluconeogenesis in diabetes of man and laboratory animals. Horm. Metab. Res. *2:* 135–141 (1970).

254 WILSON, J. E.: Brain hexokinase. A proposed relation between soluble-particulate distribution and activity *in vivo*. J. biol. Chem. *243:* 3640–3647 (1968).

255 WHITE, H. B., III: The molecular weights of glycerol-3-*p*-dehydrogenase from chicken, rabbit and honey bee. Arch. Biochem. *147:* 123–128 (1971).

256 WOLF, H. P. und LEUTHARDT, F.: Ueber die Glycerindehydrase der Leber. Helv. chim. Acta *36:* 1463–1467 (1953).

257 WOLF, H. P.; QUEISSER, W. und BECK, K.: Der initiale Phosphatabfall im Serum von Gesunden und Leberkranken nach intravenöser Verabreichung von Hexosen und Zuckeralkoholen. Klin. Wschr. *47:* 1084–1094 (1969).

258 WOODS, H. F.; EGGLESTON, L. K., and KREBS, N. A.: The cause of hepatic accumulation of fructose-1-phosphate on fructose loading. Biochem. J. *119:* 501–510 (1970).

259 YOUNG, H. L. and PACE, N.: Some physical and chemical properties of crystalline α-glycerophosphate dehydrogenase. Arch. Biochem. *75:* 125–141 (1958).

260 ZAKIM, D. and HERMAN, R. H.: Fructose metabolism. 2. Regulatory control to the triose level. J. amer. clin. Nutr. *21:* 315–319 (1968).

261 ZALITIS, J. and OLIVER, I. T.: Inhibition of glucose phosphate isomerase by metabolic intermediates of fructose. Biochem. J. *102:* 753–759 (1952).

Author's address: Dr. F. HEINZ, Medizinische Hochschule Hannover, Institut für klinische Biochemie und physiologische Chemie, Karl-Wiechert-Allee 9, *D-3 Hannover-Kleefeld* (FRG)

Progr. biochem. Pharmacol., vol. 8, pp. 57–124 (Karger, Basel 1973)

Lipid Biosynthesis

I. R. KUPKE

Medizinische Hochschule Hannover, Institut für Klinische Biochemie und Physiologische Chemie, Hannover-Kleefeld

Contents

I. Introduction

Lipid biosynthesis from carbohydrates is mainly accomplished by the conversion of its degradation products acetyl-CoA and sn-glycerol-3-phosphate (L-glycerol-3-phosphate) into complex lipids. However, it is not intended to give a complete survey in this article of the multiplicity of reaction mechanisms leading to a variety of different lipids. The main topic will be the formation of 1,2-diacyl-sn-glycerol-3-phosphate (3-sn-phosphatidate) and its activated form 1,2-diacyl-sn-glycerol-3-cytidine-diphosphate (CDP-diacylglycerol), a nucleotide intermediate serving as phosphatidyl donor for further biosynthetic processes. To date, both substances are accepted to be key intermediates on the pathway of phosphoglyceride biosynthesis.

As far as the production of long-chain acyl-CoA compounds from acetyl-CoA is concerned, the reader is referred to comprehensive reviews recently published by other authors. Therefore, only introductory remarks will be given on this subject. Biosynthesis of sterols and sterol derivatives, glycolipids including sphingolipids, plasmalogens and sulfatides will be presented only as a brief account referring to reviews lately published by others. The period from 1970 up to date will be covered by a more or less complete bibliography. The nomenclature proposed by the JUPAC-JUB Commission [130] will be used.

II. Biosynthesis of Fatty Acids

(Conversion of acetyl-CoA into acyl-CoA.) The first step on the synthetic pathway is the fixation of CO_2 by acetyl-CoA which is mediated by the biotin enzyme acetyl-CoA carboxylase (acetyl-CoA ligase [ADP],

EC 6.4.1.2) in the presence of ATP, Mg^{2+} or Mn^{2+} and bicarbonate. Malonyl-CoA is formed by this reaction [183, 184, 311].

$$\text{Enzyme-biotin} + HO-CO_2^- + ATP \xrightleftharpoons{Mg^{2+} \text{ or } Mn^{2+}} \text{enzyme-biotin-}CO_2^- + ADP + P_i. \quad (1)$$

$$\text{Enzyme-biotin-}CO_2^- + \text{acetyl-CoA} \rightleftharpoons \text{enzyme-biotin} + \text{malonyl-CoA}. \quad (2)$$

Net equation:

$$ATP + HO-CO_2^- + \text{acetyl-CoA} \xrightleftharpoons{Mn^{2+}\ Mg^{2+}} \text{malonyl-CoA} + ADP + P_i. \quad (3)$$

This reaction is considered to be the rate-limiting step in fatty acid synthesis [42, 78, 170, 198, 210]. The enzyme is activated by citrate [211] and is usually thought to be a soluble constituent of the cytoplasm. However, membrane association cannot be excluded to date [88].

The second step in fatty acid synthesis is the conversion of malonyl-CoA to palmitate catalyzed by a multi-enzyme complex. In addition, this fatty acid synthetase requires acetyl-CoA and NADPH. The enzyme is located in the soluble part of the cell [312], and it is widely distributed in unicellular organisms, plants, and animals [185]. In *Escherichia coli*, each intermediate of *de novo* fatty acid synthesis from malonyl-CoA is bound to an acyl-carrier protein [306].

Chain elongation systems from mitochondrial origin can synthetize acyl-CoA that is two carbon atoms longer than the starting compound [215] using a starting long-chain acyl-CoA, acetyl-CoA and NADH or NADPH.

Polyenoic fatty acids can be formed by desaturation systems which were found in mammalian microsomes [292]. Shortening of fatty acids is accomplished by β-α-, - and ω-oxidation.

For detailed information about the biosynthetic pathways of fatty acids and the properties of the corresponding enzyme systems so far known, the reader is referred to the reviews by PORTER *et al.*, NIKKILÄ, and SPECTOR published in 1971 in this series. Furthermore, it is suggested that the following review articles are consulted: WAKIL [313], BRESSLER [29], STUMPF [298, 299], SAMUELSSON [251], VAGELOS [307], ROUS [250], WHEREAT [322], LOWENSTEIN [180], MAJERUS and VAGELOS [192], and the articles currently published in Annual Reviews of Biochemistry. Recent publications which appeared in 1970–1972 give further information about the enzymatic properties and the reaction mode of the fatty acid synthetase [1, 21, 30, 32, 66, 99, 133, 156, 176, 177, 193, 229], and about control mechanisms of this enzyme complex [98, 178]. New observations concerning the elongation system [169, 264], acylcarrier protein [134, 262, 239], desaturation [15, 93] and biohydrogenation [144] of fatty acids, synthesis of branched chain fatty acids [90], cell compartmental studies [2, 22, 49, 65, 122, 168, 261, 315], and rate-limiting steps in fatty acid formation [204] were reported in the last 2 years. Current aspects on lipid metabolism including fatty acid synthesis were discussed during the International Symposium on Lipids in 1972 [129].

III. Biosynthesis of Sterols and Sterol Derivatives

(Conversion of acetyl-CoA into isoprenoid lipids.) An alternate pathway of acetyl-CoA to be incorporated into more complex substances is the formation of sterols including cholesterol and its esters, steroid hormones, vitamin D, carotinoids including vitamin A, and various plant steroids like digitonine and strophantine. The end-products of cholesterol metabolism are the bile acids.

A key substance for isoprenoid synthesis is mevalonic acid which is formed according to these equations:

Acetyl-CoA + acetoacetyl-CoA + $H_2O \rightarrow \beta$-hydroxy-β-methyl-glutaryl-CoA (β-HMG-CoA). (4)

β-HMG-CoA + 2NADPH + 2 $H^+ \rightarrow$ mevalonic acid + 2$NADP^+$. (5)

The β-HMG-CoA was recognized by BLOCH [25] to be a key intermediate in cholesterol synthesis. The reduction of this compound yields mevalonic acid which is catalyzed by the enzyme mevalonate: $NADP^+$ oxidoreductase (EC 1.1.1.34) [182, 231, 328].

The following equations summarize the reaction steps leading to cholesterol:

Mevalonate + 3ATP = isopentenyl-PP + 3ADP + CO_2. (6)

Isopentenyl-PP $\rightarrow$ squalene $\rightarrow$ lanosterol $\rightarrow$ cholesterol. (7)

The mechanisms involved in squalene formation were widely elucidated in the laboratories of BLOCH and LYNEN and of POPJÁK [232]. A complex cyclization process converts squalene to lanosterol [303] and, finally, further stepwise enzymatic mechanisms produce cholesterol [214].

Biosynthesis of sterols occurs in microsomes [242, 247]. Recently, SCALLEN *et al.* [257, 258] suggested that the participation of a non-catalytic sterol-carrier protein may be of key importance in the understanding of the enzymatic synthesis of sterols. This heat-stable protein appears to play a general rôle as vehicle for cholesterol and water-insoluble precursors, as was proposed by RITTER and DEMPSEY [248]. These investigators purified this protein from liver (300-fold); it binds specifically squalene and sterol precursors of cholesterol and activates the microsomal enzymatic steps of cholesterol synthesis in which these precursors participate. The non-covalent squalene- or sterol-carrier protein complex may be a component of high-density lipoproteins and may be identical with the apo-sterol carrier protein. Some characteristics of this protein have been reported [248].

An interesting observation was madc by BRADY and GAYLOR [38]. In the 10,000 *g* supernatant fraction obtained from cell-free homogenates from liver, competition between ester formation and demethylation of methyl sterol intermediates has been demonstrated. This esterification does not take place in microsomes. Incubations carried out with cultured aortic cells showed that stimulation of sterol synthesis is accompanied by an increase in protein synthesis and is followed 1 day later by an appearance of DNA synthesis [17]. Studies on mature mammalian erythrocytes revealed that these cells do not carry out the synthesis of sterols and other lipids. However, reticulocytes incorporate mevalonate into squalene and sterols and secrete cholesterol into the plasma [276].

Control of sterol biosynthesis is mainly accomplished by the enzyme mevalonate: $NADP^+$ oxidoreductase converting β-HMG-CoA to mevalonic acid [266, 273]. Recent work on this subject dealt with the rôle of cytochrome P-450 in the biosynthetic process [16], with the effect of different co-factors [18] and ubiquinone [151], and the inhibition of synthesis by carbon monoxide [80].

LOFLAND *et al.* [179] studied the regulation of plasma cholesterol levels in squirrel monkeys and concluded that the mechanism of control is related to the rate of conversion of cholesterol to bile acids, and that it depends upon the interplay between synthesis, absorption and excretion. Similar results were obtained by GRUNDY *et al.* [91]. *In vivo* experiments performed with pigs [249] gave additional quantitative information: regarding total sterol synthesis in the body, the liver contributes 67% and the small intestine only 4%.

Detailed studies on the synthesis of different sterols in aorta were recently published by DALY [54], and were reviewed by PORTMAN [234] and KRITCHEVSKY [152]. For steroid biosynthesis in the human adrenal, see the review by GRIFFITHS and CAMERON [89]. Steroid biosynthesis from mevalonate was reviewed by HANSON [100], regulation of sterol synthesis by SIPERSTEIN [274] and cholesterol turnover in man by NESTEL [209].

A comprehensive report on the different cholesterol esterifying enzymes in different organs was presented by ETO and SUZUKI [70].

IV. The Introduction of sn-Glycerol-3-Phosphate into Complex Lipids

sn-Glycerol-3-phosphate (sn-glycerol-3-P) can be incorporated into complex lipids by *two* principle reactions (fig. 1):

Type I reaction. Acylation of the primary and secondary hydroxyl group with acyl-CoA or acyl-acyl carrier protein serving as acyl-donors thus forming an ester bond. By stepwise acylation monoacyl- and diacyl-sn-glycerol-3-P (3-sn-lysophosphatidate and 3-sn-phosphatidate) are produced which may serve as immediate precursors for the biogenesis of:

1. Monoacylglycerols (monoglycerides).
2. Diacylglycerols (diglycerides).
3. Triacylglycerols (triglycerides).
4. (a) 3-sn-Lysophosphatidylcholine (lysolecithine) and 3-sn-phosphatid ylcholine (lecithin); (b) 3-sn-lysophosphatidylethanolamine and 3-sn-phosphatidylethanolamine, and (c) 3-sn-phosphatidylserine.
5. 1,2-Diacyl-sn-glycerol-3-cytidine-diphosphate (CDP-diacylglycerol) and derivatives such as 3-sn-phosphatidylinositol and 3-sn-phosphatidylserine.

Type II reaction. Phosphatidyl transfer from CDP-diacylglycerol to another sn-glycerol-3-P molecule thus forming a phosphodiester bond. 3-sn-phosphatidyl-sn-glycerol-3-P is produced as an intermediate which may serve as precursor for the biogenesis of:

1. 3-sn-Phosphatidylglycerol.
2. 3,3′-sn-Diphosphatidylglycerol (cardiolipin).

A. Type I Reaction

1. Biosynthesis of 3-sn-Lysophosphatidate and 3-sn-Phosphatidate

a) Acylation Reactions

$$\text{sn-Glycerol-3-P} + 2\ \text{acyl-CoA} \rightleftharpoons \text{3-sn-phosphatidate} + 2\ \text{CoA} \qquad (8)$$

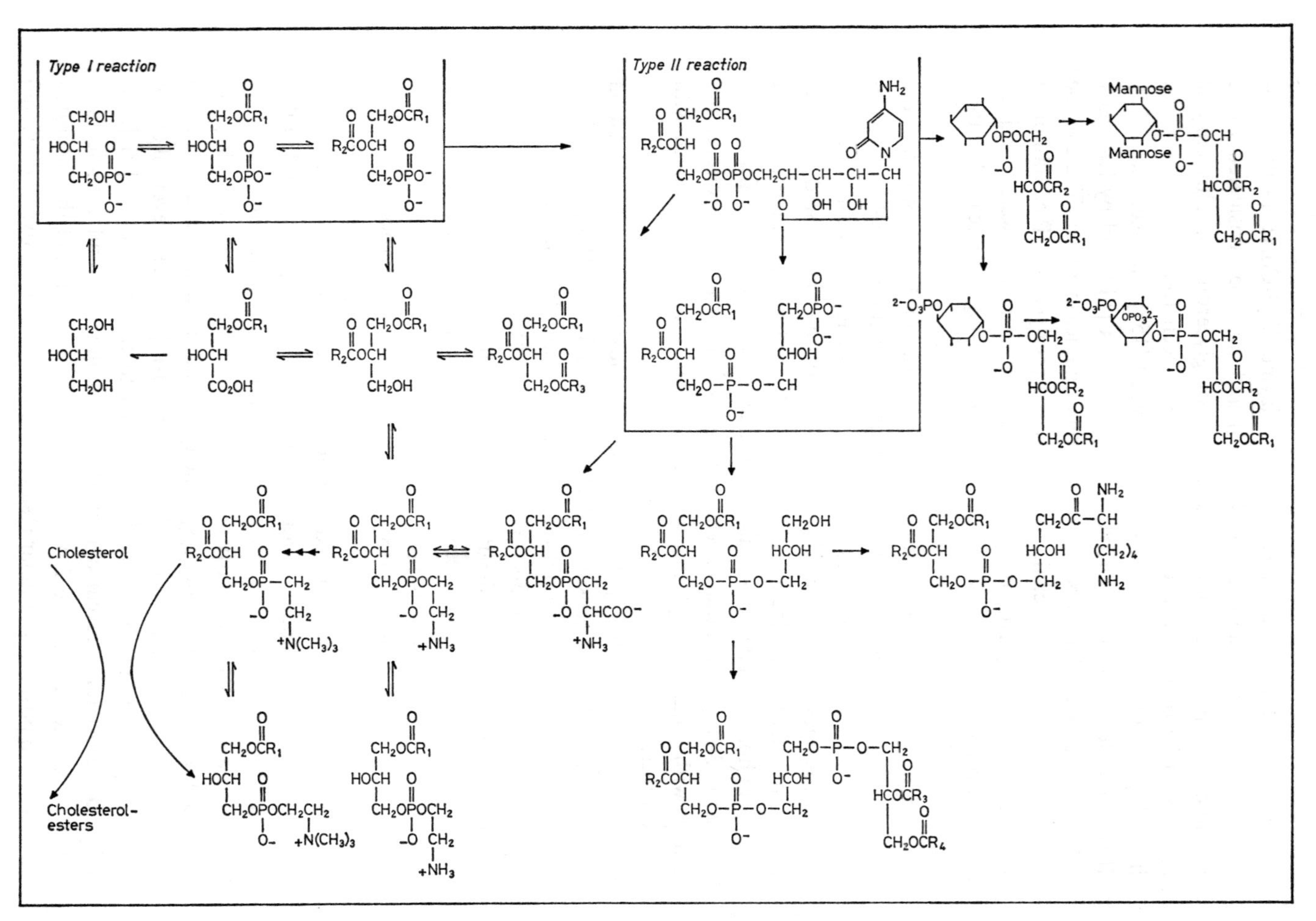

Fig. 1. Biosynthetic pathways of glycerolipids. Type I and type II reactions refer to the text. Modified according to HILL and LANDS [107] and reproduced with permission of the publisher.

This reaction is catalyzed by the enzyme acyl-CoA: L-glycerol-3-phosphate-O-acyltransferase (EC 2.3.1.15) and may be considered to be the first step in the *de novo* biosynthesis of glycerolipids. This enzyme appears to be of universal occurrence, having been found in mammalian liver [71] and brain [197], plants [43, 255], yeast [154] and bacteria [9, 85]. It was first demonstrated by Kornberg and Pricer [149] and by Kennedy [141] using a particulate subcellular fraction of guinea-pig liver. The enzyme had a preference for long-chain fatty acids and sn-glycerol-3-P could not be replaced by free glycerol or glycerol-2-P. Similar results were obtained using subcellular fractions of rat liver, guinea-pig pancreas, lactating mammary gland and hamster small intestine [125].

Compartmental studies with rat liver preparations [52] showed that the enzyme system has a bimodal intracellular localization. One-half to two-thirds of the total activity was found in the mitochondria with the rest in the microsomes [67]. Subfractions of the mitochondria showed that the acylation activity was localized mainly in the outer membrane and that 3-sn-lysophosphatidate was chiefly produced. In contrast, microsomes produced 3-sn-phosphatidate. The patterns of the reaction products obtained with total homogenates were quite similar to those obtained with microsomes.

Fallon and Lamb [71] identified 3-sn-lysophosphatidate as an acylation product in rat liver microsomes. Husbands and Lands [126] found 3-sn-phosphatidate produced in pigeon liver particles, and Possmayer *et al.* [237] reported the formation of 3-sn-phosphatidate in rat liver microsomes. The ratio acyl group: sn-glycerol-3-P of 1.3 was measured by Brandes *et al.* [39]; this ratio indicates the synthesis of a mixture of diacyl- and monoacyl derivatives in the rat liver microsomal acylation processes.

The acylation reaction was found to be a two-step process [166, 167, 228] and, therefore, further investigations were mainly concerned with the positional specificity of the acylation reactions leading to the esterification of the primary and secondary hydroxyl group of the sn-glycerol-3-P. Liver microsomal systems showed low positional specificity in the esterification of saturated and unsaturated fatty acids to sn-glycerol-3-P [105, 166, 293]. On the contrary, high positional specificity could be demonstrated in the esterification of 1- or 2-monoacyl-sn-glycerol-3-P [165]. However, slices of rat liver provided results in marked contrast to the microsomes [106]: incubation procedures revealed that 75% of monoenoic and dienoic fatty acids were esterified to the 2-position of sn-glycerol-3-P and the saturated fatty acids to position 1. Similar results were obtained by Elovson *et al.* [68]

in *in vivo* experiments with rats. The specific distribution of the fatty acids incorporated into 3-sn-phosphatidate was quite comparable to that found in liver 3-sn-phosphatidylcholines and in triacylglycerols. Incubation experiments with subcellular preparations from rat liver carried out by POSSMAYER *et al.* [237] support these observations. A rat liver particulate preparation [163] incubated with acyl-CoA and sn-glycerol-3-P produced 3-sn-lysophosphatidate which was isolated and, afterwards, acylated in a second step. The 1-isomer was the primary product when palmityl-CoA was the substrate, and the 2-isomer was the predominant product formed from oleyl-CoA. The authors conclude that this reaction may contribute to the specific distribution of fatty acids found in glycerolipids. A pronounced substrate specificity was observed in the acylation reactions in rat liver preparations [203]: in mitochondria, palmitic acid was esterified exclusively to position 1 of sn-glycerol-3-P; in microsomes, oleyl- and linoleyl-CoA were more efficient than palmityl-CoA. There is kinetic evidence that 1-palmityl-3-sn-lysophosphatidate is the precursor of 3-sn-phosphatidate rather than the product formed by the action of mitochondrial phosphatidate phosphohydrolase on 3-sn-phosphatidate. ^{14}C-linoleic acid administered intraportally to rats [12] *in vivo* was chiefly incorporated into the palmityl-linoleyl fraction of 3-sn-phosphatidate, while synthesis *de novo* of the stearyl-linoleyl fraction was very slow. ^{14}C-linoleic acid was located almost exclusively at position 2 in 3-sn-phosphatidate, indicating a high positional specificity in 3-sn-phosphatidate synthesis. The incorporation of trienoic fatty acids and particularly arachidonic acid into phospholipids occurred mainly by acyl-CoA:1-acyl-phospholipid acyltransferase reactions. This information from compartmental studies is supported by a recent report by NACHBAUR *et al.* [207]: microsomes from rat liver produce exclusively 3-sn-phosphatidate from sn-glycerol-3-P and (^{14}C) oleic acid, while 3-sn-lysophosphatidate is accumulated in the outer mitochondrial membranes. The membranes of the endothelial reticulum show enzyme acitivities comparable to that of the microsomes. The plasma membranes and the inner mitochondrial membranes do not appear to contain metabolically relevant activities.

Further observations on the intracellular localization and specificity of the acylation process were lately published by DAAE [53]: unsaturated and saturated fatty acid carnitine esters, CoA and carnitine palmityltransferase were used as a donor system for activated fatty acids with sn-(1[3]-^{3}H)glycerol-3-P as the acyl acceptor. It is shown in figure 2 that the mitochondrial preparations mainly incorporate palmitic acid carnitine ester thus producing

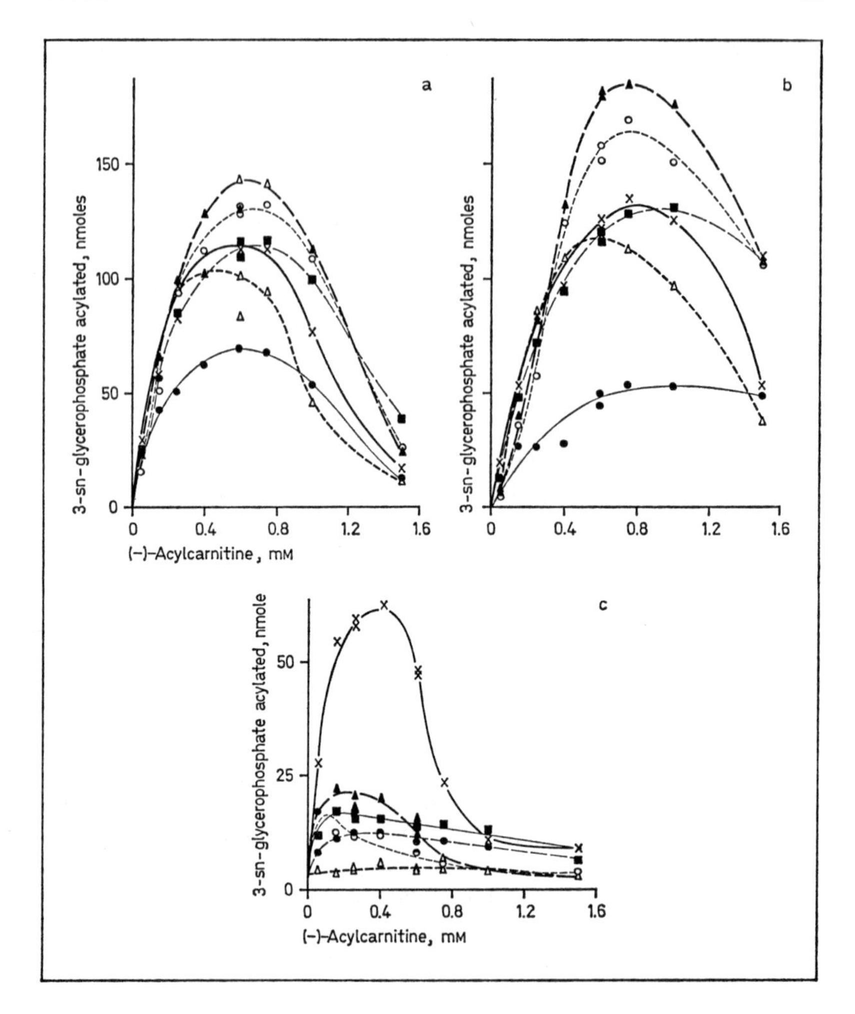

Fig. 2. Acylation of sn-glycerol-3-P with different acylcarnitines in rat liver cell subfractions. Cytoplasmic extract (a), microsomes (b) and mitochondria (c) were incubated with varying concentrations of acylcarnitines: ×——× = palmitylcarnitine; ▲——▲ = linolenylcarnitine; □-------□ = oleylcarnitine; ○.........○ = linoleylcarnitine; •-·-·-·-• = stearylcarnitine; △-·-·-·-△ = arachidonylcarnitine. ([53]; reproduced with permission of the publisher.)

3-sn-lysophosphatidate while the microsomal enzymes react with most of the carnitine esters tried forming 3-sn-phosphatidate. In microsomes, most of the acyl carnitines were acylated to the sn-glyerol-3-P at equal rates; optimal fatty acid carnitine concentration was about twice the optimal concentration found with the mitochondrial outer membrane enzyme. The author postulates that the mitochondrial system may contribute to the positional specificity observed in the intact cell.

Further enzyme characterizations. According to KUHN and LYNEN [154], one active site of this acyl-CoA:L-glycerol-3-phosphate-O-acyltransferase (EC 2.3.1.15) from yeast is responsible for both acylation steps. Moreover, LANDS and HART [167] succeeded in showing that two enzymic reactions are involved in the formation of 3-sn-phosphatidate from acyl-CoA and sn-glycerol-3-P: the synthesis of 3-sn-phosphatidate from sn-glycerol-3-P is strongly or competitively inhibited by preincubating the enzyme with thiol-binding agents as N-ethylmaleimide or *p*-chloromercuribenzoate, whereas the formation of 3-sn-phosphatidate from 3-sn-lysophosphatidate was inhibited less or was even stimulated. The existence of two separate acyl transferases is, therefore, favoured over one acyltransferase with two active sites.

Further information concerning these acylation processes was obtained by OKUYAMA *et al.* [212]: activity of 3-sn-phosphatidate as substrate of rat-liver microsomal acyltransferases was much lower than that of 3-sn-lysophosphatidate. This phenomenon was more obvious when both isomers were present in the incubation mixture. With optimal amounts of substrate, the acyltransferase reaction to the 1-position occurred at about one-tenth the rate observed for the 2-position. Oleate was esterified more rapidly than palmitate or stearate. The author suggested that 4 separate enzyme activities may account for the acylation of sn-glycerol-3-P to 3-sn-phosphatidate. However, the rate-limiting step of 3-sn-phosphatidate formation from sn-glycerol-3-P *in vivo* seems not to be the second acylation of both 3-sn-lysophosphatidate isomers because this intermediate does not accumulate. Accordingly, their results do not exclude the involvement of the first acylation in position 2 and a second one in the 1-position of the sn-glycerol-3-P. Besides, this group found that the acylation of sn-glycerol-3-P is sensitive to sulfhydryl reagents whereas the acylation of 3-sn-lysophosphatidate is not, therefore suggesting that these two reactions are catalyzed by different enzymes. 2-Acyl-sn-glycerol-3-P acyl transfer reaction, however, was not inhibited by sulfhydryl reagents.

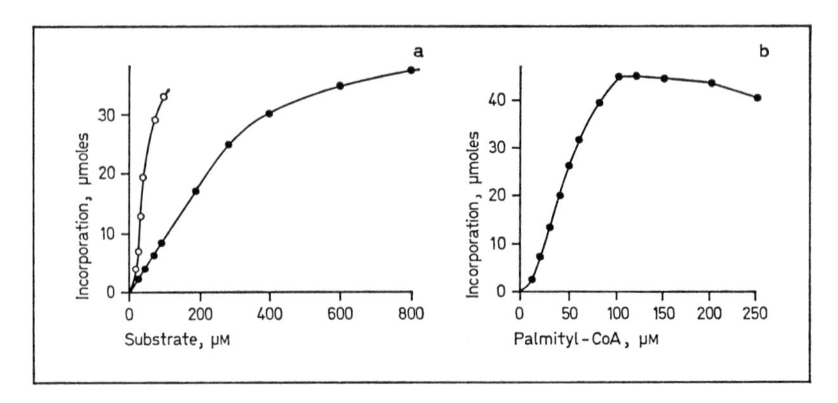

Fig. 3. Acylation of sn-glycerol-3-P by rat brain microsomes. sn-glycerol-3-P, 1.6 mM. *a* Comparison of palmityl-CoA (○) and palmitic acid (•) as substrates. *b* Incorporation as a function of palmityl-CoA concentration. ([252]; reproduced with permission of the publisher.)

OKUYAMA *et al.* [212] also studied the sn-glycerol-3-P acyltransferase activity in mitochondria. Using 3-sn-phosphatidate as substrate for the acylation system they found that the specific activities were about one-tenth of those found in microsomes (0.27 and 0.78 nmole/min/mg protein for palmityl and oleyl-CoA, respectively). Whether or not the activities were derived from microsomal contamination was not examined.

SANCHEZ *et al.* [252] reported more detailed information about the rat brain microsomal enzyme: the best acyl donors for sn-glycerol-3-P acylation were acyl-CoA containing fully saturated fatty acids with 15–18 carbons. These acyl-CoA could be replaced by Mg ATP, CoA and fatty acids (fig. 3) indicating that the endogenous fatty acids in microsomes normally are the source of fatty acids for sn-glycerol-3-P acylation rather than preformed acyl-CoA. The Michaelis constant was 0.4×10^{-3} M (table I) and it was not dependent on the acyl CoA concentration. Inhibition of the enzyme activity occurred above a certain palmityl-CoA concentration, probably because of its detergent properties, as is shown in figure 3. Very similar results were obtained by the same group [6] using rat-liver microsomes. In both cases, in brain as well as in liver microsomes, the kinetics of the respective acyl-CoA were complicated due to the micellar nature of these molecules. This is supported by the observations of ZAHLER *et al.* [332] that Michaelis kinetics are observable only at substrate levels below the critical micelle concentration (3–4 μM for palmityl-CoA). Furthermore, this enzyme loses its activity

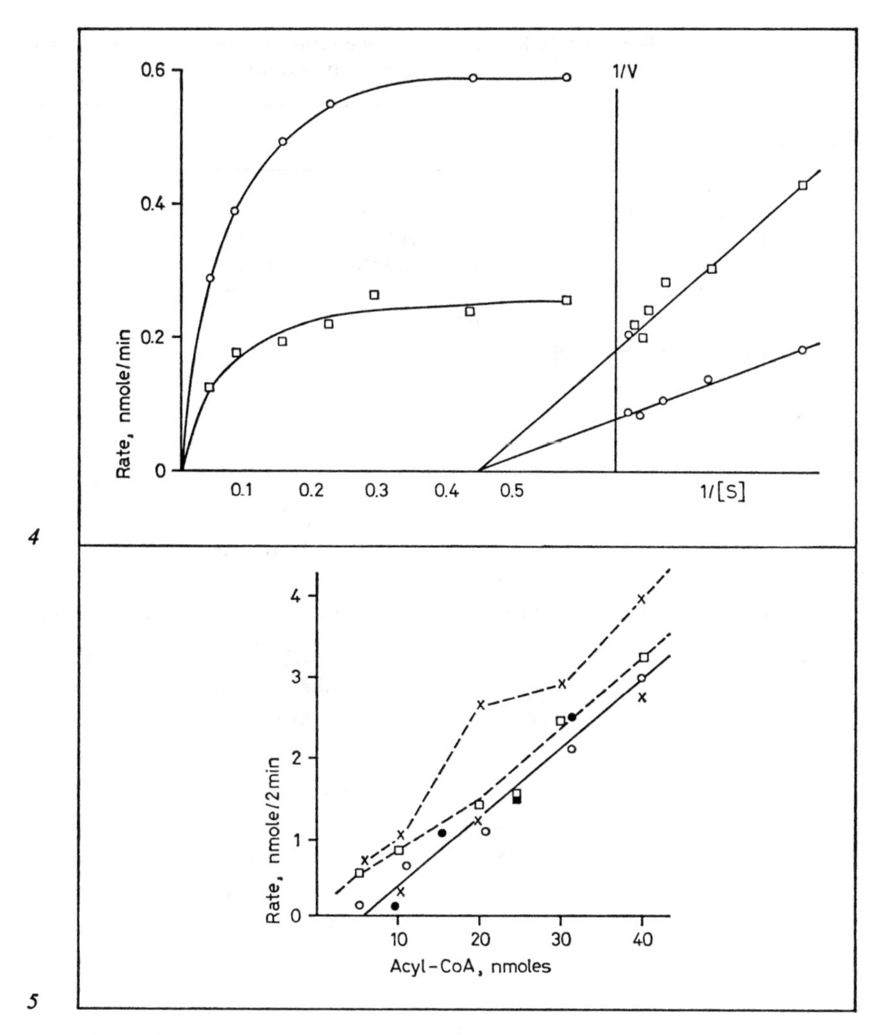

Fig. 4. Effect of sn-glycerol-3-P concentration on the rate of formation of 3-sn-phosphatidate in pigeon liver particles. Constant amount of palmityl-CoA and varied amounts of sn-glycerol-3-P added last (□——□), or varied amounts of sn-glycerol-3-P and a constant amount of palmityl-CoA added last (○——○). Abscissa: sn-glycerol-3-phosphate, in mM; 1/V = reciprocal reaction speed; 1/[S] = reciprocal substrate concentration. ([126]; reproduced with permission of the publisher.)

Fig. 5. The non-selectivity of acylation of different acyl groups in pigeon liver particles. Reaction rates of various acyl-CoA esters in the presence of bovine serum albumin. sn-Glycerol-3-P concentration constant, acyl-CoA concentrations varied; •——• = oleyl-CoA; □——□ = linoleyl-CoA; ○——○ = stearyl-CoA; ×——× = lauryl-CoA; ×-----× = palmityl-CoA. ([126]; reproduced with permission of the publisher.)

Table I. Michaelis constants K_m for the enzyme acyl-CoA:L-glycerol-3-phosphate O-acyltransferase (EC 2.3.1.15) with sn-glycerol-3-P as substrate

Enzyme source	Substrate	K_m for sn-glycerol-3-P, M
S. cerevisiae particles [154]	palmityl-CoA	1×10^{-4}
E. coli membrane preparations [9]	palmityl-CoA	8×10^{-5}
	palmityl-ACP	4×10^{-5}
Rat liver microsomes [6]	acyl-CoA (C_{15}–C_{18})	5×10^{-4}
Rat brain microsomes [252]	acyl-CoA (C_{15}–C_{18})	4×10^{-4}
Pigeon liver particles [126]	palmityl-CoA	5×10^{-5}

when phospholipids are split by phospholipase A, but this activity can be restored by the addition of micellar phospholipid in the presence of dithiothreitol [6].

HUSBANDS and LANDS [126] investigated the influence of different substrate concentrations on the reaction rate using pigeon liver particles. Figure 4 demonstrates a typical saturation phenomenon indicating an apparent K_m for sn-glycerol-3-P of 5×10^{-5} M (table I). Acyl-CoA-inhibited the enzyme at concentrations exceeding 60 μM. This inhibition can be prevented by adding protein. The enzyme was also inhibited by thiol-binding agents. Protection from thiol-binding agents is possible by preincubation of the enzyme with acyl-CoA. The amount of the thiol group protected per milligram enzyme protein was estimated by adding radioactive N-ethylmaleimide in the presence of sn-glycerol-3-P. No evidence was obtained for required sulfhydryl groups other than those protected by acyl-CoA. In the presence of bovine serum albumin, several different acyl-CoA derivatives were esterified to sn-glycerol-3-P at very simliar rates (fig. 5) indicating a non-selectivity in acylation under these conditions.

The enzyme from yeast requires acyl-CoA as an obligatory acyl donor [154], but acitivity from *E. coli* and *Clostridium butyricum* can also use the acyl group attached to acyl-carrier protein [9, 85]. Recently, VAN DEN BOSCH and VAGELOS [36] published data on acylation processes in a particulate fraction of *E. coli*. This preparation catalyzes the transfer of palmitate from palmityl-CoA and palmityl-acyl carrier protein to sn-glycerol-3-P. When a series of long-chain acyl thiol esters including myristyl, palmityl-, palmitoleyl- and cis-vaccenyl-CoA and acyl-carrier protein were tested, the palmitate esters were the most effective substrates. 3-sn-Lysophosphatidate has been identified as the initial product formed from sn-glycerol-3-P and either palmityl-CoA or palmityl-acyl carrier protein. The 3-sn-lysophosphatidate can either be acylated to 3-sn-phosphatidate or be dephosphorylated to the monoacyl-glycerol. The extent to which both reactions occur depends on experimental conditions and on the nature of the acyl donor. 3-sn-Phosphatidate synthesis from sn-glycerol-3-P was observed in the presence of both saturated or unsaturated CoA esters. When palmityl-acyl carrier protein was the only acyl donor present, the main products were 3-sn-lysophosphatidate and monoacyl-glycerol; in the additional presence of unsaturated acyl-acyl carrier protein, 3-sn-phosphatidate was formed. This indicates the possibility of 3-sn-phosphatidate biosynthesis from sn-glycerol-3-P and acyl-acyl carrier protein esters. Furthermore, the acylation of 1-acyl-sn-glycerol-3-P by acyltransferases occurred preferentially with unsaturated fatty acyl thiolesters. Especially acyl-acyl carrier protein esters were used with great selectivity for the acyl portion in this acylation reaction. The palmitate thiolester of 4′-phosphopantetheine was not used as acyl donor to sn-glycerol-3-P under conditions in which acyl thiolesters of both CoA and acyl carrier protein are active.

In contrast to bacteria, mammalian enzyme systems appear to require acyl-CoA as the acyl donor [39, 154, 192, 197].

The picture emerging from such enzyme property studies including acylation positional specificity is far from clear. This might be partly explained by the fact that different cell preparations from different organs of different species were used by the investigations reviewed here. In addition, a variety of other experimental conditions were used. However, there is strong evidence that the non-random distribution of the naturally occurring phospholipid fatty acids may partly result from acylation specificity in 3-sn-phosphatidate formation.

b) Alternate Pathways for the Biosynthesis of 3-sn-Lysophosphatidate and 3-sn-Phosphatidate

Acyldihydroxyacetone-phosphate pathway. Synthesis of 3-sn-lysophosphatidate can also be accomplished by the following reaction sequence:

$$\text{Dihydroxyacetone-P} + \text{palmityl-CoA} \rightleftharpoons \text{palmityl-dihydroxyacetone-P} + \text{CoA}. \tag{9}$$

$$\text{Palmityl-dihydroxyacetone-P} + \text{NADPH} + \text{H}^+ \rightleftharpoons \text{1-palmityl-sn-glyerol-3-P} + \text{NADP}^+. \tag{10}$$

HAJRA and AGRANOFF [95] and HAJRA *et al.* [94] first observed this pathway and characterized acylhydroxyacetone-P as an intermediate in guinea-pig mitochondria. HAJRA [96] also demonstrated an acyltransferase reaction in mitochondria and microsomes from guinea-pig liver forming acyldihydroxyacetone-P from palmityl-CoA and dihydroxyacetone-P. The highest rate of esterification was found with saturated C_{16} and C_{17} fatty acids in both subcellular fractions. In mitochondria, the acylation rate was much faster with dihydroxyacetone-P, as compared to sn-glycerol-3-P, and no competition was found between both substrates. However, this competition was present in the microsomes. The conversion of acyldihydrocyacetone-P to diacyl-sn-glycerol-3-P is mediated by a reaction sequence: a mitochondrial preparation from guinea-pig liver containing $NADP^+$ catalyzes the reduction step which is followed by another acylation using palmityl-CoA [97]. It should be noted that this system requires $NADP^+$, that it does not function with NAD^+ and that it is selective for saturated fatty acids. This might account for the preponderance of saturated fatty acids on C-1 of the glycerol of natural phospholipids [96]. Since the reduction of dihydroxyacetone-P to sn-glycerol-3-P requires NADH and that of acyl-dihydroxy-acetone-P to 1-acyl-glycerol-3-P, NADPH, AGRANOFF and HAJRA [8] tried to compare the participation of the two pathways for phospholipid synthesis by measuring the incorporation of radioactivity from tritium-labelled NADH and NADPH into C-2 of lipid glycerol. They found that the acyl-dihydroxyacetone phosphate pathway plays a significant rôle in glycerolipid synthesis in mouse liver homogenates and a clearly dominant one in Ehrlich ascites tumor cell homogenates. Apparently, this finding is related to a reported lack of L-glycerol-3-phosphate:NAD^+ oxidoreductase (EC 1.1.1.8) in tumour cells and to their high glycerol ether lipid content.

Experiments with rats kept on different diets were recently reported by RAO *et al.* [245]. Synthesis of 3-sn-lysophosphatidate from dihydroxyacetone-P and palmitate in liver microsomes was suppressed in rats either fasting or maintained on a high-fat diet. Analogous results were observed when sn-glyceraldehyde-3-P was used as the glyceride-glycerol precursor, probably because microsomes convert sn-glyceraldehyde-3-P to dihydroxyacetone-P. These studies demonstrate that 3-sn-lysophosphatidate synthesis from dihydroxyacetone-P by particulate enzymes is subject to dietary regulation.

Monoacylglycerol pathway. Another route for the formation of 3-sn-lysophosphatidate is the phosphorylation of monoacylglycerol by the monoglyceride kinase that has been described by several investigators [37, 111, 217, 227].

Monoacylglycerol + ATP $\rightleftharpoons$ monoacyl-sn-glycerol-3-P (3-sn-lyso-phosphatidate) + ADP. (11)

Using a deoxycholate-solubilized preparation from guinea-pig brain, 1-monoacylglycerol as well as 2-monoacylglycerol were found to be suitable substrates [227]. It was suggested by PARIS and CLÉMENT [217] that this type of reaction might play a rôle in the intestinal mucosa during fat absorption.

Diacylglycerol pathway. An alternate pathway for the synthesis of 3-sn-phosphatidate can be mediated by the enzyme 1,2-diglyceride kinase (EC 2.7.1.–) [110]:

1,2-Diacylglycerol + ATP $\rightleftharpoons$ 1,2-diacyl-sn-glycerol-3-P (3-sn-phosphatidate) + ADP. (12)

This enzyme has been studied in brain [110, 291] and erythrocyte ghosts [112, 113]. Diacyl-sn-glycerol-3-P was the reaction product in different organs [110, 114, 223, 240, 330,]. The reaction can be stimulated by acetylcholine [246]. Some properties of the enzyme of rat cerebral cortex, its distribution in isolated subcellular fractions and the effects of acetylcholine on its activity were recently reported by LAPETINA and HAWTHORNE [171]. The enzyme was activated by Mg^{2+}. Ca^{2+} activated it to a smaller extent but was inhibitory in the presence of optimum concentrations of Mg^{2+}. The activity was greatly increased by the addition of 1,2-diacylglycerol. Sodium deoxycholate markedly stimulated the reaction, but other detergents (Cutscum and Triton X-100) did not. The enzyme was concentrated in the supernatant and microsomal fractions from rat cerebral cortex. The distribution of the kinase in the particulate fractions resembled that of acetylcholinesterase and 5′-nucleosidase. The rate of 3-sn-phosphatidate synthesis by this route was much greater than reported rates for acylation of sn-glycerol-3-P and was also very rapid in comparison with the rates of other steps in the synthesis of 3-sn-phosphoinositides. Acetylcholine had no stimulatory effect on this kinase of isolated intact nerve-end particles or of nerve-end membranes obtained after osmotic shock.

According to HOKIN and HOKIN [113], this kinase reaction is unimportant for acylglycerol biosynthesis, but probably plays a rôle in membrane functions.

2. *Biosynthesis of Acylglycerols*

a) sn-Glycerol-3-Phosphate Pathway

The biosynthesis of 1,2-diacylglycerols can be accomplished by phosphohydrolases using either 3-sn-phosphatidates or base containing 3-snphosphoglycerides as substrates:

$$\text{3-sn-Phosphatidate} + H_2O \rightleftharpoons \text{1,2-diacylglycerol} + P_i. \quad (13)$$

This bond cleavage is mediated by the enzyme L-α-phosphatidate phosphohydrolase (EC 3.1.3.4) which was discovered in plants by KATES [139], and in animal tissues by WEISS and KENNEDY [318] and by STEIN *et al.* [288]. Detailed information on the properties of this enzyme were recently reported [202].

Another bond cleavage reaction uses a variety of phosphoglycerides as a substrate:

$$\text{3-sn-Phosphatidylcholine} + H_2O \rightleftharpoons \text{1,2-diacylglycerol} + \text{phosphorylcholine}. \quad (14)$$

This reaction is catalyzed by the enzyme phosphatidylcholine choline-phosphohydrolase (EC 3.1.4.3) (phospholipase C). 3-sn-Lysophosphatidate can also be split by this enzyme in *Bacillus cereus* [58]. In *E. coli*, mono-acylglycerol formation was recently found to occur in a very specific manner [36]: in a particulate fraction using palmityl-acyl carrier protein as the only acyl donor for the acylation of 3-sn-glycerol-3-P, 3-sn-lysophosphatidate and monoacylglycerol were identified as the main products. A great selectivity for the acyl portion in this acylation reaction was observed. The dephos-phorylation process forming monoacylglycerol does not require divalent cations, and it is not affected by NaF, an agent which causes 70% inhibition of the 3-sn-phosphatidylglycerol-3-P phosphatase. Both phosphatases appear to depend on sulfhydryl groups for enzymatic activity. The pH optimum was found to be 7.0 for both substrates, 3-sn-lysophosphatidate and 3-sn-phosphatidate. ÅKESSON [12] observed in *in vivo* experiments with rats that labelled linoleic acid was rapidly incorporated into liver 3-sn-phosphatidate and 1,2-diacylglycerols. This labelling was almost exclusively located at position 2 in 3-sn-phosphatidate indicating a high positional specificity in 3-sn-phosphatidate synthesis. For further information the reader is referred to the reviews by HILL and LANDS [107] and by LENNARZ [173].

b) Monoacylglycerol Pathway

In addition, acylglycerol biosynthesis, excluding hydrolysis of 3-sn-phosphatidates and derivatives, is catalyzed by a multi-enzyme complex termed triglyceride synthetase (EC class 2.3.1) which is tightly bound to subcellular fractions, and which includes, among others, the following reactions:

$$\text{Monoacylglycerol} + \text{acyl-CoA} \rightleftharpoons \text{1,2-diacylglycerol} + \text{CoA}. \quad (15)$$

This reaction is catalyzed by the enzyme monoglyceride acyltransferase (EC 2.3.1-type). Both isomers of the substrate may be used.

$$1,2\text{-Diacylglycerol} + \text{acyl-CoA} \rightleftharpoons \text{triacylglycerol} + \text{CoA}. \quad (16)$$

This reaction is catalyzed by the enzyme acyl-CoA: 1,2-diglyceride-O-acyl-transferase (EC 2.3.1.20). Mechanisms (15) and (16) are termed 'monoacylglycerol pathway' for acylglycerol formation.

The occurrence of these enzymes in different species, organs and subcellular compartments as well as enzyme properties including the positional specificity in the acylation processes were comprehensively reviewed by HÜBSCHER in 1970 [125]; therefore, the reader may consult this article. However, recent observations are reported here: RAO and JOHNSTON [243, 244] described an approximately 70-fold purified preparation containing acyl-CoA synthetase, monoacylglycerol transacylase, diacylglycerol transacylase and lipid. This triglyceride synthetase appeared to reside not only on membranes but also in a lipid-rich medium probably essential for optimal activity. Furthermore, a microsomal fraction synthetized triacylglycerols using free palmitate. This reaction took place after CoA was bound to the enzyme complex using ATP. In order to get more information about the intracellular localization of the synthetase, the same group [260] prepared a brush border and a microsomal fraction from rat and hamster intestinal mucosa. The results suggest that synthetase activity is chiefly located in the microsomes, and that activity detected in the brush border fraction might be due to microsomal contamination.

In contrast, GALLO and TREADWELL [77] demonstrated that in rat intestine diacylglycerol formation via the monoacylglycerol pathway is mostly attributed to the microsomal fraction and triacylglycerol synthesis to the brush border membranes. Diacylglycerols produced either in microsomes or in the brush border preparation were identified as the 1,2-isomer. These results were not due to cross-contamination of the prepared fractions. Further investigations reported by ÅKESSON [10] on the acylation of diacylglycerols with labelled oleic acid in pig-liver revealed a stereospecific incorporation of radioactivity into position 1 of 1,2-diacylglycerols. In studies using a microsomal fraction from pig liver, labelled oleic acid, ATP and CoA, this group [302] pointed out that 1,2-diacylglycerol was rapidly formed when 2-monoacylglycerol was the acceptor indicating a stereospecific acylation at position 1. Rac-1-mono-oleylglycerol was much less effective as an acyl acceptor than the 2-isomer. In this case the labelled oleic acid was incorporated mainly into triacylglycerols, without any accumulation in 1,2-diacylglycerols. It was concluded from the positional distribution of radioactivity in triacylglycerols that the positional isomers of mono-oleylglycerol were incorporated without prior acyl migration.

c) Operation of Both Pathways

It is well known from studies on the positional specificity of the fatty acids in acylglycerols that the unsaturated fatty acids predominate in the 2-position. This was supported by data lately published by RAJU and REISER [241]: in an *in vitro* system from rat liver, biosynthesis of diacyl- and triacylglycerols from preformed fatty acids and sn-glycerol-3-P was found to occur by incorporation of unsaturated fatty acids preferentially into the 2-position. Furthermore, it is well established that monoacylglycerols are the predominant acceptor for fatty acids in the resynthesis of triacylglycerols in intestinal mucosa [145], and in other tissues, and it was also found that 2-monoacylglycerol is more effective than rac-1-monoacylglycerol [302].

Similar metabolic patterns observed between adipose tissue and the intestinal mucosa including triacylglycerol synthesis and the presence and subcellular distribution of certain enzymes raised the question as to whether the monoacylglycerol pathway for the formation of triacylglycerols exists in adipose tissue [263]. Using subcellular fractions and whole cell preparations of white and brown adipose tissue, SCHULTZ and JOHNSTON [263] obtained conclusive evidence for the existence of the monoacyl pathway in these tissues. The monoacyl and sn-glycerol-3-P pathways were found to be operational mainly in microsomes. In whole cell preparations, both pathways were approximately of the same order of magnitude. In microsomes, the monoacylglycerol pathway was approximately one-half of the activity of the sn-glycerol-3-P pathway. The authors presume that the quantitation and the relationship of both pathways might have importance in certain genetically related obese conditions, and that the presence of the monoacylglycerol pathway in brown adipose tissue may be of significance in the process of thermogenesis.

d) Compartmental and Stereochemical Studies

Investigations on diacylglycerol synthesis in hamster intestinal microsomes [131] revealed that the stereochemical configuration of the diacylglyerols produced by the sn-glyerol-3-P and the monoacylglyerol pathway is the sn-1,2 configuration.

In a recent report by BRECKENRIDGE and KUKSIS [26], biosynthesis of diacylglycerols from 2-monoacylglycerols and free fatty acids was described in everted sacs of rat intestinal mucosa. It was shown that the main products of synthesis were the sn-1,2-diacylglycerols (60%), but sn-2,3-diacylglycerols (40%) were also formed in significant amounts. The total yield and proportions of the isomeric diacylglycerols recovered appeared to vary with the nature of the monoacylglycerol and the complexity of the free fatty acid mixture supplied.

KRUYFF *et al.* [153] found that the rate of triacylglycerol synthesis in rat-liver microsomes from 1,2-diacylglycerols was the same regardless of the fatty acid composition of the diacylglyerol used. Similarly, ÅKESSON *et al.* [11] observed diacylglycerols to be randomly utilized for triacylglycerol biosynthesis in rat liver.

In another study concerning the subcellular distribution of the endogenously synthetized lipids in isolated adipose tissue cells from rats, ANGEL [14] pointed out that at least two forms of compartmentation of these lipids occurred. The first, termed 'structural', refers to localization of lipids to organelle fractions. The second type of compartmentation, termed 'chemical', concerns the intracellular segregation of a specific lipid class. Using labelled glucose as precursor for lipid formation, the following relative specific activities of organelle triacylglycerol were measured: mitochondria $\gg$ microsomes $>$ liposomes $>$ soluble supernatant $>$ bulk lipid. It is suggested from these data that a significant proportion of newly synthetized lipid is transferred from mitochondrial membranes into the storage vacuole by direct lipid-lipid interaction.

3. Biosynthesis of Base Containing Diacylglycerols

The synthesis of 3-sn-phosphatidylcholine is known to occur via a number of metabolic pathways. One pathway is the successive methylation of 3-sn-phosphatidylethanolamine as the sole mode of 3-sn-phosphatidylcholine synthesis in bacteria [28]. In mammalian systems, both the methylation pathway and the synthesis of 3-sn-phosphatidylcholine from CDP choline and 1,2-diacylglycerols have been demonstrated [28, 317]. Original studies concerned with the biosynthetic pathways leading to phosphoglycerides were mainly undertaken by KENNEDY an his coworkers. Comprehensive reviews were presented by KENNEDY [142], HILL and LANDS [107], and LENNARZ [173].

a) Incorporation of Bases

In 1952, KORNBERG and PRICER [149a], observed that phosphorylcholine can be incorporated into liver lipids as a unit. In 1956, WEISS *et al.* [317]

found CTP to be a co-factor in phosphorylcholine incorporation into 3-sn-phosphatidylcholine (lecithin), and they suggested the following pathway for its biosynthesis:

$$\text{Phosphorylcholine} + \text{CTP} \rightleftharpoons \text{CDP-choline} + \text{PP}_i. \quad (17)$$

$$\text{CDP-choline} + 1,2\text{-diacylglycerol} \rightleftharpoons 3\text{-sn-phosphatidylcholine} + \text{CMP}. \quad (18)$$

Reaction mechanism (17) is catalyzed by the enzyme cholinephosphate cytidyltransferase (CTP: cholinephosphate cytidyltransferase, EC 2.7.7.15) which has a maximal activity at pH 7.2 requiring divalent cations (Mg^{2+} or Mn^{2+}), CTP and phosphorylcholine. CTP can be replaced by dCTP.

Reaction mechanism (18) is mediated by the enzyme CDP-choline: 1,2-diglyceride cholinephosphotransferase (EC 2.7.8.2) having maximal activity at pH 8.6 in the presence of 1,2-diacylglycerol. Both, CDP- and dCDP-choline are active with this enzyme, and it is markedly inhibited by Ca^{2+}. Among different 1,2-diacylglycerols tried, mostly 1,2-diolein was found to stimulate synthesis. Animal organ preparations which catalyze the formation of 3-sn-phosphatidylcholine, also catalyze the synthesis of 3-sn-phosphatidylethanolamine alternatively using CTP or dCTP:

$$\text{Phosphorylethanolamine} + \text{CTP} \rightleftharpoons \text{CDP-ethanolamine} + \text{PP}_i. \quad (19)$$

$$\text{CDP-ethanolamine} + 1,2\text{-diacylglycerol} \rightleftharpoons 3\text{-sn-phosphatidylethanolamine} + \text{CMP}. \quad (20)$$

This reaction is catalyzed by the enzyme CDP-ethanolamine: 1,2-diglyceride ethanolaminephosphotransferase (EC 2.7.8.1).

In animal tissues [34], incorporation of L-serine into phosphoglycerides takes place via a Ca^{2+}-stimulated exchange reaction starting from 3-sn-phosphatidylethanolamine which is derived from 1,2-diacylglycerol:

$$3\text{-sn-Phosphatidylethanolamine} + \text{L-serine} \rightleftharpoons 3\text{-sn-phosphatidylserine} + \text{ethanolamine}. \quad (21)$$

This reaction is catalyzed by the enzyme phosphatidylethanolamine-L-serine phosphatidyltransferase, the 'exchange enzyme' which was found to be almost entirely microsomal in rat liver [60]. This reaction was detected in rat-liver [34, 124], in the house fly *Musca domestica* [51], and in the protozoon *Tetrahymena pyriformis* [59]. In animal tissues, reaction (21) can be followed by a decarboxylation process leading to net production of 3-sn-phosphatidylethanolamine and CO_2 according to the 'decarboxylation cycle' which was postulated by KENNEDY [143]. This phosphatidyldecarboxylase was found to be located in rat-liver mitochondria [60].

In a mutant of *Neurospora crassa* [267], 3-sn-phosphatidylserine was synthetized from CDP-diacylglycerol and L-serine by a particulate subfraction which was then converted into 3-sn-phosphatidylethanolamine by decarboxylation. Fungi appear to synthetize 3-sn-phosphatidylethanolamine in a manner similar to that found in gram-negative bacteria and animal tissues.

In order to study the significance of the exchange mechanism, SPITZER [282] recently incubated liver microsomes with labelled bases and, afterwards, the displacement of the radioactive bases from the phosphoglycerides was determined. He found that the serine from 3-sn-phosphatidylserine could be almost completely replaced by bases, whereas choline from 3-sn-phosphatidylcholine could not be displaced by other bases. The finding that bases can displace the serine of 3-sn-phosphatidylserine suggests that the level of serine and/or 3-sn-phosphatidylserine may be a control point in phosphoglyceride synthesis via the base displacement mechanism. Using phosphorylcholine and its phosphonate analogues as precursors for 3-sn-phosphatidylcholine synthesis in rat liver and kidney, BJERVE [24] observed that the phosphonate substrates were incorporated as a unit into the phosphonate analogues of 3-sn-phosphatidylcholine; phosphorylcholine behaved as a strong inhibitor of this reaction. From radioactivity distribution studies in the various reaction products it was suggested that most of the choline incorporation into 3-sn-phosphatidylcholine *in vivo* occurs by the same mechanism as phosphorylcholine incorporation.

ABDEL-LATIF and SMITH [3] administered labelled choline, glycerol and orthophosphate to rats during the stage of active myelination, and then prepared brain membranes of synaptosomes, microsomes and an intermediate fraction. All subfractions incorporated the 3 labelled precursors with the following rate: microsomes $>$ intermediate fraction $>$ synaptosomes. It was concluded from these data that at least part of the phosphoglycerides can be synthetized at the synapse and that this synthesis is probably independent of the cell body of the neurone.

b) N-Methylation

An alternate pathway for the biosynthesis of 3-sn-phosphatidylcholine in liver microsomes can occur by stepwise N-methylation of 3-sn-phosphatidylethanolamine with S-adenosylmethionine which was first suggested by BREMER and GREENBERG [27] and by WILSON *et al.* [326]:

3-sn-Phosphatidylethanolamine + 3-S-adenosylmethionine → 3-sn-phosphatidylcholine + 3-S-adenosylhomocysteine. (22)

This pathway was also found to be operational in *N. crassa* and in erythrocytes. Studies with mutants revealed that at least two N-methyltransferases are involved in the formation of 3-sn-phosphatidylcholine. Furthermore, the principle species of 3-sn-phosphatidylcholine formed by this pathway is one containing arachidonic acid [107].

Experiments with *Saccharomyces cerevisiae* carried out by STEINER and LESTER [289] showed that particulate, cell-free preparations from yeast incorporated the labelled methyl group of S-adenosyl-L-methionine into 3-sn-phosphatidylcholine, 3-sn-phosphatidyl-N-methylethanolamine, 3-sn-phosphatidyl-N,N-dimethylethanolamine, and ergosterol. When particles were prepared from yeast grown in the presence of choline, they had a much lower rate of incorporation, as compared to yeast in the absence of choline. Both preparations were qualitatively similar, incorporation occurred at linear rates, and both preparations reached steady state levels of the methylation intermediates. A difference in the amount of enzyme in both preparations was suggested to be an explanation for the observed differences. Presumably, this results from the repression of enzyme synthesis in the presence of choline. Similar data were obtained by GLENN and AUSTIN [81] using liver microsomes from rats kept on a normal and on a choline-deficient diet: the methylation of 3-sn-phosphatidylethanolamine to 3-sn-phosphatidylcholine was significantly increased in male rats after 21 h on a choline-deficient diet, whereas this methylation was increased in female rats only after they had been on a choline-deficient diet for longer periods. The different 3-sn-phosphatidylcholine subspecies were methylated at different rates. There is no explanation for this sex difference.

Compartmental studies in *T. pyriformis* reported by SMITH and LAW [277] revealed that the methylation process was located in the microsomal fraction. Exogenous 3-sn-phosphatidyl-monomethylethanolamine was accepted as substrate by the enzyme but not 3-sn-phosphatidylethanolamine or 3-sn-phosphatidyldimethylethanolamine. CDP-choline: 1,2-diglyceride cholinephosphotransferase activity, however, was localized in the mitochondria. The enzyme was twice as active with the optimal concentrations of Mg^{2+} and Mn^{2+}; the enzymatic activity was inhibited by low concentrations of Ca^{2+} or Hg^{2+}. Incorporation of the intact choline molecule was demonstrated when the cells were grown on (Me-^{14}C) choline. In contrast to other organisms, phosphonic acids were not methylated.

In a mutant of *N. crassa* having a block in the methylation pathway, SHERR and BYK [267] found that submerged cultures incorporated 94% of (Me-^{14}C) choline and 89% of (1,2- $^{14}C_2$) choline into 3-sn-phosphatidylcholine compared to the other lipid classes. This indicated that choline is incorporated as a whole molecule by an alternate pathway.

Developmental investigations with rats undertaken by WEINHOLD [316] showed that labelled choline was also incorporated into 3-sn-phosphatidylcholine of liver as a unit, and that only little methylation of 3-sn-phosphatidylethanolamine occurred. Liver slices from fetal and young rats incorporated about 8–10 times more choline into phosphorylcholine than did liver slices from adults. Adult levels were reached 10 days after birth. The incorporation of radioactivity from labelled methionine into 3-sn-phosphatidylcholine by fetal liver slices was much lower than that by adult liver slices and increased to twice adult levels by 10–12 days after birth. Adult levels were reached by 25 days after birth. The results are discussed in view of phosphoglyceride requirement in membrane formation and of phosphoglyceride supply to the serum and bile.

A different pathway for the formation of saturated 3-sn-phosphatidylcholine may exist in lung. AKINO *et al.* [13] incubated lung slices with (^{3}H) glycerol and (^{14}C)palmitate; the ratio $^{14}C/^3H$ was much higher in 3-sn-phosphatidylcholine than in either 3-sn-phosphatidate or 1,2-diacylglycerol, suggesting that palmitate was incorporated into 3-sn-phosphatidylcholine by pathways other than *de novo* synthesis via 3-sn-phosphatidate and 1,2-diacylglycerol. It was postulated from these and other data that a transacylation between two molecules of 3-sn-lysophosphatidate can be a principal pathway for the formation of saturated 3-sn-phosphatidylcholine species in lung. Further support for a deacylation-reacylation mechanism in lung was given by VEREYKEN *et al.* [308]: using labelled glycerol as precursor for phosphoglyceride synthesis in rats, it was found that in lung as well as in liver *de novo* synthesis of 3-sn-phosphatidylcholine primarily contributes to the biosynthesis of linoleic acid-containing molecules of 3-sn-phosphatidylcholine. Acylation of monoacyl derivatives of these phosphoglycerides was suggested to play an important rôle particularly in the formation of arachidonic acid containing phosphoglyceride species. In lung, deacylation-reacylation processes may contribute significantly to the formation of dipalmityl-3-sn-phosphatidylcholine, a major constituent of lung pulmonary surfactant. In strong contrast to liver, lung microsomes incorporated palmitate into the 2-position of 3-sn-phosphatidylcholine. Comparison of the composition of 3-sn-phosphatidylethanolamine and 3-sn-phosphatidylcholine from lung suggested that obviously methylation does not play an important rôle in the formation of dipalmitoyl-3-sn-phosphatidylcholine.

c) Lecithin: Cholesterol Acyltransferase Reaction

Another mechanism leading to the formation of 3-sn-lysophosphatidylcholine in human plasma was first described by Glomset *et al.* [82–84]:

3-sn-Phosphatidylcholine + cholesterol → 3-sn-lysophosphatidylcholine + cholesterolester. (23)

This reaction is catalyzed by the enzyme lecithin: cholesterol acyltransferase (EC class 2.3.1). Activity was found to be sufficient to esterify 5–10% of the free cholesterol in plasma per hour. These data are consistent with results obtained by Kupke [162]: dog plasma incubated with ^{3}H-cholesterol for 8 h esterified approximately 30% of the added cholesterol. Esterification occurred to some extent in a linear way. ^{3}H-activity was mostly carried by the α_1 lipoproteins; however, there was a considerable shift of labelling to the α_2, and chiefly to the β-fraction when plasma from hypercholesteremic animals was used. The distribution of ^{3}H activity between the lipoprotein fractions was not affected by the addition of the alkaloid nicotine *in vitro*.

Glomset proposed that the acyltransferase reaction is the major source of plasma cholesterolesters in man, and that mainly fatty acids from the 2-position of 3-sn-phosphatidylcholine are transferred to cholesterol.

Abdulla *et al.* [4] obtained evidence that the transferase reaction is also operational in rabbit and human aorta, and that activity increases in human fatty atherosclerotic lesions and in atherosclerotic aortas from cholesterol-fed rabbits. Similarly, Portman *et al.* [235] observed in long-term hyperlipaemia and atherosclerosis in squirrel monkeys that the transferase activity in plasma was elevated *in vitro*. Simultaneously, a rapid equilibration of plasma 3-sn-lysophosphatidylcholine and that of liver and aortic intima plus inner media was seen. In a previous report, Portman and Alexander [233] described that in atherosclerotic aortas from squirrel monkeys 3-sn-lysophosphatidylcholine was increased, and that linoleic acid was incorporated predominantly into 3-sn-phosphatidylcholine when linoleyl-^{14}C CoA was the substrate. The extent of this reaction was dependent on the concentration of 3-sn-lysophosphatidylcholine. These authors concluded from these results that the concentrations of 3-sn-lysophosphatidylcholine which are higher in atherosclerotic than in control aortic tissues, could be a factor controlling rates of fatty acid incorporation into 3-sn-phosphatidylcholine. The increased amount of 3-sn-lysophosphatidylcholine in atherosclerotic aortas [233] is consistent with data obtained by KUPKE [162] from experiments with dogs fed an atherogenic diet: significantly elevated sterolester contents in aorta and coronary artery probably indicate an augmented transferase activity.

ABDULLA and ADAMS [5] found sulfhydryl groups to be essential for enzyme activity. Ultracentrifugal studies on this enzyme from rat plasma [300] showed its presence in the plasma-residual protein fraction of density 1.210. After starvation of the rats overnight, arachidonate was preferably transferred, at the expense of linoleate.

FIELDING *et al.* [73] found in density-gradient centrifugation followed by chromatographical procedures that the activation of the transferase by high-density lipoproteins is associated with a specific protein co-factor, and that another high-density lipoprotein partially inhibits the effect of the activator protein. Inhibition of the enzyme by *p*-chloromercuriphenyl sulfonate [128] reduced the exchange of sphingomyeline but had no effect on lecithin exchange. Further studies on the enzyme activity of human serum by FIELDING *et al.* [74] revealed maximal enzyme activity at a lecithin-cholesterol molar ratio of 4 in the presence of a specific protein co-factor. Cholesterolesters, at proportions similar to those present in high-density lipoproteins, reduced enzyme activity to about 20% of that obtained in the presence of only cholesterol and 3-sn-phosphatidylcholine as lipid components. This lower rate was similar to that found in the native lipoprotein. Triacylglycerol inhibited the enzyme under the same conditions, but to a lesser extent than did cholesterol esters. 3-sn-lysophosphatidylcholine diminished enzyme activity only in the absence of albumin.

In vitro experiments with rat liver slices and heated rat serum resulted in a significant amount of newly synthetized cholesterolester in the serum indicating that the liver is a source of plasma lecithin cholesterol acyltransferase [195].

The current information about this transferase system gives further support to the hypothesis proposed by GLOMSET [84], namely, that esterified cholesterol may be only a by-product, and that the principle physiological rôle of this reaction is connected with the exchanges in lipoprotein phosphoglyceride. Presumably, the conversion of 3-sn-phosphatidylcholine into 3-sn-lysophosphatidylcholine and the subsequent loss of 3-sn-lysophosphatidylcholine from lipoproteins have important effects on lipoprotein structure and metabolism. Furthermore, 3-sn-lysophosphatidylcholine formed by the reaction is probably metabolized rapidly by cells of the liver and other tissues.

d) Localization of the Phosphoglyceride Synthetizing Enzyme Systems in the Cell

As far as phosphoglyceride synthesis in mitochondria is concerned, contradictionary observations are reported. WILGRAM and KENNEDY [324], WIRTZ and ZILVERSMIT [327], GURR *et al.* [92] and MCMURRAY and DAWSON [201] published data supporting the suggestion that mitochondria do not contain phosphoglyceride synthetizing enzymes. These authors gave evidence that the phosphoglycerides are synthetized in the endoplasmic reticulum and then exchanged with mitochondria. However, STOFFEL and SCHIEFER [294] and KAISER [135] showed that mitochondria are able to synthetize their own phosphoglycerides. It was also shown by STOFFEL and SCHIEFER that considerable acyltransferase activities are located in the outer membrane of rat liver mitochondria. Studies by SARZALA *et al.* [254] with rat liver strongly indicate that mitochondria do contain the acyltransferases involved in the conversion of both isomeric 3-sn-lysophosphatidylcholines into 3-sn-phosphatidylcholine and of 1-acyl-3-sn-lysophosphatidate into 3-sn-phosphatidate. These mechanisms were bound to the outer mitochondrial membrane fraction. In contrast to microsomes, mitochondria and mitochondrial subfractions did not catalyze to any extent the conversion of 1,2-diacylglycerols with (^{14}C) choline into 3-sn-phosphatidylcholine or that of 1,2-diacylglycerols with (^{14}C) fatty acids into triacylglycerols. Incubation of submitochondrial fractions with labelled sn-glycerol-3-P and fatty acids in the presence of outer membranes showed that sn-glycerol-3-P was converted into 3-sn-phosphatidate. It was concluded from isotope distribution that the uptake of fatty acids into these phosphoglycerides proceeds mainly via acylation of their monoacyl derivatives.

In order to elucidate whether mitochondria are able to synthetize their own phosphoglycerides, SAUNER and LÉVY [256] incubated mitochondria with labelled glycerol, orthophosphate and palmitic acid, thus demonstrating that phosphoglycerides are only labelled when palmitic acid was present. It was concluded from this observation that mitochondria cannot synthetize phosphoglycerides *de novo* but are able only to exchange exogenous fatty acids with mitochondrial phosphoglyceride fatty acids. The authors, therefore, believe that mitochondrial phosphoglycerides are synthetized elsewhere in the cell and are then transferred to the mitochondria. Incubation of microsomes labelled *in vivo*, with mitochondria showed transfer of phosphoglycerides between the microsomes and both mitochondrial membranes. Exchange of the microsomal phosphoglyceride with that of the inner membrane was the same when the incubation was carried out with intact mitochondria or with mitochondria devoid of outer membranes. The degree of transfer was 3-sn-phosphatidylethanolamine ≥ 3-sn-phosphatidylcholine > 3-sn-phosphatidylinositol. For the outer membrane phosphoglycerides, the transfer was about 13–20% for 3-sn-phosphatidylethanolamine, 15% for 3-sn-phosphatidylcholine and 8% for 3-sn-phosphatidylinositol. For the inner membrane, it was 34% for 3-sn-phosphatidylethanolamine, 25% for 3-sn-phosphatidylcholine and 4–8% for 3-sn-phosphatidylinositol. These results suggest that the whole phosphoglyceride molecule is transferred from microsomes to mitochondrial membranes. Furthermore, it was shown that there is the same degree of transfer to each of the two mitochondrial membranes, and that the transfer of phosphoglycerides to the inner membrane does not depend on the presence of the outer membrane. These data are consistent with results reported by WILLIAMS and BYGRAVE [325]: incorporation of labelled orthophosphate into either rat-liver microsomes or mitochondria alone was negligible, and maximal incorporation into the major phosphoglycerides occurred only when mitochondria, microsomes and cytosol were present together in the reaction medium. Optimal incorporation required NADH, fructose-1,2-diphosphate and CMP. There is conclusive evidence that the bulk of mitochondrial phosphoglycerides are not synthetized *de novo* by mitochondria. Further information concerning the utilization of the sn-glycerol-3-P and monoacylglycerol pathways for 3-sn-phosphatidylcholine biosynthesis in the hamster intestine were reported by JOHNSTON *et al.* [131]: it was demonstrated that the diacylglycerols synthetized via the monoacylglycerol pathway are not incorporated into 3-sn-phosphatidylcholine, while the diacylglycerols synthetized via the sn-glycerol-3-P pathway can serve as precursor for 3-sn-phosphatidylcholine formation. These data suggest that both pathways are located at different sites in the microsomes and that the bound intermediates are not transported from one site to the other. Alternatively, the microsomal fraction used for the experiments in reality consists of different subcellular particles carrying different activities.

e) Miscellaneous

Further information about the fatty acid composition, distribution and pairing of fatty acids in photosynthetic and nonphotosynthetic plants was presented by DEVOR and MUDD [61, 62, 63]: in spinach leaf chloroplasts and microsomes and in cauliflower inflorescence, the saturated fatty acids were predominantly esterified to the 1-position of 3-sn-phosphatidylcholine and unsaturated fatty acids to position 2. In the predominant species of cauliflower, linolenic acid comprised 22% of the total, linolenic and oleic acid 19% and linolenic and palmitic acids 37%. In general, the fatty acid pairing in the 1,2-diacylglycerol species of cauliflower 3-sn-phosphatidylcholine was found to be similar to that of animals. However, the variation from random is in no case as great as it is in the animal tissues studied.

KANOH [137] compared the mode of utilization of biosynthetically labelled 1,2-diacylglycerol species in the biosynthesis of 3-sn-phosphatidylcholine and 3-sn-phosphatidylethanolamine in rat liver microsomes. When each molecular species was incubated separately, no obvious difference in incorporation was found. However, relatively higher utilization of hexaenoic 1,2-diacylglycerol was found to occur in 3-sn-phosphatidylethanolamine formation in contrast to 3-sn-phosphatidylcholine biosynthesis when incorporation of various 1,2-diacylglycerols was studied.

Further information about the biosynthesis of various molecular species of phosphoglycerides [35, 155, 205, 213, 224, 285] and about phenobarbital-induced alterations in 3-sn-phosphatidylcholine and triacylglycerol synthesis in hepatic endoplasmic reticulum [331] was recently reported.

4. Biosynthesis of CDP-Diacylglycerols

Cytidine diphosphate-diacylglycerols (CDP-diglycerides) are lipid-soluble nucleotides that may be regarded as activated forms of 3-sn-phosphatidates. CDP-diacylglycerols have been identified as precursors of 3-sn-phosphatidylinositol [222] 3-sn-phosphatidylglycerol [147, 285] and 3-sn-phosphatidylglycerol-P [147, 285] in animal tissues, and 3-sn-phosphatidylserine [136], 3-sn-phosphatidylethanolamine [136] 3-sn-phosphatidylglycerol [41], 3-sn-phosphatidylglycerol-P [41] and 3-sn-diphospatidylglycerol (cardiolipin) [283] in *E. coli*. In *E. coli*, and possibly other biological systems, CDP-diacylglycerols may constitute the sole nucleotide intermediate for the biogenesis of phosphoglycerides.

The occurrence of CDP-diacylglycerol as an intermediate in the synthesis of 3-sn-phosphatidylinositol using a particulate fraction of guinea-pig kidney had already been observed by AGRANOFF *et al.* in 1958 [7]. They suggested that the reaction apparently takes place according to this equation:

$$\text{CDP-choline} + \text{3-sn-phosphatidate} \rightarrow \text{CDP-diacylglycerol} + \text{choline}. \quad (24)$$

In 1960, PAULUS and KENNEDY [222] reported that incubation of inositol and CDP-^{14}C-choline with a chick-liver particulate preparation lead to an exchange of inositol with 3-sn-phosphatidylinositol without a measurable release of choline.
Accordingly, they postulated a reaction as follows:

$$\text{3-sn-Phosphatidate} + \text{CTP} \rightarrow \text{CDP-diacylglycerol} + \text{PP}_i. \quad (25)$$

This reaction is catalyzed by the enzyme CTP:1,2-diacyl-L-glycerol-3-phosphate-cytidyltransferase (EC 2.7.7). PETZOLD and AGRANOFF [225] proposed an additional pathway for the synthesis of CDP-diacylglycerol in guinea-pig liver:

$$\text{CDP-diacylglycerol} + \text{inositol} \rightleftharpoons \text{CMP} + \text{phosphatidylinositol}. \quad (26)$$

However, this reversal of the synthesis of 3-sn-phosphatidylinositol appeared to be unlikely.

In 1966, CARTER and KENNEDY [33] presented evidence for the enzymatic formation of CDP-diacylglycerol in microsomal preparations from guinea-pig liver according to reaction (25). The incorporation of labelled CTP into CDP-diacylglycerol was dependent upon added 3-sn-phosphatidate. Incorporation of CMP was very slight. Net synthesis of CDP-diacylglycerol was measured. The synthesis was specific for CTP and showed an absolute requirement for added divalent cations. Manganese was twice as effective as magnesium. The pH optimum was found to be 7.5. Very low enzyme concentrations had only little effect, increasing concentrations lead to a linear elevation of the amount of products, and very high concentrations even caused an inhibition. Accordingly, it was presumed that the stabilization of the magnesium-phosphatidate complex by the lipoprotein of the microsomal enzyme preparation is needed for activity, and at very low concentrations of enzyme this effect may not be obtained. Pyrophosphate in a concentration of 4 mM was found to inhibit by 90% the incorporation of labelled CTP into CDP-diacylglycerol. The enzyme activity was mostly located in microsomes of guinea-pig liver, and the following relative specific activities were measured in guinea-pig organs: liver 74, intestinal mucosa 25, kidney 9, brain 8, and heart 2. Further interesting observations were reported by PETZOLD and AGRANOFF [225]: 3-sn-phosphatidate prepared from egg-yolk was more effective than synthetic lecithins. Under the same experimental conditions the amount of newly synthetized CDP-diacylglycerol from different 3-sn-phosphatidates was as follows: natural 2.0 nmoles; dihexanoyl-glycerol-P (GP) 1.4 nmoles; dipalmityl-GP, 1.1 nmoles; distearyl-GP,

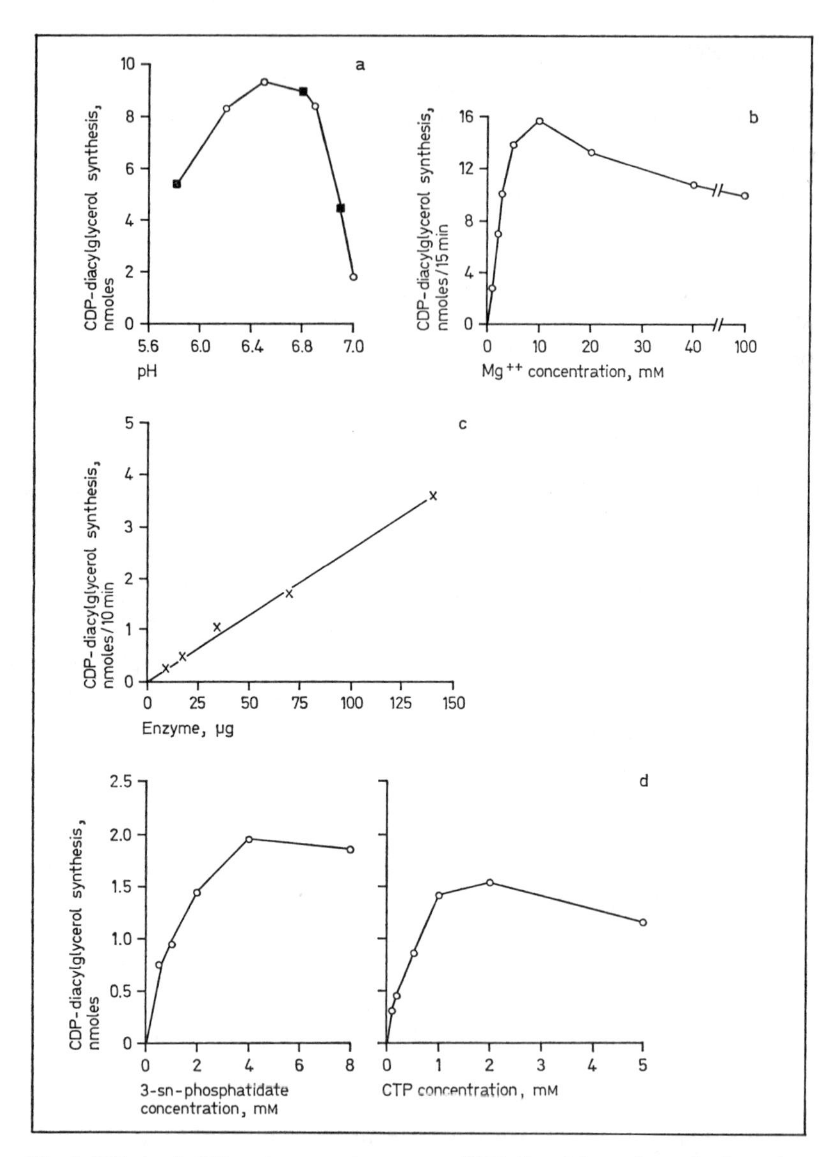

Fig. 6. Effect of different parameters upon CDP-diacylglycerol synthesis. *a* Influence of pH; the incubation mixture contained phosphate buffer, 3-sn-phosphatidate, CTP, enzyme protein, and $MgCl_2$. Incubation was carried out for 30 min at 37 °C. *b* Influence of Mg^{2+} concentration; imidazol buffer of pH 6.5 was used. The assay was carried out for 15 min at 37 °C. *c* Influence of enzyme concentration; the assay mixture was identical with *a* except that the concentration of $MgCl_2$ was 20 mM. Incubation was for 10 min at 37 °C. *d* Influence of substrate concentration; the concentration of 3-sn-

Table II. Kinetic data for the enzyme CTP:1,2-diacyl-L-glycerol-3-phosphate-cytidyltransferase (EC 2.7.7) [40]. Substrate 1 was varied, substrate 2 was constant

Substrate 1	Substrate 2	K_m, M
CTP	3-sn-phosphatidate (2×10^{-3} M)	7×10^{-4}
3-sn-Phosphatidate	CTP (10^{-3} M)	2×10^{-3}

1.1 nmoles; dieicosanyl-GP, 0.6 nmoles; no added lipid, 0.6 nmoles. Synthesis was inhibited by palmityl-CoA.[1]

Detailed studies about further enzyme properties were performed with particulate preparations from *E. coli* by CARTER [40]: CTP was incorporated into lipids in the presence of exogenous 3-sn-phosphatidate and Mg^{2+} and CDP-diacylglycerol was identified as the product. The influence of various parameters on the synthesis of CDP-diacylglycerol is clearly demonstrated in some figures (for detailed experimental conditions see reference [40]). Figure 6a and b show that the reaction is optimal at a pH of 6.5 and Mg^{2+} concentration of 5–10 mM. Figure 6c indicates that the initial rate of synthesis is a linear function of enzyme concentration over a wide range. As is shown in figure 6d, increasing substrate concentrations cause an increased synthetic rate which is slightly inhibited at higher substrate concentrations. Kinetic data are shown in table II.

Magnesium can have two separate effects that inhibit the enzyme reaction (figure 6b): the formation of insoluble Mg^{2+}-3-sn-phosphatidate prevents the reaction with the enzyme; a direct effect of Mg^{2+} on the enzyme protein may prevent the penetration of the lipid substrate to the active site. The authors assume that 3-sn-phosphatidate apparently protects the membrane-bound enzyme against this effect of Mg^{2+}.

Recently, SCHEGGET *et al.* [259] reported that the deoxy analogue of CTP, dCTP, is incorporated by rat liver mitochondria yielding dCDP-

1 CDP-diacylglycerol synthesis was also observed in cell-free extracts of *S. cerevisiae* by HUTCHISON and CRONAN [127].

phosphatidate was varied with CTP constant at 1 mmole (left) and that of CTP was varied with 3-sn-phosphatidate constant at 2 mmole (right). Other conditions as in *a*. Incubation was for 20 min at 37°C; ([40] reproduced with permission of the publishers.)

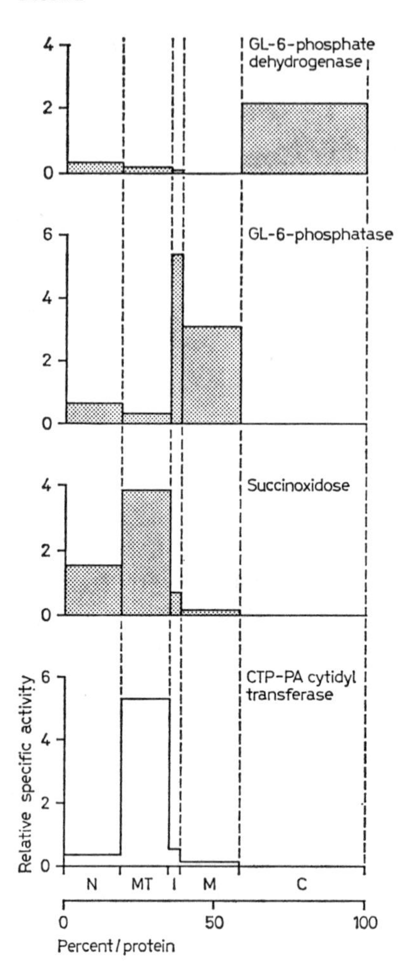

Fig. 7. Distribution patterns of CTP: 1,2-diacyl-L-glycerol-3-phosphate-cytidyl-transferase (EC 2.7.7) and marker enzymes among subfractions of rat liver homogenate (N = nuclear; MT = mitochondrial; I = 'intermediate'; M = microsomal; and C = cytosol). [Experimental conditions see reference 310]. (Reproduced with permission of the publisher.)

diacylglycerol. The apparent K_m for dCTP was 4×10^{-5} M. Synthesis was inhibited by rCTP, whereas the incorporation of the rCDP moiety of rCTP into rCDP-diacylglycerol was inhibited by dCTP. It is suggested from these results that the synthesis of CDP-diacylglycerol and dCDP-diacylglycerol is catalyzed by the same enzyme. A particulate enzyme preparation from

Micrococcus cerificans [200] showed an absolute requirement for K^+ and a nonionic detergent. It was the most active enzyme preparation which was ever obtained from any mammalian source.

In view of the central rôle of CDP-diacylglycerols in phosphoglyceride biosynthesis, VORBECK and MARTIN [310] determined its intracellular site of formation. Rat liver and bovine heart homogenates were fractionated by differential centrifugation and the composition of the fractions was evaluated by marker enzymes and electron-microscopy.

Figure 7 presents the relative specific activities in subfractions of a rat liver homogenate. In liver as well as in heart, the highest relative specific activities were associated with the mitochondrial fractions and closely paralleled the distribution pattern for succinoxidase, a mitochondrial marker. These data are compatible with results obtained by PETZOLD and AGRANOFF [225] using chick-brain preparations. Further support is given by observations by ZBOROWSKI and WOJTCZAK [335] concerning CTP-stimulated 3-sn-phosphatidate synthesis in rat liver mitochondria. On the contrary, CARTER and KENNEDY [33], using guinea-pig liver, reported the greatest activity in the microsomal fraction. Whether this divergency is due to species differences or differences in preparation of the subcellular fractions remains to be evaluated.

5. Biosynthesis of 3-sn-Phosphatidylinositol

The original work on *de novo* synthesis of 3-sn-phosphatidylinositol was done by PAULUS and KENNEDY [222]. Using ^{32}P-labelled sn-glycerol-3-P and diacylglycerol-3-P as precursors for 3-sn-phosphatidylinositol formation in a chick liver particulate fraction, they found an absolute requirement for CTP and Mn^{2+}. The optimal pH was 8. The following pathway was postulated:

$$\text{CDP-diacylglycerol} + \text{inositol} \rightarrow \text{3-sn-phosphatidylinositol} + \text{CMP}. \quad (27)$$

HOLUB *et al.* [116] identified over 27 molecular species of different 3-sn-phosphatidylinositols. In further studies using labelled phosphate for 3-sn-phosphatidylinositol synthesis in rat liver, this group observed an active synthesis of the monoenoic and dienoic 3-sn-phosphatidylinositols by way of the 3-sn-phosphatidate, followed by a deacylation-reacylation cycle, as claimed for other rat liver phosphoglycerides [117]. It was suggested that rat liver 3-sn-phosphatidylinositols have a heterogenous metabolism *in vivo* and that more than one mechanism must contribute to their biosynthesis and degradation.

For further information about these lipids see the review by HILL and LANDS [107].

6. Biosynthesis of 3-sn-Phosphatidylserine

Net synthesis of 3-sn-phosphatidylserine was found in a soluble fraction of *E. coli* according to this equation [136]:

CDP-diacylglycerol + L-serine → 3-sn-phosphatidylserine + CMP. (28)

This reaction is catalyzed by the enzyme CDP-1,2-diacylglycerol: L-serine diacylglycerol transferase. The enzyme has a pH optimum of 7.5 and an apparent K_m for L-serine of 8×10^{-4} M. No divalent cations are required, and no inhibition by hydrazine, KF or N-ethylmalimide occurs. Similar observations were reported by PATTERSON and LENNARZ [220] using ghosts of *Bacillus megaterium*. CDP-diacylglycerol has also been shown to be directly involved in 3-sn-phosphatidyl serine synthesis in *S. cerevisiae* [290]. However, there is no evidence that 3-sn-phosphatidylserine formation according to reaction (28) occurs in animals including the protozoon *T. pyriformis* [59].

In animal tissues [34], incorporation of L-serine into phospholipids takes place via an exchange mechanism starting from 3-sn-phosphatidylethanolamine which is derived from 1,2-diacylglycerol (see IV A, section 3a).

B. Type II Reaction

1. Biosynthesis of 3-sn-Phosphatidylglycerol and Derivatives

a) Biosynthesis of 3-sn-Phosphatidylglycerol

3-sn-Phosphatidylglycerol has been shown to be a major component of the phospholipids of green leaves [61, 62, 196], algae and bacteria, but only a minor component of mammalian lipids [87, 191, 297]. KIYASU *et al.* [147] first demonstrated that this lipid was synthetized in chicken and rat liver particles by reactions (29) and (30). 3-sn-Phosphatidyl synthesis was shown to occur by the same reactions in rat brain mitochondria [236, 284] and in *E. coli* [136]. In contrast to *E. coli*, the transferase of liver catalyzing reaction (29) does not require divalent cations. A pH optimum between 6.2 and 8.5 was determined [147]. Biosynthesis occurs as follows:

CDP-diacylglycerol + sn-glycerol-3-P
→ 3-sn-phosphatidyl-sn-glycerol-3-P + CMP. (29)

3-sn-phosphatidyl-sn-glycerol-3-P → 3-sn-phosphatidylglycerol + P_i. (30)

Table III. Kinetic data for the synthesis of 3-sn-phosphatidylglycerol from sn-glycerol-3-P by rat-liver mitochondria [259]

Substrate	K_m, M
dCDP-diacylglycerol	2×10^{-5}
rCDP-diacylglycerol	7×10^{-5}

This reaction sequence leads to the formation of a phospho-diester bond (figure 1). PATTERSON and LENNARZ [221] demonstrated that 3-sn-phosphatidylglycerol synthesis in ghosts of *Bacillus* sp. proceeds via CDP-diacylglycerol requiring Mg^{2+} and a non-ionic detergent. Furthermore, SCHEGGET *et al.* [259] observed that dCDP-diacylglycerol substituted for rCDP-diacylglycerol in the synthesis of 3-sn-phosphatidylglycerol from sn-glycerol-3-P by mitochondrial preparations from rat liver. Kinetic data are given in table III.

Recently, FANG and MARINETTI [72] published information concerning the involvement of phospholipids in the dynamics of mitochondrial function. Rat-liver mitochondria incubated with labelled P_i, ATP and glycerol synthetized 3-sn-phosphatidylphosphate. The incorporation of P_i was stimulated by succinate, glycerol and NAD^+, and this stimulation was inhibited by oligomycin, atractyloside and dinitrophenol. Glycerol and NAD^+ also stimulated labelling of 3-sn-phosphatidylglycerol-P from labelled ATP and labelled glycerol, and this stimulation was inhibited by oligomycin but not by atractyloside. The authors conclude from these results that NAD^+ and glycerol act on the outer surface of the inner mitochondrial membrane and/or on the outer mitochondrial membrane. Moreover, they postulate that the stimulating effects of NAD^+ and glycerol are not primarily related to their redox properties but rather appear to be mediated by some allosteric action on the mitochondrial membrane.

OLSON and BALLOU [216] presented the first complete structural characterization of 3-sn-phosphatidyl-1′-(3′-acyl)-sn-glycerol, a new phospholipid from *Salmonella typhimurium.* This lipid comprises 2% of the phospholipids in this organism, 3-sn-phosphatidylethanolamine 83%, 3-sn-phosphatidylglycerol 11%, and 3,3′-sn-diphosphatidylglycerol 4%. The two glycerols of the molecule are of opposite configuration with respect to their attachments to the central P. The fatty acid composition resembles that of total lipid fatty acids.

b) Formation of 3,3′-sn-Diphosphatidylglycerol (Cardiolipin)

The most abundant phospholipids found in rat liver mitochondria are 3-sn-phosphatidylcholine, 3-sn-phosphatidylethanolamine and 3,3′-sn-diphosphatidylglycerol (cardiolipin) [294]. Cardiolipin may be of special importance in the process of mitochondrial biogenesis since, unlike 3-sn-phosphatidylcholine and 3-sn-phosphatidylethanolamine, it appears to be uniquely mitochondrial, present only in small amounts in extra-mitochondrial structures [60].

STANACEV *et al.* [283] first suggested that 3,3′-sn-diphosphatidylglycerol synthesis in cell-free preparations of *E. coli* can occur by the following pathway:

3-sn-Phosphatidylglycerol + CDP-diacylglycerol
→ 3,3′-sn-diphosphatidylglycerol + CMP. (31)

This reaction is catalyzed by the enzyme CDP-diacylglycerol-phosphatidylglycerol-phosphatidyltransferase and may be stimulated by CDP-diacylglycerol. This was indirectly supported by further data obtained by DAVIDSON and STANACEV [55] who studied mitochondria and microsomes from guinea-pig liver, adding labelled sn-glycerol-3-P and exogenous CDP-diacylglycerol. They found that only with mitochondria and a system generating CDP-diacylglycerol, was 3,3′-sn-diphosphatidylglycerol synthetized. It was concluded from these results that 3,3′-sn-diphosphatidylglycerol synthetase has high specificity for the endogenously formed CDP-diacylglycerol, and that biosynthesis of polyphosphoglycerides in mitochondria is highly compartmentalized. The biosynthesis of 3,3′-sn-diphosphatidylglycerol is associated with a compartment capable of generating CDP-diacylglycerol, or permeable to the endogenously generated CDP-diacylglycerol, but not permeable to the exogenously added CDP-diacylglycerol. However, biosynthesis of 3-sn-phosphatidylglycerol is associated with a compartment permeable to the exogenously added CDP-diacylglycerol. Apparently, the compartment synthetizing 3,3′-sn-diphosphatidylglycerol is permeable to newly synthetized 3-sn-phosphatidylglycerol. Similarly, HOSTELLER *et al.* [118] found 3,3′-sn-diphosphatidylglycerol formation from CDP-diacylglycerol and sn-glycerol-3-P using rat-liver mitochondria. The reaction required Mg^{2+} and the rate of 3,3′-sn-diphosphatidylglycerol synthesis was slow, probably explaining the fact that this reaction mechanism was not previously observed. Furthermore, STANACEV *et al.* [287] could demonstrate for the first time that both 3-sn-phosphatidylglycerol and CDP-diacylglycerol directly participate in the enzymatic formation of 3,3′-sn-diphosphatidylglycerol according to reaction mechanism (31) using isolated rat-liver mitochondria. A recent report by HOSTELLER *et al.* [119] gives more detailed information about the subcellular and submitochondrial localization of the biosynthesis of 3,3′-sn-diphosphatidylglycerol and related phospholipids in rat liver. 3,3′-sn-diphosphatidylglycerol synthesis was found exclusively in the inner mitochondrial membrane. 3-sn-Phosphatidylglycerol was predominantly formed in the inner membrane although a small contribution was noted in the outer membrane. CDP- and dCDP-diacylglycerol synthesis occurred primarily in the microsomal fraction. 3,3′-sn-Diphosphatidylglycerol synthesis from 3-sn-phosphatidylglycerol required CDP-diacylglycerol or dCDP-diacylglycerol and either Mg^{2+}, Mn^{2+} or Co^{2+}. The reaction was strongly inhibited by non-ionic detergents such as Triton X-100.

An alternate pathway for 3,3′-sn-diphosphatidylglycerol synthesis was proposed by STANACEV and STUHNE-SEKALEC [286]:

2 3-sn-phosphatidylglycerol = 3, 3′-sn-diphosphatidylglycerol + glycerol. (32)

This reaction mechanism of intermolecular transphosphatidylation should not require CDP-diacylglycerol; 3-sn-phosphatidylglycerol is the donor and the acceptor of the 3-sn-phosphatidyl group. Evidence for the occurrence of this reaction in bacteria was provided by DE SIERVO *et al.* [270], POLONOVSKI *et al.* [230], SHORT and WHITE [268] and HIRSCHBERG and KENNEDY [109].

In a very recent report, HOSTELLER *et al.* [120] described studies using intact mitochondria and inner membranes of rat liver, and a particulate fraction of *E. coli* for 3,3′-sn-diphosphatidylglycerol synthesis from labelled precursors. In mitochondria, 3,3′-sn-diphosphatidylglycerol synthesis from 3-sn-phosphatidyl (1′-^{14}C)glycerol was supported when CDP- and dCDP-diacylglycerol were added exogenously. The maximum rate observed for CDP-diacylglycerol was about twofold greater than for dCDP-diacylglycerol. Optimal concentrations of CDP-diacylglycerol caused a threefold stimulation of 3,3′-sn-diphosphatidylglycerol formation. (^{14}C)Glycerol was not produced in considerable amounts under these conditions. In mitochondria, 3-sn-(2-^{3}H)phosphatidyl(1′-^{14}C)glycerol was converted in the presence of unlabelled CDP-diacylglycerol to 3,3′-sn-diphosphatidylglycerol having a nearly identical ^{3}H:^{14}C ratio. These data confirm that 3,3′-sn-diphosphatidylglycerol is synthetized from 3-sn-phosphatidylglycerol and CDP-diacylglycerol in mitochondria.

On the contrary, in membrane preparations from *E. coli* [120] the 3,3′-diphosphatidylglycerol synthetized at low concentrations of CDP-diacylglycerol had a $^{3}H:{}^{14}C$ ratio of 2.0 relative to its precursor 3-sn-(2-^{3}H) phosphatidyl(1′-^{14}C) glycerol and significant amounts of (1′-^{14}C)glycerol were released during incubation. These findings indicate that in *E. coli* 3,3′-sn-diphosphatidylglycerol is formed from two molecules of 3-sn-phosphatidylglycerol. However, the authors presented evidence suggesting that the CDP-diacylglycerol pathway may be operational in *E. coli* at higher concentrations of CDP-diacylglycerol.

From the information on 3,3′-sn-diphosphatidylglycerol biosynthesis reviewed here, it can be postulated that pathway 1 occurs in mammalian tissues, and pathway 2 in *E. coli.* Whether mechanism (32) is functioning in animals remains to be evaluated. However, there is evidence that mechanism (31) might contribute to 3,3′-sn-diphosphatidylglycerol formation in *E. coli.*

Furthermore, results described by DAVIDSON and STANACEV [55] concerning 3,3′-sn-diphosphatidylglycerol formation together with previously reported observations on mitochondrial capability to synthetize 3-sn-phosphatidate and 3-sn-lysophosphatidate [115, 215, 292, 306, 313), CDP-diacylglycerol [29, 299] 3-sn-phosphatidyl-sn-glycerol-3-P and 3-sn-phosphatidylglycerol [42, 88, 170, 312] have established mitochondrial autonomy in the biosynthesis of all known polyglycerol phosphatidates.

C. Control of Phosphoglyceride Biosynthesis

MARINETTI and his co-workers [64, 69, 194] observed that CTP and CMP inhibited the incorporation of ^{14}C-glycerol into 3-sn-phosphatidate in rat organs, and the incorporation by the acylglycerols was stimulated. They suggested that this effect was due to the hydrolysis of CDP-diacylglycerol to produce CDP and diacylglycerol. More recently, POSSMAYER and MUDD [238] found CTP to be the most effective co-factor both for the inhibition of sn-glycerol-3-P acylation and for the stimulation of 3-sn-phosphatidyl-inositol synthesis in rat brain preparations; acylglycerol production was only moderately elevated. These data indicate that the effect of the cytidine nucleotides is due to a direct inhibition of 3-sn-phosphatidate production, an obligatory intermediate in *de novo* diacylphosphoglyceride and acylglycerol synthesis. Since 3-sn-phosphatidate is a key intermediate located at a critical branch point in lipid metabolism not accumulating in most tissues, the factors controlling its synthesis will have a marked influence on the formation of diacyl lipids. Presumably, the site of the cytidine nucleotide effect is prior to the action of phosphatidate phosphohydrolase.

V. Biosynthesis of Glycolipids Including Sphingolipids

Members of this lipid class contain N-acylsphingosin as an alcohol instead of diacylglycerol. Except the sphingomyelines, the others are glycosides, therefore also called glycolipids. About 60 naturally occurring sphingolipid long-chain bases have been identified or proposed. A review is available on their nomenclature, distribution, metabolism, biologic properties, chemistry and methods of characterization [138]. Recently, this topic was reviewed comprehensively by LENNARZ [174] STOFFEL [295], WIEGANDT [323] and by VOLK and ARONSON [309].

VI. Biosynthesis of Plasmalogens

These are O-alkyl-phospholipids which have alkyl groups linked to the glycerol moiety of the lipid by a vinyl ether bond. The reader is referred to an extensive review of the chemistry of ether bonds in glycerolipids, including historical aspects, occurrence, biological effects and metabolism by SNYDER and HOKMAN [278] and GOLDFINE [86]. New aspects concerning plasmalogen biosynthesis were recently reported [57, 76, 279, 280, 281, 296, 329].

VII. Selected Aspects on Lipid Biosynthesis in Mammalian Organs

Carbohydrates and also ethanol are excellent precursors for lipid synthesis. Furthermore, epidemiological studies have suggested that dietary sugars and ethanol are responsible for the increasing incidence of diseases related to disorders in carbohydrate and lipid metabolism. Therefore, some aspects concerned with lipid formation from fructose [79, 157–159] as one of the main dietary carbohydrate constituents, and from acetate [160–162] as a direct degradation product of ethanol will be discussed in this chapter. If fructose is used as precursor, the glycerol moiety of the lipids is chiefly produced; acetate, however, is mainly incorporated into the fatty acid components, sterols and sterol derivatives.

A. Lipid Biosynthesis from Fructose in Liver

It was reported by MACDONALD [187–190] and HERMAN *et al.* [104] that fructose raises the level of serum triacylglycerols and that this hexose seems to be a better triacylglycerol precursor than glucose or glucose-containing compounds. Similarly, ZAKIM *et al.* [333] observed the highest triacylglycerol concentrations in serum and liver of rats fed a fructose diet, and a transient increase of hepatic sn-glycerol-3-P after the intravenous injection of fructose to rats [334]. Moreover, BAILY *et al.* [19] found an increase of enzymes associated with lipogenesis in liver. Feeding rats with glucose and fructose lead to an enhanced lipogenesis in liver and adipose tissue; however, there was no difference between glucose and fructose, as was reported by COHEN and TEITELBAUM [48]. On the contrary, BAR-ON and STEIN [20] found that in liver more fructose than glucose was converted to triacylglycerols and secretion of triacylglycerols into the serum was higher after fructose than after glucose administration to Triton-treated rats. As feeding fructose did not induce lipoprotein-lipase activity in adipose tissue, the egress of triacylglycerols from the serum and, thus, the homeostatic regulation of triacylglycerol levels may be impaired, leading to its accumulation in the bloodstream.

Further evidence for increased rates of triacylglycerol synthesis in the liver due to sucrose and fructose diets, and to an elevated hepatic glycerol-3-P dehydrogenase activity leading to an elevated sn-glycerol-3-P supply was obtained by MUKHERJEE *et al.* [206]. In spite of low sterol degradation rates in the liver of rats fed on fructose, the serum cholesterol level was not significantly altered because hepatic cholesterol formation was diminished due to a relative increase in the rate of hepatic fatty acid synthesis from acetate. HILL [108] reported that fructose feeding of rats led to an elevation of serum triacylglycerols and caused an accumulation of liver triacylglycerols in fed but not in fasted mature rats. Recently, CHEVALIER *et al.* [45] found that dietary fructose or sucrose increased serum triacylglycerol levels in mature but not in weanling rats, and TOPPING and MAYES [304] showed that liver from rats perfused with fructose-containing blood raised the secretion of triacylglycerols in very low-density lipoproteins, and that lipogenesis was enhanced in the livers treated with fructose plus bovine insuline. MARUHAMA [199] observed in patients with a fructose-induced hypertriglyceridaemia, who had had a myocardial infarction, that the incorporation of fructose carbon into serum triacylglycerol-glycerol increased linearly. However, the formation of triacylglycerol-glycerol and fatty acids was much greater after

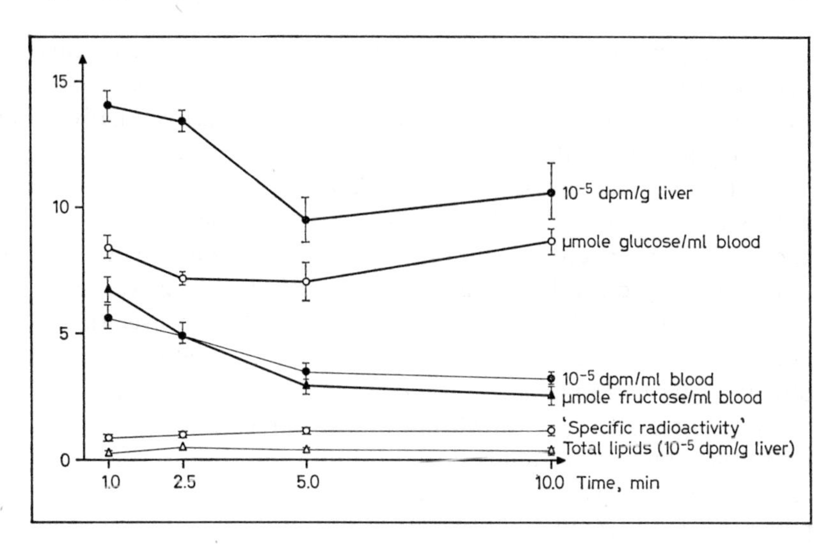

Fig. 8. Time-course of selected parameters in blood and liver after fructose administration. 20 μCi (U-^{14}C) fructose + 555 μmole carrier fructose were infused intraportally to male rats. After 1,2.5, 5 and 10 min mixed blood from the tail and liver samples were taken from the animals. 'Specific radioactivity': total radioactivity of the blood is related to the corresponding fructose concentrations. The data for 1 min are unpublished; the data for 2.5, 5 and 10 min are taken from KUPKE and LAMPRECHT [157]. dpm = Disintegrations per minute.

fructose than after glucose ingestion. The fact that fructose obviously is converted into lipids more rapidly than glucose may be possibly due to the by-passing of the second phosphorylation of the hexose molecule and to the direct breakdown of the fructose-1-phosphate into the immediate precursors for sn-glycerol-3-P [164].

However, a hitherto undescribed response of almost all patients with carbohydrate-induced hypertriglyceridaemia and normal subjects to oral fructose was a transient decrease in serum triacylglycerols as was observed by KAUFMANN *et al.* [140]. Serum-free fatty acid levels also decreased in patients as well as in normal subjects in response to fructose.

Absorption studies in man carried out by COOK [50] showed that 80–90% of fructose is absorbed in the jejunum as fructose. Therefore, experiments previously performed [79, 157–159] administering uniformly labelled ^{14}C-fructose intraportally to rats may be regarded as fairly physiological ones if metabolic aspects due to anaesthesia are neglected. An amount of carrier fructose (555 μmoles) was chosen which did not affect considerably intermediary metabolic concentrations in the liver, thus avoiding enhanced lipid formation by an elevated supply of the respective precursors [157]. In short-term (10-min) experiments, the highest radioactivity derived from fructose appeared in the liver after 1 min and it decreased rapidly up to 5 min (fig. 8). This time-course of ^{14}C-labelling in the blood was almost paralleled by the corresponding fructose concentrations. This observation and a nearly unchanged 'specific radioactivity' (total radioactivity as related to fructose) suggested that ^{14}C activity in the blood is chiefly carried by unchanged fructose which is successively extracted by the liver (fig. 8) and other organs. The same response of blood fructose and also of glucose was observed later by HEINZ and JUNGHÄNEL [103] in similar experiments.

After 1 min, the approximate distribution of radioactivity between some liver metabolites was as follows: fructose 30%, fructose-1-phosphate 55%, glucose 12%, sn-glycerol-3-P 2%, total lipids 1% [unpublished data]. The percentage radioactivity measured in the fructose-1-phosphate is consistent with the amount of fructose converted into fructose-1-phosphate in similar experiments [103]. The relative small ^{14}C incorporation into lipids (fig. 8) reached maximal values after 2.5 min, nearly holding this level up to 10 min; this also indicates that obligatory lipid degradation is counteracted by the continued uptake of ^{14}C-fructose from the blood.

Figures 9 and 10 show the ^{14}C-distribution between the different lipid fractions within the experimental period. Total radioactivity is mainly comprised by the phosphoglycerides (96%) whereas the highest specific radioactivities were found after 2.5 min in the diacylglycerols. It was postulated from this result that the newly synthetized diacylglycerols have a very high turnover rate at low concentrations (1.23% of the total lipids). Furthermore, it is shown in figure 9 that up to 5 min specific ^{14}C-activity of the phosphoglycerides decreases at the expense of the diacylglycerols, probably indicating that primarily 3-sn-phosphatidate was formed which is hydrolyzed to diacylglycerols. This reaction was clearly demonstrated by HÜBSCHER [125] in *in vitro* experiments using phosphatidate phosphohydrolase from the particle-free supernatant of rat liver for 3-sn-phosphatidate hydrolysis producing diacylglycerols (fig. 11). In acylation studies *in vivo* using labelled linoleic acid, ÅKESSON [12] also observed that in rat-liver, 3-sn-phosphatidates were rapidly labelled, and that the maximum was found after 10 sec. The maximum in diacylglycerols was at 30–60 sec, while radioactivity in the other glycerolipids rose throughout the experimental period.

After 5 min (fig. 9), specific radioactivity of the diacylglycerols de-

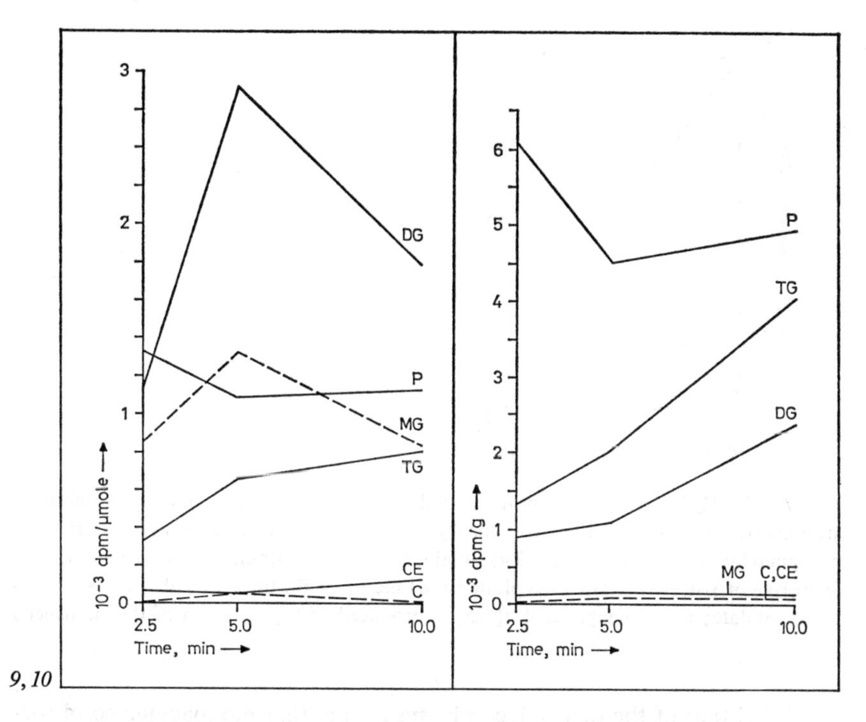

Fig. 9. Specific radioactivity of different liver lipids after the intraportal infusion of (U-^{14}C) fructose to rats ([157]; reproduced with permission of the publisher.) DG = diacyglycerols; P = phospholipids; MG = monoacyglycerols; TG = triacyglycerols; CE = cholesterol esters; C = cholesterol; dpm = disintegrations per minute.

Fig. 10. Total radioactivity of different liver lipids after the intraportal infusion of (U-^{14}C)fructose to rats. For the phospholipids, the ordinate scale has to be multiplied by 10 ([157]; reproduced by the permission of the publisher.) Abbreviations as for figure 9.

creases in favour of the phosphoglycerides. It is very likely that this demonstrates base containing phosphoglyceride formation via diacylglycerols. Similarly, Elovson *et al.* [68] found that in rat-liver microsomes incorporation of labelled precursors into diacylglycerols occurs more rapidly than into the corresponding 3-sn-phosphatidylcholines, suggesting the existence of a normal precursor-product relationship between the specific species of these classes of glycerolipids.

The time-course for triacylglycerol synthesis (fig. 9) also agrees well with the known mechanisms about triacylglycerol formation via diacylglycerols. Moreover, the specific radioactivity of the monoacylglycerols

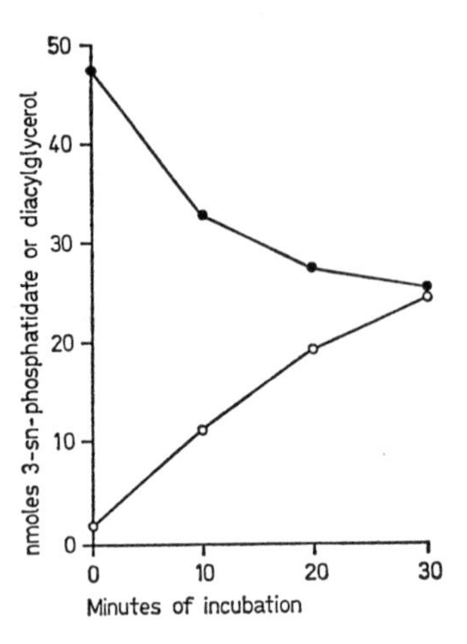

Fig. 11. Hydrolysis of membrane-bound 3-sn-phosphatidate. A mitochondrial preparation of rat liver containing biosynthetically formed 3-sn-phosphatidate was incubated with phosphatidate hydrolase (EC 3,1.3.4) obtained from the patricle-free supernatant. The hydrolysis of 3-sn-phosphatidate is plotted as a function of the time of incubation. • = 3-sn-Phosphatidate; ○ = diacylglycerol. ([125]; reproduced with permission of the publisher.)

paralleled that of the diacylglycerols, suggesting that monoacylglycerol formation in the liver may be also mediated by the sn-glycerol-3-P pathway although total ^{14}C-incorporation into these lipids (fig. 10) is very low and in the range of sterols and sterolesters.

Analysis of the glycerol and fatty acid components showed that on the average 90% of the radioactivity was incorporated into the glycerol moiety of the acylglycerols.

In order to follow the introduction of the trioses dihydroxyacetone-phosphate and D-glyceraldehyde formed from fructose-1-phosphate by the aldolase reaction into glycerolipids, (1-^{14}C) and (6-^{14}C) fructose were administered to the rats [158]. ^{14}C-incorporation into the glycerol of the acylglycerols was nearly independent of the site of labelling in the fructose molecule. Also, after (6-^{14}C) fructose, the radioactivity of the phosphoglycerides was about 70% of the value found after (1-^{14}C) fructose. Furthermore, 74% of the total radioactivity of the acylglycerol-glycerol was found in carbon 3 [79]. It was concluded from these results that the unphosphorylated triose D-glyceraldehyde derived from fructose-l-phosphate cleavage is metabolized to sn-glycerol-3-P to a considerable extent via glyceraldehyde-3-P formed by the triokinase reaction. Possibly, this D-glyceraldehyde might also be introduced into lipid glycerol by the reduction to glycerol which is phosphorylated to sn-glycerol-3-P [79].

B. Lipid Biosynthesis from Acetate in Arteries

Arterial tissues have been shown to synthetize lipids actively from various precursors [44, 218, 272, 320, 338] modifying this pattern under different influences, such as hypoxia [75 150,] air pollutants [321], stress [226], atherogenic diets [47, 56, 121, 123, 181, 337] and in atherosclerotic human arterial segments [46]. PARKER *et al.* [218] found some similarities between the mode of phosphoglyceride synthesis in liver and aorta from humans and rabbits. For current aspects on arterial lipid metabolism see JONES [132].

However, the results available were obtained under very different experimental conditions, using primarily aortas from various species. Little information exists about coronary arteries. Therefore, dog aortas and coronary arteries were simultaneously perfused with the animals' own serum using pulsatile pressures for 8 h by KUPKE [160, 161, 162]. (2-^{14}C) acetate was added to the perfusion fluid (serum) at a very high specific radioactivity thus avoiding a considerably increased acetate supply which may cause enhanced lipid formation. This experimental design was used in order to approach physiological conditions as near as possible *in vitro*. The examination of lipid metabolism in three aortic layers and in coronary arteries showed that the intima preparation containing some layers of adjacent media and the coronary arteries incorporated identical amounts of ^{14}C-acetate, whereas the media and the adventitia also containing some medial layers incorporated only 20–50% of label, as compared to the intima preparation [160]. Nevertheless, ^{14}C-distribution between the various lipids examined was nearly identical in the intima preparation and in the media (fig. 12). Diacyl- and triacylglycerols were mostly produced in the adventitia and in the coronary arteries; labelling of diacylglycerols and sterolesters was highest in the coronary arteries and synthesis of phosphoglycerides mainly occurred in the aortic media and intima. Interestingly, the pattern of the percent ^{14}C incorporation into the various lipids roughly reflects the corresponding lipid contents in the tissues. It is suggested that the arterial lipid concentrations, which are comprised of endogenous and exogenous lipids, are regulated by this relative synthetic rate; that is, the interrelationship between the newly synthetized lipids.

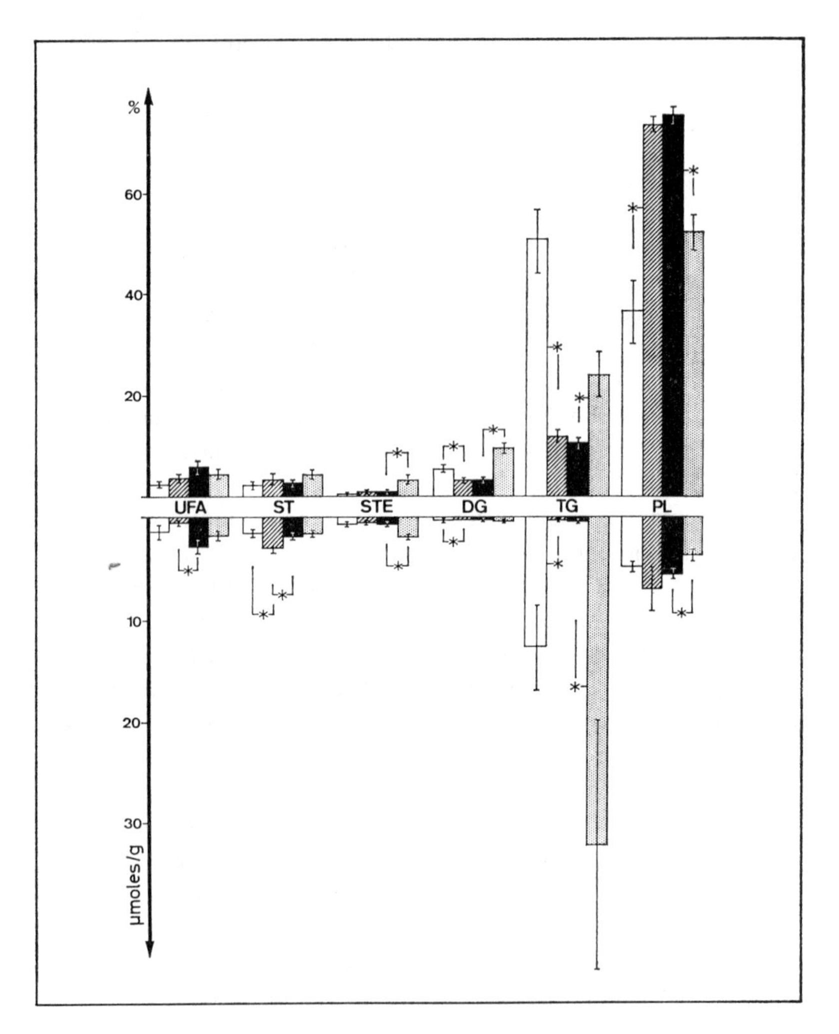

Fig. 12. Lipid pattern of the normal dog aorta and coronary artery. The percentage of ^{14}C incorporation, as related to the total lipid radioactivity, is compared with the corresponding lipid content in each tissue. □ = Aortic adventitia; ▨ = aortic media; ■ = aortic intima (the intima and adventitia contained several layers of adjacent media); ⁙ = coronary arteries. Upright line with bars $\pm s_{\bar{x}}$. UFA = unesterified fatty acids; ST = free sterols; STE = sterolesters; DG = diacylglycerols; TG = triglycerols; PL = phospholipids. * = significant ([160]; reproduced with permission of the publisher.)

Moreover, 99% of the acetate label was found in the acylglycerol fatty acids, and 95% in the phosphoglyceride fatty acids. This difference was highly significant, indicating that radioactivity was also incorporated by the phosphoglyceride constituents other than glycerol.

Feeding the animals on an atherogenic diet [160] caused a severe hyperlipaemia but no morphological changes in the arteries and no alterations in ^{14}C incorporation by the total lipids of the vessels was observed. However, early changes in some lipid classes were seen: the sterolester content and its labelling were markedly increased in the aortic media and intima. This result is consistent with data recently reported by HOWARD [123], for atherosclerotic rabbit aortas, and by CHOBANIAN and MANZUR [46] for human fatty streak lesions. Moreover, percent radioactivity of the free sterols was also markedly increased in the intima preparation but it was significantly diminished in the coronary arteries. Similarly, one dog of this group which developed severe atherosclerotic lesions in both vessels had also a drastically reduced sterol formation in the aortic layers [162]. It is postulated from these observations that in coronary arteries the ATP supply necessary for sterol synthesis is exhausted at an earlier stage of atherogenesis, as compared to aorta. Incorporation of ^{14}C-acetate into the acylglycerol-glycerol was abolished, and it was significantly diminished in the phosphoglycerides. It is suggested that acetate metabolism via sn-glycerol-3-P is inhibited by this diet.

Arteries from another group of normal animals were perfused with serum containing the alkaloid nicotine [161]. ^{14}C-incorporation into free sterols was reduced in the aorta, and it was unchanged in the coronary arteries; however, ^{14}C incorporation by the sterolesters was significantly elevated in the aortic intima preparation and in the coronary arteries. In the coronary arteries, an enhanced labelling of the triacylglycerols was also observed. These findings may be interpreted in terms of an impaired oxidative metabolism in the cell by nicotine leading to alterations in arterial lipid metabolism comparable to changes caused by hypoxia or by atherosclerosis.

It is suggested from these and other results that arteries having a different morphological structure and physiological function such as aorta and coronary arteries, are probably furnished with inherent biochemical factors including mechanisms of lipid biosynthesis *in situ* which predispose dog coronary arteries for atherogenesis, as compared to aorta. Comparison of normal vessels and their response to an atherogenic diet and to nicotine *in vitro* support this conclusion [162].

C. Lipid Biosynthesis in Various Mammalian Organs

Miscellaneous aspects concerning lipid formation in mammary cells [146], in adipose tissue [101, 172, 219], in brain [265], in skin [319], in bone cells [64] and in liver [31, 175] were recently reported. Comparative aspects of lipogenesis in mammalian tissues were discussed by SHRAGO *et al.* [269].

D. Factors Influencing Lipogenesis

Different aspects of lipid formation were lately examined: dietary [186, 301] and hormonal effects [23, 208, 275, 336], the influence of unesterified fatty acid [148] and temperature [271], relations to carbohydrate [102, 219, 314, 315] and to protein metabolism [305].

VIII. Acknowledgements

I wish to thank Mrs. J. KREYZI and Mrs. J. VON DER HEYDE for their kind assistance in preparing the manuscript.

IX. References

1 AAS, M.: Long-chain acyl-CoA synthetase in rat liver stimulation by high salt concentrations. Biochim. biophys. Acta *202:* 250–258 (1970).
2 AAS, M.: Organ and subcellular distribution of fatty acid activating enzymes in the rat. Biochim. biophys. Acta *231:* 32–47 (1971).
3 ABDEL-LATIF, A. A. and SMITH, J. P.: *In vivo* incorporation of choline, glycerol and orthophosphate into lecithin and other phospholipids of subcellular fractions of rat cerebrum. Biochim. biophys. Acta *218:* 134–140 (1970).
4 ABDULLA, Y. H.; ORTON, C. C., and ADAMS, C. W. M.: Cholesterol esterification by transacylation in human and experimental atheromatous lesions. J. Atheroscler. Res. *8:* 967–973 (1968).
5 ABDULLA, Y. H. and ADAMS, C. W. M.: Sulfhydryl dependence of arterial lecithin.: cholesterol transacylase. Atherosclerosis *12:* 319–320 (1970).
6 ABOU-ISSA, H. M. and CLELAND, W. W.: Studies on the microsomal acylation of L-glycerol-3-phosphate. II. The specificity and properties of the rat liver enzyme. Biochim. biophys. Acta *176:* 692–698 (1969).
7 AGRANOFF, B. W.; BRADLEY, R. M., and BRADY, R. D.: The enzymatic synthesis of inositol phosphatide. J. biol. Chem. *233:* 1077–1083 (1958).
8 AGRANOFF, B. W. and HAJRA, A. K.: The acyl dihydroxyacetonephosphate pathway

for glycerolipid biosynthesis in mouse liver and Ehrlich ascites tumor cells. Proc. nat. Acad. Sci., Wash. *68:* 411–415 (1971).

9 AILHAUD, G. P. and VAGELOS, P. R.: Palmityl-acyl carrier protein as acyl donor for complex lipid biosynthesis in *Escherichia coli.* J. biol. Chem. *241:* 3866–3869 (1966).

10 ÅKESSON, B.: The acylation of diglycerols in pig liver. Europ. J. Biochem. *9:* 406–414 (1969).

11 ÅKESSON, B.; ELOVSON, J., and ARVIDSON, G.: Initial incorporation into rat liver glycerolipids of intraportally injected ($9,10^{-3}$ H_2) palmitic acid. Biochim. biophys. Acta *218:* 44–56 (1970).

12 ÅKESSON, B.: Initial esterification and conversion of intraportally injected ($1\text{-}^{14}C$) linoleic acid in rat liver. Biochim. biophys. Acta *218:* 57–70 (1970).

13 AKINO, T.; ABE, M., and ARAI, T.: Studies on the biosynthetic pathways of molecular species of lecithin by rat lung slices. Biochim. biophys. Acta *248:* 274–281 (1971).

14 ANGEL, A.: Studies on the compartmentation of lipid in adipose cells. I. Subcellular distribution, composition, and transport of newly synthesized lipid: liposomes. J. Lipid Res. *11:* 420–432 (1970).

15 APPLEBY, R. S.; SAFFORD, R., and NICHOLS, B. W.: The involvment of lecithin and monogalactosyl diglyceride in linoleate synthesis by green and blue-green algae. Biochim. biophys. Acta *248:* 205–211 (1971).

16 ATKIN, S. D.; PALMER, E. D.; ENGLISH, P. D.; MORGAN, B.; CAWTHORNE, M. A., and GREEN, J.: The role of cytochrome P-450 in cholesterol biogenesis and catabolism. Biochem. J. *128:* 237–242 (1972).

17 AVIGAN, J.; BHATHENA, S. J.; WILLIAMS, C. D., and SCHREINER, M. E.: *In vitro* biosynthesis of lipids, proteins and deoxyribonucleic acid in aortic tissue and in cultured aortic cells. Biochim. biophys. Acta *270:* 279–287 (1972).

18 AXELROD, L. R. and GOLDZIEHER, J. W.: The effect of cofactors on steroid biosynthesis in normal ovarian tissue. Biochim. biophys. Acta *202:* 349–353 (1970).

19 BAILEY, E.; TAYLOR, C. B., and BARTLEY, W.: Effect of dietary carbohydrate on hepatic lipogenesis in the rat. Nature, Lond. *217:* 471–472 (1968).

20 BAR-ON, H. and STEIN, Y.: Effect of glucose and fructose administration on lipid metabolism in the rat. J. Nutr. *94:* 95–105 (1968).

21 BAR-TANA, J.; ROSE, G., and SHAPIRO, B.: The purification and properties of microsomal palmitoylcoenzyme-A synthetase. Biochem. J. *122:* 353–362 (1971).

22 BARTH, C.; SLADEK, M., and DECKER, K.: The subcellular distribution of short-chain fatty acyl-CoA synthetase activity in rat tissues. Biochim. biophys. Acta *248:* 24–33 (1971).

23 BERGER, C. K. and FOÀ, P. P.: The effect of 2-deoxy-glucose and 3-methyl glucose on insulin-stimulated lipogenesis in the chick embryo heart. Horm. Metab. Res. *3:* 98–102 (1971).

24 BJERVE, K. S.: Lecithin biosynthesis in the rat studied with phosphate analogues of phosphorylcholine. Biochim. biophys. Acta *270:* 348–363 (1972).

25 BLOCH, K.: Biological Synthesis of cholesterol.
Harvey Lect. Ser. *48:* 68–88 (1952).

26 BRECKENRIDGE, W. C. and KUKSIS, A.: Stereochemical course of diacylglycerol formation in rat intestine. Lipids *7:* 256–258 (1972).

27 BREMER, J. and GREENBERG, D. M.: Biosynthesis of choline *in vitro*. Biochim. biophys. Acta *37:* 173–175 (1960).

28 BREMER, J.; FIGARD, P. H., and GREENBERG, D. M.: The biosynthesis of choline and its relation to phospholipid metabolism. Biochim. biophys. Acta *43:* 477–488 (1960).

29 BRESSLER, R.: Physiological-chemical aspects of fatty acid oxydation; in WAKIL Lipid metabolism (Academic Press, New York 1970).

30 BROWN, O. R.; HOWITT, H. F.; STEES, J. L., and PLATNER, W. S.: Effects of hyperoxia on composition and rate of synthesis of fatty acids in *E. coli*. J. Lipid Res. *12:* 692–698 (1971).

31 CAPUZZI, D. M. and MARGOLIS, S.: Metabolic studies in isolated rat liver cells. I. Lipid synthesis. Lipids *6:* 601–608 (1971).

32 CAREY, E. M.; HANSEN, H. J. M., and DILS, R.: Fatty acid biosynthesis. IX. Fatty acid synthesis from enzymatically and chemically acetylated rabbit mammary gland fatty acid synthetase. Biochim. biophys. Acta *260:* 527–540 (1972).

33 CARTER, J. R. and KENNEDY, E. P.: Enzymatic synthesis of cytidine diphosphate diglyceride. J. Lipid Res. *7:* 678–683 (1966).

34 BORKENHAGEN, L. F.; KENNEDY, E. P., and FIELDING, L.: Enzymatic formation and decarboxylation of phosphatidylserine. J. biol. Chem. *236:* 28–30 (1961).

35 BOSCH, H. VAN DEN; SLOTBOOM, A. J., and DEENEN, L. L. M. VAN: The conversion of unsaturated 2-acyl-sn-glycero-3-phosphorylcholines into major molecular species of phosphatidylcholines. Biochim. biophys. Acta *176:* 632–634 (1969).

36 BOSCH, H. VAN DEN and VAGELOS, P. R.: Fatty acyl-CoA and fatty acyl-acyl carrier protein as acyl donors in the synthesis of lysophosphatidate and phosphatidate in *Escherichia coli*. Biochim. biophys. Acta *218:* 233–248 (1970).

37 BRADBEER, C. and STUMPF, P. K.: Fat metabolism in higher plants. XIII. Phosphatidic acid synthesis and diglyceride phosphokinase activity in mitochondria from peanut cotyledons. J. Lipid Res. *1:* 214–220 (1959).

38 BRADY, D.R. and GAYLOR, J. L.: Enzymic formation of esters of methyl sterol precursors of cholesterol. J. Lipid Res. *12:* 270–276 (1971).

39 BRANDES, R.; OLLEY, J., and SHAPIRO, B.: Assay of glycerol phosphate acyltransferase in liver particles. Biochem. J. *86:* 244–247 (1963).

40 CARTER, J. R.: Cytidine triphosphate. Phosphatidic acid cytidyltransferase in *Escherichia coli*. J. Lipid Res. *9:* 748–754 (1968).

41 CHANG, Y. Y. and KENNEDY, E. P.: Biosynthesis of phosphatidyl glycerophosphate in *Escherichia coli*. J. Lipid Res. *8:* 447–455 (1967).

42 CHANG, H.-C.; SEIDMAN, I.; TEEBOR, G., and LANE, M. D.: Liver acetyl-CoA carboxylase and fatty acid synthetase. Relative activities in the normal state and in hereditary obesity. Biochem. biophys. Res. Commun. *28:* 682–686 (1967).

43 CHENFAE, G. M.: Phosphatidic acid and glyceride synthesis by particles from spinach leaves. Plant Physiol. *40:* 235–243 (1965).

44 CHERNICK, S. S. and CHAIKOFF, I. L.: The metabolism of arterial tissue. II. Lipid synthesis. The formation *in vitro* of fatty acids and phospholipids by rat artery with C^{14} and P^{32} as indicators. J. biol. Chem. *179:* 113–118 (1949).

45 CHEVALIER, M.; WILEY, J. H., and LEVEILLE, G. A.: The age-dependent response of serum triglycerides to dietary fructose. Proc. Soc. exp. Biol. Med. *139:* 220–222 (1972).

46 CHOBANIAN, A. V. and MANZUR, F.: Metabolism of lipid in the human fatty streak lesion. J. Lipid Res. *13:* 201–206 (1972).

47 CLAIR, R. W. ST.; LOFLAND, H. B., and CLARKSON, T. B.: Influence of atherol sclerosis on the composition, synthesis and esterification of lipids in aortas of squirre-monkeys. J. Atheroscler. Res. *10:* 193–206 (1969).

48 COHEN, A. M. and TEITELBAUM, A.: Effect of glucose, fructose, sucrose and starch on lipogenesis in rats. Life Sci. *7:* 23–29 (1968).

49 CONNETT, R. J.; WITTELS, B., and BLUM, J. B.: Metabolic pathways in *Tetrahymena.* II. Compartmentalization of acyl coenzyme A and structure of the glycolytic and gluconeogenetic pathways. J. biol. Chem. *247:* 2657–2661 (1972).

50 COOK, G. C.: Absorption products of D(–)fructose in man. Clin. Sci. *37:* 675–687 (1969).

51 CRONE, H. D.: The calcium-stimulated incorporation of ethanolamine and serine into the phospholipids of the housefly *Musca domestica.* Biochem. J. *104:* 695–704 (1967).

52 DAAE, L. N. W. and BREMER, J.: The acylation of glycerophosphate in rat liver. A new assay procedure for glycerophosphate acylation. Studies on its subcellular and submitochondrial localization and determination of the reaction products. Biochim. biophys. Acta *210:* 92–104 (1970).

53 DAAE, L. N. W.: The mitochondrial acylation of glycerophosphate in rat liver. Fatty acid and positional specificity. Biochim. biophys. Acta *270:* 23–31 (1972).

54 DALY, M. M.: Biosynthesis of squalene and sterols by rat aorta. J. Lipid Res. *12:* 367–375 (1971).

55 DAVIDSON, J. B. and STANACEV, N. Z.: Biosynthesis of cardiolipin in mitochondria isolated from guinea pig liver. Biochem. biophys. Res. Commun. *42:* 1191–1199 (1971).

56 DAY, A. J. and GOULD-HURST, P. R. S.: Cholesterolesterase activity of normal and atherosclerotic rabbit aorta. Biochim. biophys. Acta *116:* 169–171 (1966).

57 DEBUCH, H.; MÜLLER, J., and FÜRNISS, H.: The synthesis of plasmalogenes at the time of myelination in the rat. IV. The incorporation of 14 C-labelled O-(1-alkyl-sn-glycero-3-phosphoryl)-ethanolamine. A direct precursor of plasmalogenes. Z. Physiol. Chem. *352:* 984–990 (1971).

58 DEENEN, L. L. M. VAN: Some structural investigations on phospholipids from membranes. J. amer. Oil Chem. Soc. *43:* 296–304 (1966).

59 DENNIS, E. A. and KENNEDY, E. P.: Enzymic synthesis and decarboxylation of phosphatidylserine in *Tetrahymena pyriformis.* J. Lipid Res. *11:* 394–403 (1970).

60 DENNIS, E. A. and KENNEDY, E. P.: Intracellular sites of lipid synthesis and the biogenesis of mitochondria. J. Lipid Res. *13:* 263–267 (1972).

61 DEVOR, K. A. and MUDD, J. B.: Structural analysis of phosphatidylcholine of plant tissue. J. Lipid Res. *12:* 396–402 (1971).

62 DEVOR, K. A. and MUDD, J. B.: Biosynthesis of phosphatidylcholine by enzyme preparations from spinach leaves. J. Lipid Res. *12:* 403–411 (1971).

63 DEVOR, K. A. and MUDD, J. B.: Control of fatty acid distribution in phosphatidylcholine of spinach leaves. J. Lipid Res. *12:* 412–419 (1971).

64 DIRKSEN, T. R.; MARINETTI, G. V., and PECK, W. H.: Lipid metabolism in bone and bone cells. 1. The *in vitro* incorporation of (14C)glycerol and (14C)glucose into lipids of bone cell cultures. Biochim. biophys. Acta *202:* 67–79 (1970).

65 DONALDSON, W. E.; WITT-PEETERS, E. M., and SCHOLTE, H. R.: Fatty acid synthesis in rat liver. Relative contributions of the mitochondrial, microsomal and non-particulate systems. Biochim. biophys. Acta *202:* 35–42 (1970).

66 DUGAN, R. E.; SLAKEY, L. L., and PORTER, J. W.: Stereospecificity of the transfer of hydrogen from reduced nicotineamide adenine dinucleotide phosphate to the acyl chain in the dehydrogenase-catalyzed reactions of fatty acid synthesis. J. biol. Chem. *245:* 6312–6316 (1970).

67 EIBL, H.; HILL, E. E., and LANDS, W. E. M.: The subcellular distribution of acyl-transferases which catalyze the synthesis of phosphoglycerides. Europ. J. Biochem. *9:* 250–258 (1969).

68 ELOVSON, J.; AKESSON, B., and ARVIDSON, G.: Positional specificity of liver 1,2-diglyceride biosynthesis *in vivo*. Biochim. biophys. Acta *176:* 214–217 (1969).

69 ERBLAND, J. F.; BROSSARD, M. V., and MARINETTI, G. V.: Controlling effects of ATP, Mg^{2+} and CTP in the biosynthesis of lipids. Biochim. biophys. Acta *137:* 23–32 (1967).

70 ETO, Y. and SUZUKI, K.: Cholesterol ester metabolism in the brain. Properties and subcellular distribution of cholesterol-esterifying enzymes and cholesterol ester hydrolases in adult rat brain. Biochim. biophys. Acta *239:* 293–311 (1971).

71 FALLON, H. J. and LAMB, R. G.: Acylation of sn-glycerol-3-phosphate by cell fractions of rat liver. J. Lipid Res. *9:* 652–660 (1968).

72 FANG, M. and MARINETTI, G. V.: The effects of pyridine nucleotides and glycerol on lipid phosphorylation in rat liver mitochondria. Biochim. biophys. Acta *202:* 91–105 (1970).

73 FIELDING, C. J.; SHORE, V. G., and FIELDING, P. E.: A protein cofactor of lecithin. Cholesterol acyltransferase. Biochem. biophys. Res. Commun. *46:* 1493–1498 (1972).

74 FIELDING, C. J.; SHORE, V. G., and FIELDING, P. E.: Lecithin: cholesterol acyl-transferase. Effects of substrate composition upon enzyme activity. Biochim. biophys. Acta *270:* 513–518 (1972).

75 FILIPOVIC, I. and BUDDECKE, E.: Increased fatty acid synthesis of arterial tissue in hypoxia. Europ. J. Biochem. *20:* 587–592 (1971).

76 FRIEDBERG, S. J.; HEIFETZ, A., and GREEN, R. C.: Studies on the mechanism of O-alkyl lipid synthesis. Biochemistry *11:* 297–301 (1972).

77 GALLO, L. and TREADWELL, C. R.: Localization of the monoglyceride pathway in subcellular fractions of rat intestinal mucosa. Arch. Biochem. *141:* 614–621 (1971).

78 GANGULY, J.: Studies on the mechanism of fatty acid synthesis. VII. Biosynthesis of fatty acids from malonyl-CoA. Biochim. biophys. Acta *40:* 110–118 (1960).

79 GERICKE, C.; RAUSCHENBACH, P.; KUPKE, I. und LAMPRECHT, W.: Ueber die Biosynthese von Lipiden aus Fructose in Leber. III. Biosynthese des Glyceridglycerins aus (6-^{14}C)Fructose. Z. Physiol. Chem. *349:* 1055–1061 (1968).

80 GIBBONS, G. F. and MITROPOULOS, K. A.: Inhibition of cholesterol synthesis by carbon monoxide. Accumulation of lanosterol and 24,25-dihydrolanosterol. Biochem. J. *127:* 315–317 (1972).

81 GLENN, J. L. and AUSTIN, W.: The conversion of phosphatidyl ethanolamines to lecithins in normal and choline-deficient rats. Biochim. biophys. Acta *231:* 153–160 (1971).

82 GLOMSET, J. A.; PARKER, F.; TJADEN, M., and WILLIAMS, R. H.: The esterification

in vitro of free cholesterol in human and rat plasma. Biochim. biophys. Acta *58:* 398–406 (1962).

83 GLOMSET, J. A.: Further studies of the mechanism of the plasma cholesterol esterification reaction. Biochim. biophys. Acta *70:* 389–395 (1963).

84 GLOMSET, J. A.: The plasma lecithin:cholesterol acyltransferase reaction. J. Lipid Res. *9:* 155–167 (1968).

85 GOLDFINE, H.: Acylation of glycerol-3-phosphate in bacterial extracts. Stimulation by acyl carrier protein. J. biol. Chem. *241:* 3864–3866 (1966).

86 GOLDFINE, H.: Lipid chemistry and metabolism. Annu. Rev. Biochem. *37:* 303–330 (1968).

87 GRAY, G. M.: The isolation of phosphatidyl glycerol from rat-liver mitochondria. Biochim. biophys. Acta *84:* 35–40 (1964).

88 GREGOLIN, C.; RYDER, E.; WARNER, R. C.; KLEINSCHMIDT, A. K.; CHANG, H.-C., and LANS, M. D.: Liver acetyl coenzyme A carboxylase. II. Further molecular characterization. J. biol. Chem. *243:* 4236–4245 (1968).

89 GRIFFITHS, K. and CAMERON, E. H. D.: Steroid biosynthetic pathways in the human adrenal; in Advances in steroid biochemistry and pharmacology, vol. 2, pp. 223–265 (Academic Press, London 1970).

90 GRIGOR, M. R.; DUNCKLEY, G. G., and PURVES, H. D.: The synthesis of the branched-chain fatty acids of rat skin surface lipid. Biochim. biophys. Acta *218:* 389–399 (1971).

91 GRUNDY, S. M.; AHRENS, E. H., and DAVIGNON, J.: The interaction of cholesterol absorption and cholesterol synthesis in man. J. Lipid Res. *10:* 304–315 (1969).

92 GURR, M. I.; PROTTEY, C. and HAWTHORNE, J. N.: The phospholipids of liver-cell fractions. II. Incorporation of (^{32}P)orthophosphate *in vivo* in normal and regenerating rat liver. Biochim. biophys. Acta *106:* 357–370 (1965).

93 GURR, M. I.: The biosynthesis of polyunsaturated fatty acids in plants. Lipids *6:* 266–273 (1971).

94 HAJRA, A. K.; SEGUIN, E. B., and AGRANOFF, B. W.: Rapid labeling of mitochondrial lipids by labeled orthophosphate and adenosine triphosphate. J. biol. Chem. *243:* 1609–1616 (1968).

95 HAJRA, A. K. and AGRANOFF, B. W.: Acyl dihydroxyacetone phosphate. Characterization of a ^{32}P-labeled lipid from guinea pig liver mitochondria. J. biol. Chem. *243:* 1617–1622 (1968).

96 HAJRA, A. K.: Biosynthesis of acyl dihydroxyacetone phosphate in guinea pig liver mitochondria. J. biol. Chem. *243:* 3458–3465 (1968).

97 HAJRA, A. K. and AGRANOFF, B. W.: Reduction of palmitoyl dihydroxyacetone phosphate by mitochondria. J. biol. Chem. *243:* 3542–3555 (1968).

98 HANSEN, H. J. M.; CAREY, E. M., and DILS, R.: Fatty acid biosynthesis. VII. Substrate control of chain length of products synthesized by rat liver fatty acid synthetase. Biochim. biophys. Acta *210:* 400–410 (1970).

99 HANSEN, H. J. M.; CAREY, E. M., and DILS, R.: Fatty acid biosynthesis. VIII. The fate of malonyl-CoA in fatty acid biosynthesis by purified enzymes from lactating mammary gland. Biochim. biophys. Acta *248:* 391–405 (1971).

100 HANSON, R. J.: The application of multiply labelled mevalonates in terpene and steroid biosynthesis; in BRIGGS Advances in steroid biochemistry and pharmacology, vol. 1, pp. 51–72 (Academic Press, London 1970).

101 HANSON, R. W.; PATEL, M. S.; JOMAIN-BAUM, M., and BALLARD, F. J.: Regulation of lipogenesis and lipolysis in adipose tissue. Role of mitochondria in metabolism of pyruvate and lactate by rat adipose tissue. Metabolism *20:* 27–42 (1971).
102 HASSELBLATT, A.: Interrelations between lipid and carbohydrate metabolism. Arch. exp. Path. Pharmakol. *269:* 331–346 (1971).
103 HEINZ, F. and JUNGHÄNEL, J.: Levels of metabolites in rat liver following the application of fructose. Z. Physiol. Chem. *350:* 859–866 (1969).
104 HERMAN, R. H.; ZAKIM, D., and STIFEL, F. B.: Effect of diet on lipid metabolism in experimental animals and man. Fed. Proc. *29:* 1302–1308 (1970).
105 HILL, E. E. and LANDS, W. E. M.: Incorporation of long-chain and polyunsaturated acids into phosphatidate and phosphatidylcholine. Biochim. biophys. Acta *152:* 645–648 (1968).
106 HILL, E. E.; HUSBANDS, D. R., and LANDS, W. E. M.: The selective incorporation of ^{14}C-glycerol into different species of phosphatidic acid, phosphatidylethanolamine, and phosphatidylcholine. J. biol. Chem. *243:* 4440–4451 (1968).
107 HILL, E. E. and LANDS, W. E. M.: Phospholipid metabolism; in WAKIL Lipid metabolism (Academic Press, New York 1970).
108 HILL, P.: Effect of fructose on rat lipids. Lipids *5:* 621–627 (1970).
109 HIRSCHBERG, C. B. and KENNEDY, E. P.: Mechanism of the enzymatic synthesis of cardiolipin in *Escherichia coli.* Proc. nat. Acad. Sci., Wash. *69:* 648–651 (1972).
110 HOKIN, M. R. and HOKIN, L. E.: The synthesis of phosphatidic acid from diglyceride and adenosine triphosphate in extracts of brain microsomes. J. biol. Chem. *234:* 1381–1386 (1959).
111 HOKIN, L. E. and HOKIN, M. R.: Synthesis of a new phosphatide from monoglyceride and adenosine triphosphate. Biochim. biophys. Acta *37:* 176–177 (1960).
112 HOKIN, L. E. and HOKIN, M. R.: Diglyceride kinase and phosphatidic acid phosphatase in erythrocyte membranes. Nature, Lond. *189:* 836–837 (1961).
113 HOKIN, L. E. and HOKIN, M. R.: Diglyceride kinase and other pathways for phosphatidic acid synthesis in the erythrocyte membrane. Biochim. biophys. Acta *67:* 470–484 (1963).
114 HOKIN, M. R. and HOKIN, L. E.: The synthesis of phosphatidic acid and protein-bound phosphorylserine in salt gland homogenates. J. biol. Chem. *239:* 2116–2122 (1964).
115 HOLMES, W. L. and BORTZ, W. M. (ed.): Biochemistry and pharmacology of free fatty acids (Karger, Basel 1971).
116 HOLUB, B. J.; KUKSIS, A., and THOMPSON, W.: Molecular species of mono-, di-, and triphosphoinositides of bovine brain. J. Lipid Res. *11:* 558–564 (1970).
117 HOLUB, B. J. and KUKSIS, A.: Differential distribution of orthophosphate-^{32}P and glycerol-^{14}C among molecular species of phosphatidylinositols of rat liver *in vivo.* J. Lipid Res. *12:* 699–705 (1971).
118 HOSTELLER, K. Y.; BOSCH, H. VAN DEN, and DEENEN, L. L. M. VAN: Biosynthesis of cardiolipin in liver mitochondria. Biochim. biophys. Acta *239:* 113–119 (1971).
119 HOSTELLER, K. Y. and BOSCH, H. VAN DEN: Subcellular and submitochondrial localization of the biosynthesis of cardiolipin and related phospholipids in rat liver. Biochim. biophys. Acta *260:* 380–386 (1972).
120 HOSTELLER, K. Y.; BOSCH, H. VAN DEN, and DEENEN, L. L. M. VAN: The mechanism

of cardiolipin biosynthesis in liver mitochondria. Biochim. biophys. Acta *260:* 507–513 (1972).

121 HOWARD, A. N.; BOWYER, D. E., and GRESHAM, G. A.: Cholesterol metabolism in normal and atherosclerotic aortas. Circulation *35:* suppl. 11 (1967).

122 HOWARD, C. F.: Synthesis of fatty acids in outer and inner membranes of mitochondria. J. biol. Chem. *245:* 462–468 (1970).

123 HOWARD, C. F.: Lipogenesis from glucose-2-^{14}C and acetate-1-^{14}C in aorta. J. Lipid Res. *12:* 725–730 (1971).

124 HÜBSCHER, G.: Metabolism of phospholipids. VI. The effect of metal ions on the incorporation of L-serine into phosphatidylserine. Biochim. biophys. Acta *57:* 555–561 (1962).

125 HÜBSCHER, G.: Glyceride metabolism; in WAKIL Lipid metabolism, pp. 280–370 (Academic Press, New York 1970).

126 HUSBANDS, D. R. and LANDS, W. E. M.: Phosphatidate synthesis by sn-glycerol-3-phosphate acyltransferase in pigeon liver particles. Biochim. biophys. Acta *202:* 129–140 (1970).

127 HUTCHISON, H. T. and CRONAN, J. E.: The synthesis of cytidine diphosphate diglyceride by cell-free extracts of yeast. Biochim. biophys. Acta *164:* 606–608 (1968).

128 ILLINGWORTH, D. R. and PORTMAN, O. W.: Exchange of phospholipids between low and high density lipoproteins of squirrel monkeys. J. Lipid Res. *13:* 220–227 (1972).

129 International Symposium on Lipids: Proceedings of the Biochemical Society. Biochem. J. *128:* 1–68 (1972).

130 IUPAC-IUB Commission on Biochemical Nomenclature: The nomenclature of lipids. Biochim. biophys. Acta *152:* 1–9 (1968).

131 JOHNSTON, J. M.; PAULTAUF, F.; SCHILLER, C. M., and SCHULTZ, L. D.: The utilization of the alpha-glycerolphosphate and monoglyceride pathways for phosphatidylcholine biosynthesis in the intestine. Biochim. biophys. Acta *218:* 124–133 (1970).

132 JONES, R. J. (ed.): Atherosclerosis. Proc. 2nd Int. Symp. (Springer, Berlin 1970).

133 JOSHI, V. C.; PLATE, C. A., and WAKIL, S. J.: Studies on the mechanism of fatty acid synthesis. XXIII. The acyl binding sites of the pigeon liver fatty acid synthetase. J. biol. Chem. *245:* 2857–2867 (1970).

134 JOSHI, V. C. and WAKIL, S. J.: Studies on the mechanism of fatty acid synthesis. XXVI. Purification and properties of malonyl-coenzyme A-acyl-carrier protein transacylase of *Escherichia coli.* Arch. Biochem. *143:* 493–505 (1971).

135 KAISER, W.: Incorporation of phospholipid precursors into isolated rat liver mitochondria. Europ. J. Biochem. *8:* 120–127 (1969).

136 KANFER, J. N. and KENNEDY, E. P.: Metabolism and function of bacterial lipids. II. Biosynthesis of phospholipids in *Escherichia coli.* J. biol. Chem. *239:* 1720–1726 (1964).

137 KANOH, H.: Biosynthesis of lecithins and phosphatidyl ethanolamines from various radioactive 1,2-diglycerides in rat liver microsomes. Biochim. biophys. Acta *218:* 249–258 (1970).

138 KARLSSON, K.-A.: Sphingolipid long chain bases. Lipids *5:* 878–891 (1970).

139 KATES, M.: Hydrolysis of lecithin by plant plastid enzymes. Canad. J. Biochem. *33:* 575–589 (1955).

140 KAUFMANN, N. A.; KAPITULNIK, J., and BLONDHEIM, S. H.: Studies in carbohydrate-induced hypertriglyceridemia. Aspects of fructose metabolism. Israel J. Med. *6:* 80–85 (1970).

141 KENNEDY, E. P.: Synthesis of phosphatides in isolated mitochondria. J. biol. Chem. *201:* 399–412 (1953).

142 KENNEDY, E. P.: Metabolism of lipids. Annu. Rev. Biochem. *26:* 119–145 (1957).

143 KENNEDY, E. P.: The Harvey Lectures. Series 57, pp. 143–171 (Academic Press, New York 1962).

144 KEPLER, C. R.; TUCKER, W. P., and TOVE, S. B.: Biohydrogenation of unsaturated fatty acids. V. Stereospecificity of proton addition and mechanism of action of linoleic acid Δ^{12}-cis,Δ''-trans-isomerase from *Butyrivibrio fibrisolvens*. J. biol. Chem. *246:* 2765–2771 (1971).

145 KERN, F., jr. and BORGSTRÖM, B.: Quantitative study of the pathways of triglyceride synthesis in hamster intestinal mucosa. Biochim. biophys. Acta *98:* 520–531 (1965).

146 KINSELLA, J. E.: Biosynthesis of lipids from (2-^{14}C)acetate and D(–)-β-hydroxy(1,3-^{14}C)butyrate by mammary cells from bovines and rat. Biochim. biophys. Acta *210:* 28–38 (1970).

147 KIYASU, J. Y.; PIERINGER, R. A.; PAULUS, H., and KENNEDY, E. P.: The biosynthesis of phosphatidylglycerol. J. biol. Chem. *238:* 2293–2298 (1963).

148 KOHOUT, M.; KOHOUTOVA, B., and HEIMBERG, M.: The regulation of hepatic triglyceride metabolism by free fatty acids. J. biol. Chem. *246:* 5067–5074 (1971).

149 KORNBERG, A. and PRICER, W. E., jr.: Enzymatic esterification of α-glycerophosphate by long chain fatty acids. J. biol. Chem. *204:* 345–357 (1953).

149a KORNBERG, A. and PRICER, W. E.: Studies on the enzymatic synthesis of phospholipids. Fed. Proc. *11:* 242 (1952).

150 KRESSE, H.; FILIPOVIC, I. und BUDDECKE, E.: Gesteigerte ^{14}C-Incorporation in die Triacylglycerine (Triglyceride) des Arteriengewebes bei Sauerstoffmangel. Z. Physiol. Chem. *350:* 1611–1618 (1969).

151 KRISHNAIAH, K. V. and RAMASARMA, T.: Regulation of hepatic cholesterolgenesis by ubiquinone. Biochim. biophys. Acta *202:* 332–342 (1970).

152 KRITCHEVSKY, D.: Cholesterol metabolism in aorta and in tissue culture (review). Lipids *7:* 305–309 (1972).

153 KRUYFF, B. DE; GOLDE, L. M. G. VAN, and DEENEN, L. L. M. VAN: Utilization of diacylglycerol species by cholinephosphotransferase, ethanolaminephosphotransferase and diacylglycerol acyltransferase in rat liver microsomes. Biochim. biophys. Acta *210:* 425–435 (1970).

154 KUHN, N. J. and LYNEN, F.: Phosphatidic acid synthesis in yeast. Biochem. J. *94:* 240–246 (1965).

155 KUKSIS, A.; BRECKENRIDGE, W. C.; MARAI, L., and STACHNYK, O.: Molecular species of lecithins of rat heart, kidney and plasma. J. Lipid Res. *10:* 25–33 (1969).

156 KUMAR, S.; DORSEY, J. A.; MUESING, R. A., and PORTER, J. W.: Comparative studies of the pigeon liver fatty acid synthetase complex and its subunits. J. biol. Chem. *245:* 4732–4744 (1970).

157 KUPKE, I. und LAMPRECHT, W.: Ueber die Biosynthese von Lipiden aus Fructose in Leber. I. Einbau von uniform markierter (^{14}C)Fructose in Leberlipide. Z. Physiol. Chem. *348:* 17–26 (1967).

158 Kupke, I. und Lamprecht, W.: Ueber die Biosynthese von Lipiden aus Fructose in Leber. II. Einbau von spezifisch markierter (^{14}C)Fructose in Leberlipide. Z. Physiol. Chem. *348:* 929–935 (1967).

159 Kupke, I.: Interrelationships between carbohydrate and lipid metabolism. Angiologica, Basel *6:* 179–181 (1969).

160 Kupke, I. R.: Biosynthesis of lipids in perfused dog aorta and coronary artery. I. Incorporation of (2-^{14}C)acetate into the lipids of three aortic layers and of the coronary artery in normal and hyperlipemic dogs. J. molec. Cell. Cardiol. *4:* 11–26 (1972).

161 Kupke, I. R.: Biosynthesis of lipids in perfused dog aorta and coronary artery. II. Incorporation of (2-^{14}C)acetate into lipids of two aortic layers and of the coronary artery under the influence of nicotine. J. molec. Cell. Cardiol. *4:* 27–38 (1972).

162 Kupke, I. R.: Biosynthesis of lipids in perfused dog aorta and coronary artery. III. Incorporation of (2-^{14}C)acetate into sterols and uptake of (^{3}H)-cholesterol in three aortic layers and in coronary artery of normal and hyperlipemic dogs and under the influence of nicotine. J. molec. Cell. Cardiol. *4:* 255–268 (1972).

163 Lamb, R. G. and Fallon, H. F.: The formation of monoacylglycerophosphate from sn-glycerol-3-phosphate by a rat liver particulate preparation. J. biol. Chem. *245:* 3075–3083 (1970).

164 The Lancet (ed.): Glucose *versus* fructose. Lancet *ii:* 1178–1179 (1968).

165 Lands, W. E. M. and Merkl, I.: Metabolism of glycerolipids. III. Reactivity of various acylesters of coenzyme A with α'-acyl-glycerophosphorylcholine, and positional specificities in lecithin synthesis. J. biol. Chem. *238:* 898–904 (1963).

166 Lands, W. E. M. and Hart, P.: Metabolism of glycerolipids. V. Metabolism of phosphatidic acid. J. Lipid Res. *5:* 81–87 (1964).

167 Lands, W. E. M. and Hart, P.: Metabolism of glycerolipids. VI. Specificities of acyl coenzyme A: phospholipid acyltransferases. J. biol. Chem. *240:* 1905–1911 (1965).

168 Landiscrina, C.; Gnoni, G. V., and Quagliariello, E.: Mechanisms of fatty acid synthesis in rat liver microsomes. Biochim. biophys. Acta *202:* 405–414 (1970).

169 Landiscrina, C.; Gnoni, G. V., and Quagliariello, E.: Fatty acid biosynthesis. The physiological role of the elongation system present in microsomes and mitochondria of rat liver. Europ. J. Biochem. *29:* 188–196 (1972).

170 Lane, M. D. and Moss, J.: Regulation of fatty acid synthesis in animal tissues. Metab. Pathways *5:* 23–54 (Academic Press, New York 1971).

171 Lapetina, E. G. and Hawthorne, J. N.: The diglyceride kinase of rat cerebral cortex. Biochem. J. *122:* 171–179 (1971).

172 Lefebvre, P.: Le métabolisme de la cellule adipeuse. Les phénomènes de lipogenèse et de lipolyse. Rev. Méd., Liège *25:* 18–26 (1970).

173 Lennarz, W. J.: Lipid metabolism. Annu. Rev. Biochem. *39:* 359–388 (1970).

174 Lennarz, W. J.: Bacterial lipids; in Wakil Lipid metabolism, pp. 155–184 (Academic Press, New York 1970).

175 Liberti, J. P. and Jezyk, P. F.: Lipid biosynthesis in rat liver slices. Effects of ions, ATP and substrate concentration on glycerol incorporation. Biochim. biophys. Acta *210:* 221–229 (1970).

176 Lin, C. Y. and Kumar, S.: Pathway for the synthesis of fatty acids in mammalian tissues. J. biol. Chem. *247:* 604–610 (1972).

177 LIPPEL, K.; LLEWELLYN, A., and JARETT, L.: Palmitoyl-CoA synthetase activity in isolated rat epididymal fat cells in the absence and presence of various lipolytic and anti-lipolytic compounds. Biochim. biophys. Acta *231:* 48–51 (1971).

178 LIPPEL, K.: Regulation of rat liver acyl-CoA synthetase activity. Biochim. biophys. Acta *239:* 384–392 (1971).

179 LOFLAND, H. B., jr.; CLARKSON, T. B.; CLAIR, R. W. ST., and LEHNER, N. D. M.: Studies on the regulation of plasma cholesterol levels in squirrel monkeys of two genotypes. J. Lipid Res. *13:* 39–47 (1972).

180 LOWENSTEIN, J. J. (ed.): Methods in enzymology. XIV. Lipids (Academic Press, New York 1969).

181 DI LUZIO, N. R. and O'NEAL, R. M.: The rapid development of arterial lesions in dogs fed an 'infarct-producing' diet. Exp. molec. Path. *1:* 122–132 (1962).

182 LYNEN, F.; AGRANOFF, B. W.; EGGERER, H.; HENNING, U. und MÖSLEIN, E .M.: γ,γ-Dimethyl-allyl-pyrophosphat und Geranyl-pyrophosphat, biologische Vorstufen des Squalens. Angew. Chem. *71:* 657–663 (1959).

183 LYNEN, F.; MATSUHASHI, M.; NUMA, S., and SCHWEIZER, E.: The cellular control of fatty acid synthesis at the enzymatic level; in GRANT The control of lipid metabolism, pp. 43–56 (Academic Press, New York 1963).

184 LYNEN, F.: The role of biotin-dependent carboxylations in biosynthetic reactions. Biochem. J. *102:* 381–400 (1967).

185 LYNEN, F.: Enzyme systems for fatty acid synthesis. Biochem. J. *128:* 1–2 (1972).

186 MADAPPALLY, M. M.; PAQUET, R. J.; MEHLMAN, M. A., and TOBIN, R. B.: Gluconeogenic and lipogenic enzyme activities in growing chicks fed high fat and high carbohydrate diets. J. Nutr. *101:* 755–760 (1971).

187 MACDONALD, I.: Influence of fructose and glucose on serum lipid levels in men and pre- and postmenopausal women. Amer. J. clin. Nutr. *18:* 369–372 (1966).

188 MACDONALD, I.: Dietary carbohydrates in normolipemia. Amer. J. clin. Nutr. *20:* 185–190 (1967).

189 MACDONALD, I.: Physiological role of dietary carbohydrates. Wld Rev. Nutr. Diet. *8:* 143–183 (1967).

190 MACDONALD, I.: Serum-fructose levels after sucrose or its constituent monosaccharides. Lancet *i:* 841–843 (1968).

191 MACFARLANE, M. G.: Phosphatidylglycerols and lipoamino acids. Adv. Lipid Res. *2:* 91–125 (1964).

192 MAJERUS, P. W. and VAGELOS, P. R.: Fatty acid biosynthesis and the role of the acyl carrier protein. Adv. Lipid Res. *5:* 2–33 (1967).

193 MARCELL, Y. L. and SUZUE, G.: Kinetic studies on the specificity of long chain acyl coenzyme A synthetase from rat liver microsomes. J. biol. Chem. *247:* 4433–4436 (1972).

194 MARINETTI, G. V.; ERBLAND, J. F., and BROSSARD, M.: Biosynthesis of phospholipids and glycerides; in DAWSON and RHODES Metabolism and physiological significance of lipids, pp. 71–93 (Wiley, London 1964).

195 MARSH, J. B.: Secretion of lecithin cholesterol acyl transferase by rat liver. Fed. Proc. *29:* 673 (1970).

196 MARSHALL, M. O. and KATES, M.: Biosynthesis of phosphatidylglycerol by cell-free preparations from spinach leaves. Biochim. biophys. Acta *260:* 558–570 (1972).

197 MARTENSSON, E. and KAUFER, J.: The conversion of L-glycerol-^{14}C-3-phosphate into phosphatidic acid by solubilized preparation from rat brain. J. biol. Chem. *243:* 497–501 (1968).

198 MARTIN, D. B. and VAGELOS, P. R.: The mechanism of tricarboxylic acid cycle regulation of fatty acid synthesis. J. biol. Chem. *237:* 1787–1792 (1962).

199 MARUHAMA, Y.: Conversion of ingested carbohydrate-^{14}C into glycerol and fatty acids of serum triglyceride in patients with myocardial infarction. Metabolism *19:* 1085–1093 (1970).

200 MCCAMAN, R. E. and FINNERTY, W. R.: Biosynthesis of cytidine diphosphate-diglyceride by a particulate fraction from *Micrococcus cerificans.* J. biol. Chem. *243:* 5074–5080 (1968).

201 MCMURRAY, W. C. and DAWSON, R. M. C.: Phospholipid exchange reactions within the liver cell. Biochem. J. *112:* 91–108 (1969).

202 MICHEL, M. P.; BRINDLEY, D. N., and HÜBSCHER, G.: Properties of phosphatidate phosphohydrolase. Europ. J. Biochem. *18:* 214–220 (1971).

203 MONROY, G. and PULLMAN, M. E.: A substrate and position specific acylation of glycerol 3-phosphate by rat liver mitochondria. Fed. Proc. *31:* 453 (1972).

204 MOSS, J. and LANE, M. D.: The biotin-dependent enzymes. Adv. Enzymol. *35:* 321–442 (1971).

205 MUDD, J. B.; GOLDE, L. M. G. VAN, and DEENEN, L. L. M. VAN: Utilization of molecular species of diglycerides in the synthesis of lecithin. Biochim. biophys. Acta *176:* 547–556 (1969).

206 MUKHERJEE, S.; BASU, M., and TRIVEDI, K.: Effect of low dietary levels of glucose, fructose, and sucrose on rat lipid metabolism. J. Atheroscler. Res. *10:* 261–272 (1969).

207 NACHBAUR, J.; COLBEAU, A., and VIG'NAIS, P. M.: Acylation of sn-glycerol-3-phosphate by subcellular fractions of rat liver. C. R. Acad. Sci. *272:* 1015–1018 (1971).

208 NAKAMURA, M. and IBAYASHI, H.: Lipogenesis and hormones. Clin. Endocr. *19:* 501–508 (1971).

209 NESTEL, P. J.: Cholesterol turnover in man. Adv. Lipid Res. *8:* 1–39 (1970).

210 NUMA, S.; MATSUHASHI, M., and LYNEN, F.: On disorders of fatty acid synthesis in hunger and alloxane diabetes. I. Fatty acid synthesis in the liver of normal and fasting rats. Biochem. Z. *334:* 203–217 (1961).

211 NUMA, S.; BORTZ, W. M. and LYNEN, F.: Regulation of fatty acid synthesis at the acetyl-CoA-carboxylation step. Adv. Enzyme Reg. *3:* 407–423 (1965).

212 OKUYAMA, H.; EIBL, H., and LANDS, W. E. M.: Acyl coenzyme A:2-acyl-sn-glycerol-3-phosphate acyltransferase activity in rat liver microsomes. Biochim. biophys. Acta *248:* 263–273 (1971).

213 OKUYAMA, H. and LANDS, W. F. M.: Variable selectivities of acyl coenzyme A:mono-acyl-glycerophosphate acyltransferase in rat liver. J. biol. Chem. *247:* 1414–1423 (1972).

214 OLSON, J. A., jr.; LINDBERG, M., and BLOCH, K.: On the demethylation of lanosterol to cholesterol. J. biol. Chem. *226:* 941–956 (1957).

215 OLSON, J. A.: Lipid metabolism. Annu. Rev. Biochem. *35:* 559–598 (1966).

216 OLSON, R. V. and BALLOU, C. E.: Acyl phosphatidylglycerol. A new phospholipid from *Salmonella typhimurium.* J. biol. Chem. *246:* 3305–3313 (1971).

217 PARIS, R. and CLÉMENT, G.: Biosynthesis of lysophosphatidic acid from ATP and

1-monoolein by subcellular particles of intestinal mucosa. Proc. Soc. exp. Biol. Med. *131:* 363–365 (1969).

218 PARKER, F.; SCHIMMELBUSCH, W., and WILLIAMS, R. H.: The enzymatic nature of phospholipid synthesis in normal rabbit and human aorta. Diabetes *13:* 182–188 (1964).

219 PATEL, M. S.; JOMAIN-BAUM, M.; BALLARD, F. J., and HANSON, R. W.: Pathway of carbon flow during fatty acid synthesis from lactate and pyruvate in rat adipose tissue. J. Lipid Res. *12:* 179–191 (1971).

220 PATTERSON, P. H. and LENNARZ, W. J.: Biosynthesis and incorporation of phospholipids into the cytoplasmic membrane of *Bacillus megaterium*. Fed. Proc. *28* (2): 403 (1969).

221 PATTERSON, P. H. and LENNARZ, W. J.: Studies on the membranes of bacilli. I. Phospholipid biosynthesis. J. biol. Chem. *246:* 1062–1072 (1971).

222 PAULUS, H. and KENNEDY, E. P.: The enzymatic synthesis of inositol monophosphatide. J. biol. Chem. *235:* 1303–1311 (1960).

223 PAYSANT-DIAMENT, M. and POLONOVSKI, J.: *In vitro* incorporation of P-32 into the globular phospholipids of human blood. Bull. Soc. Chim. biol. *42:* 337–350 (1960).

224 PENNINGTON, R. J. and WORSFELD, M.: Biosynthesis of lecithin by skeletal muscle. Biochim. biophys. Acta *176:* 774–782 (1969).

225 PETZOLD, C. L. and AGRANOFF, B. W.: The biosynthesis of cytidine diphosphate diglyceride by embryonic chick brain. J. biol. Chem. *242:* 1187–1191 (1967).

226 PICK, R. and KATZ, L. N.: Effect of environmental changes on the development of hypercholesterolaemia and atherosclerosis in cholesterol oil fed cockerels. Ischem. Heart Dis. *5:* 635–638 (1969).

227 PIERINGER, R. A. and HOKIN, L. E.: Biosynthesis of lysophosphatidic acid from monoglyceride and adenosine triphosphate. J. biol. Chem. *237:* 653–658 (1962).

228 PIERINGER, R. A. and HOKIN, L. E.: Biosynthesis of phosphatidic acid from lysophosphatidic acid and palmityl coenzyme A. J. biol. Chem. *237:* 659–663 (1962).

229 PLATE, C. A.; JOSHI, V. C., and WAKIL, S. J.: Studies on the mechanism of fatty acid synthesis. XXIV. The acetyl- and malonyltransacylase activities of the pigeon liver fatty acid synthetase. J. biol. Chem. *245:* 2868–2875 (1970).

230 POLONOVSKI, J.; WALD, R.; PAYSANT, M.; RAMPINI, C. et BARBU, E.: Métabolisme du phosphatidylglycérol et du cardiolipide chez *Staphylococcus aureus*. Ann. Inst. Pasteur *120:* 589–598 (1971).

231 POPJÁK, G. and CORNFORTH, J. W.: The biosynthesis of cholesterol. Adv. Enzymol. *22:* 281–335 (1960).

232 POPJÁK, G. and CORNFORTH, J. W.: Substrate stereochemistry in squalene biosynthesis. The 1st Ciba Medical Lecture. Biochem. J. *101:* 553–568 (1966).

233 PORTMAN, O. W. and ALEXANDER, M.: Lysophosphatidylcholine concentrations and metabolism in aortic intima plus inner media. Effect of nutrionally induced atherosclerosis. J. Lipid Res. *10:* 158–165 (1969).

234 PORTMAN, O. W.: Arterial composition and metabolism. Esterified fatty acids and cholesterol. Adv. Lipid Res. *8:* 41–114 (1970).

235 PORTMAN, O. W.; SOLTYS, P.; ALEXANDER, M., and OSUGA, T.: Metabolism of lysolecithin *in vivo*. Effects of hyperlipemia and atherosclerosis in squirrel monkeys. J. Lipid Res. *11:* 596–604 (1970).

236 POSSMAYER, F.; BALAKRISHAN, G., and STRICKLAND, K. P.: The incorporation of labelled glycerophosphoric acid into the lipids of rat brain preparations. III. On the biosynthesis of phosphatidyl glycerol. Biochim. biophys. Acta *164:* 79–87 (1968).

237 POSSMAYER, F.; SCHERPHOF, G. L.; DUBBELMAN, T. M. A. R.; GOLDE, L. M. G. VAN, and DEENEN, L. L. M. VAN: Positional specificity of saturated and unsaturated fatty acids in phosphatidic acid from rat liver. Biochim. biophys. Acta *176:* 95–110 (1969).

238 POSSMAYER, F. and MUDD, J. B.: The regulation of sn-glycerol-3-phosphate acylation by cytidine nucleotides in rat brain cerebral hemispheres. Biochim. biophys. Acta *239:* 217–233 (1971).

239 PRESCOTT, D. J. and VAGELOS, P. R.: Acyl carrier protein. XIV. Further studies on β-ketoacyl carrier protein synthetase from *E. coli.* J. biol. Chem. *245:* 5484–5490 (1970).

240 PURY, G. G. DE and COLLINS, F. D.: Uptake of phosphorus-32 by brain microsomal phosphatidic acid in rats deficient in essential fatty acids. Nature, Lond. *198:* 788–789 (1963).

241 RAJU, P. K. and REISER, R.: Fatty acid specificity in the biosynthesis of diglycerides and triglycerides in rat liver *in vitro*. Biochim. biophys. Acta *202:* 212–215 (1970).

242 RAMSEY, R. B.; JONES, J. P.; NAQVI, S. H. M., and NICHOLAS, H. J.: The biosynthesis of cholesterol and other sterols by brain tissue. I. Subcellular biosynthesis *in vitro.* Lipids *6:* 154–161 (1971).

243 RAO, G. A. and JOHNSTON, J. M.: Purification and properties of triglyceride synthetase from the intestinal mucosa. Biochim. biophys. Acta *125:* 465–473 (1966).

244 RAO, G. A. and JOHNSTON, J. M.: Studies of the formation and utilization of bound CoA in glyceride biosynthesis. Biochim. biophys. Acta *144:* 25–33 (1967).

245 RAO, G. A.; SORRELS, M. F., and REISER, R.: Dietary regulation of phosphatidic acid synthesis from dihydroxyacetone phosphate and fatty acid by rat liver microsomes. Lipids *6:* 88–92 (1971).

246 REDMAN, C. M. and HOKIN, L. E.: Stimulation of the metabolism of phosphatidylinositol and phosphatidic acid in brain cytoplasmic fractions by low concentrations of cholinergic agents. J. Neurochem. *11:* 155–163 (1964).

247 RILLING, H. C.: Biosynthesis of presqualene pyrophosphate by liver microsomes. J. Lipid Res. *11:* 480–485 (1970).

248 RITTER, M. C. and DEMPSEY, M. E.: Specificity and role in cholesterol biosynthesis of a squalene and sterol carrier protein. J. biol. Chem. *246:* 1536–1539 (1971).

249 ROMSOS, D. R.; ALLEE, G. L., and LEVEILLE, G. A.: *In vivo* cholesterol and fatty acid synthesis in the pig intestine. Proc. Soc. exp. Biol. Med. *137:* 570–573 (1971).

250 ROUS, S.: The origin of hydrogen in fatty acid synthesis. Adv. Lipid Res. *9:* 73–118 (1971).

251 SAMUELSSON, B.: Structures, biosynthesis, and metabolism of prostaglandins; in WAKIL Lipid metabolism, pp. 107–153 (Academic Press, New York 1970).

252 SÁNCHEZ, E. DE JIMÉNEZ and CLELAND, W. W.: Studies of the microsomal acylation of L-glycerol-3-phosphate. I. The specificity of the rat brain enzyme. Biochim. biophys. Acta *176:* 685–691 (1969).

253 is eliminated.

254 SARZALA, M. G.; GOLDE, L. M. G. VAN; KRUYFF, B. DE, and DEENEN, L. J. M. VAN: The intramitochondrial distribution of some enzymes involved in the biosynthesis of rat-liver phospholipids. Biochim. biophys. Acta *202:* 106–119 (1970).

255 SASTRY, P. S. and KATES, M.: Biosynthesis of lipids in plants. II. Incorporation of glycerophosphate-32-P into phosphatides by cell-free preparations from spinach leaves. Canad. J. Biochem. *44:* 459–467 (1966).

256 SAUNER, M.-T. and LÉVY, M.: Study of the transfer of phospholipids from the endoplasmic reticulum to the outer and inner mitochondrial membranes. J. Lipid Res. *12:* 71–75 (1971).

257 SCALLEN, T. E.; SCHUSTER, M. W., and DHAR, A. K.: Evidence for a non-catalytic carrier protein in cholesterol synthesis. J. biol. Chem. *246:* 224–230 (1971).

258 SCALLEN, T. J.; SCHUSTER, M. W.; DHAR, A. K., and SKRDLANT, H. B.: Studies on the enzymatic synthesis of cholesterol. Use of a liver acetone powder. Lipids *6:* 162–165 (1971).

259 SCHEGGET, J. TER; BOSCH, H. VAN DEN; BAAK, M. A. VAN; HOSTELLER, K. Y., and BORST, P.: The synthesis and utilization of dCDP-diglyceride by a mitochondrial fraction from rat liver. Biochim. biophys. Acta *239:* 234–242 (1971).

260 SCHILLER, C. M.; DAVID, J. S. K., and JOHNSTON, J. M.: The subcellular distribution of triglyceride synthetase in the intestinal mucosa. Biochim. biophys. Acta *210:* 489–495 (1970).

261 SCHOLTE, H. R.; WIT-PEETERS, E. M., and BAKKER, J. C.: The intracellular and intramitochondrial distribution of short-chain acyl-CoA synthetases in guinea pig heart. Biochim. biophys. Acta *231:* 479–486 (1971).

262 SCHULZ, H. and WAKIL, S. J.: Studies on the mechanism of fatty acid synthesis. XXV. On the mechanism of β-ketoacyl acyl-carrier protein reductase from *Escherichia coli*. J. biol. Chem. *246:* 1895–1901 (1971).

263 SCHULTZ, F. M. and JOHNSTON, J. M.: The synthesis of higher glycerides via the monoglyceride pathway in hamster adipose tissue. J. Lipid Res. *12:* 132–138 (1971).

264 SEUBERT, W.; LAMBERTS, I.; KRAMER, R., and OHLY, B.: On the mechanism of malonyl-CoA-independent fatty acid synthesis. I. The mechanism of elongation of long chain fatty acid by acetyl-CoA. Biochim. biophys. Acta *164:* 498–517 (1968).

265 SHAH, S. N.; PETERSON, N. A., and MCKEAN, C. M.: Cerebral lipid metabolism in experimental hyperphenylalaninaemia. Incorporation of ^{14}C-labelled glucose into total lipids. J. Neurochem. *17:* 279–284 (1970).

266 SHAPIRO, D. J. and RODWELL, V. W.: Regulation of hepatic 3-hydroxy-3-methyl-glutaryl coenzyme A reductase and cholesterol synthesis. J. biol. Chem. *246:* 3210–3216 (1971).

267 SHERR, S. and BYK, CH.: Choline and serine incorporation into the phospholipids of *Neurospora crassa*. Biochim. biophys. Acta *239:* 243–247 (1971).

268 SHORT, ST. A. and WHITE, D. C.: Biosynthesis of cardiolipin from phosphatidyl-glycerol in *Staphylococcus aureus*. J. Bact. *109:* 820–826 (1972).

269 SHRAGO, E.; GLENNON, J. A., and GORDON, E. S.: Comparative aspects of lipo-genesis in mammalian tissues. Metabolism *20:* 54–62 (1971).

270 SIERVO, A. J. DE and SALTON, M. R. J.: Biosynthesis of cardiolipin in the membranes of *Micrococcus lysodeikticus*. Biochim. biophys. Acta *239:* 280–292 (1971).

271 SINENSKY, M.: Temperature control of PL biosynthesis in *Escherichia coli.* J. Bact. *106:* 449–455 (1971).
272 SIPERSTEIN, M. D.; CHAIKOFF, I. L. and CHERNICK, S. S.: Significance of endogenous cholesterol in atherosclerosis. Synthesis in arterial tissue. Science *113:* 747–749 (1951).
273 SIPERSTEIN, M. D. and GUEST, M. J.: Studies on the site of the feedback control of cholesterol synthesis. J. clin. Invest. *39:* 642–652 (1960).
274 SIPERSTEIN, M. D.: Regulation of cholesterol biosynthesis in normal and malignant tissues; in HORECKER and STADTMAN Current topics in cellular regulation, vol. 2, pp. 65–100 (Academic Press, New York 1970).
275 SKOSEY, J. L.: Adrenocorticotropin stimulation of 2-ketoglutarate oxidation by isolated rat epididymal adipose tissue. J. biol. Chem. *245:* 510–518 (1970).
276 SLOVITER, H. A.; JANIC, V., and CIRCOVIC, D.: Biosynthesis of cholesterol in rabbit reticulocytes and its exchange with plasma. J. Lipid Res. *11:* 82–86 (1970).
277 SMITH, J. D. and LAW, J. H.: Phosphatidylcholine biosynthesis in *Tetrahymena pyriformis*. Biochim. biophys. Acta *202:* 141–152 (1970).
278 SNYDER, F.: in HOLMAN Progress in the chemistry of fats and other lipids, vol. 10, part. 3, pp. 287–335 (Pergamon Press, Oxford 1969).
279 SNYDER. F.; BLANK, M. L.; MALONE, B., and WYKLE, R. L.: Identification of L-alkyldihydroxyactone phosphate, O-alkyl-dihydroxyacetone, and diacyl-glyceryl ethers after enzymatic synthesis. J. biol. Chem. *245:* 1800–1805 (1970).
280 SNYDER, F.; BLANK, M. L., and MALONE, B.: Requirement of cytidine derivatives in the biosynthesis of O-alkyl phospholipids. J. biol. Chem. *245:* 4016–4018 (1970).
281 SNYDER, F.; RAINEY, W. T ; BLANK, M. L., and CHRISTIE, W. H.: The source of oxygen in the ether bond of glycerolipids. ^{18}O studies. J. biol. Chem. *245:* 5853–5856 (1970).
282 SPITZER, H. L.: The biosynthesis of phospholipids by base displacement (abstract). Fed. Proc. *31:* 454 (1972).
283 STANACEV, N. Z.; CHANG, Y.-Y., and KENNEDY, E. P.: Biosynthesis of cardiolipin in *Escherichia coli.* J. biol. Chem. *242:* 3018–3019 (1967).
284 STANACEV, N. Z.; ISAAK, D. C., and BROOKES, K. B.: The enzymatic synthesis of phosphatidylglycerol in sheep brain. Biochim. biophys. Acta *152:* 806–808 (1968).
285 STANACEV, N. Z.; STUHNE-SEKALEC, L.; BROOKES, K. B., and DAVIDSON, J. B.: Intermediary metabolism of phospholipids. The biosynthesis of phosphatidylglycerophosphate and phosphatidylglycerol in heart mitochondria. Biochim. biophys. Acta *176:* 650–653 (1969).
286 STANACEV, N. Z. and STUHNE-SEKALEC, L.: On the mechanism of enzymatic phosphatidylation. Biosynthesis of cardiolipin catalyzed by phospholipase D. Biochim. biophys. Acta *210:* 350–352 (1970).
287 STANACEV, N. Z.; DAVIDSON, J. B.; STUHNE-SEKALEC, L., and DOMAZET, Z.: The mechanism of the biosynthesis of cardiolipin in mitochondria. Biochem. biophys. Res. Commun. *47:* 1021–1027 (1972).
288 STEIN, Y.; STEIN, O., and SHAPIRO, B.: Enzymic pathways of glyceride and phospholipid synthesis in aortic homogenates. Biochim. biophys. Acta *70:* 33–42 (1963).
289 STEINER, M. R. and LESTER, R. L.: *In vitro* study of the methylation pathway of phosphatidylcholine synthesis and the regulation of this pathway in *Saccharomyces cerevisiae*. Biochemistry *9:* 63–69 (1970).

290 STEINER, M. R. and LESTER, R. L.: *In vitro* studies of phospholipid biosynthesis in *Saccharomyces cerevisiae*. Biochim. biophys. Acta *260:* 222–243 (1972).

291 STICKLAND, K. P.: Phosphorylation of diglycerides by rat brain. Canad. J. Biochem. *40:* 247–259 (1962).

292 STOFFEL, W.: Biosynthesis of polyenoic fatty acids. Biochem. biophys. Res. Commun. *6:* 270–273 (1961).

293 STOFFEL, W.; THOMAS, M. E. DE und SCHIEFER, H. G.: Die enzymatische Acylierung von Lysophosphatidsäure, gesättigtem und ungesättigtem Lysolecithin. Z. Physiol. Chem. *348:* 882–890 (1967).

294 STOFFEL, W. and SCHIEFER, H. G.: Biosynthesis and composition of phosphatides in outer and inner mitochondrial membranes. Z. Physiol. Chem. *349:* 1017–1026 (1968).

295 STOFFEL, W.: Sphingolipids. Annu. Rev. Biochem. *40:* 51–82 (1971).

296 STOFFEL, W. and LEKIM, D.: Studies on the biosynthesis of plasmalogens. Z. Physiol. Chem. *352:* 501–511 (1971).

297 STRICKLAND, E. H. and BENSON, A. A.: Neutron activation paper chromatographic analysis of phosphatides in mammalian cell fractions. Arch. Biochem. *88:* 344–348 (1960).

298 STUMPF, P. U.: Metabolism of fatty acids. Annu. Rev. Biochem. *38:* 159–212 (1969).

299 STUMPF, P. K.: Fatty acid metabolism in plants; in WAKIL Lipid metabolism, pp. 79–106 (Academic Press, New York 1970).

300 SUGANO, M.: Phosphatidylcholine-cholesterol acyltransferase in the ultracentrifugal residual protein fraction of rat plasma. Biochem. J. *122:* 469–475 (1971).

301 SULLIVAN, A. C.; MILLER, O. N.; WITTMAN, J. S., and HAMILTON, J. G.: Factors influencing the *in vivo* and *in vitro* rates of lipogenesis in rat liver. J. Nutr. *101:* 265–272 (1971).

302 SUNDLER, R. and AKESSON, B.: The acylation of monoglycerol isomers by pig liver microsomes. Biochim. biophys. Acta *218:* 89–96 (1970).

303 TAMELEN, E. E. VAN: Bioorganic chemistry of sterols and acyclic terpene terminal epoxides. Accounts chem. Res. *1:* 111–120 (Amer. Chem. Soc., Wash., 1968).

304 TOPPING, D. L. and MAYES, P. A.: The immediate effects of insulin and fructose on the metabolism on the perfused liver. Biochem. J. *126:* 295–311 (1972).

305 TROPP, B. E.; MEADE, L. C., and THOMAS, P. J.: Consequences of expression of the 'relaxed' genotype of the RC gene. Lipid synthesis. J. biol. Chem. *245:* 855–858 (1970).

306 VAGELOS, P. R.; MAJERUS, P. W.; ALBERTS, A. W.; LARRABEE, A. R., and AILHAUD, G. P.: Structure and function of the acyl carrier protein. Fed. Proc. *25:* 1485–1494 (1966).

307 VAGELOS, P. R.: Regulation of fatty acid biosynthesis. Current topics in cellular regulation, vol. 4, pp. 119–166 (Academic Press, New York 1971).

308 VEREYKEN, J. M.; MONTFOORT, A., and GOLDE, L. M. G. VAN: Some studies on the biosynthesis of the molecular species of phosphatidylcholine from rat lung and phosphatidylcholine and phosphatidylethanolamine from rat liver. Biochim. biophys. Acta *260:* 70–81 (1972).

309 VOLK, B. W. and ARONSON, ST. M.: Sphingolipids, sphingolipidoses and allied disorders (Plenum Press, New York 1972).

310 VORBECK, M. L. and MARTIN, A. P.: Glycerophosphatide biogenesis. I. Subcellular localization of cytidine triphosphate: phosphatidic acid cytidyl transferase. Biochem. biophys. Res. Commun. *40:* 901–908 (1970).

311 WAITE, M. and WAKIL, S. J.: Studies on the mechanism of action of acetyl coenzyme A carboxylase. II. On the mechanism of action of enzyme-bound biotin. J. biol. Chem. *238:* 81–90 (1963).

312 WAKIL, S. J.: Mechanism of fatty acid synthesis. J. Lipid Res. *2:* 1–24 (1961).

313 WAKIL, S. J.: Fatty acid metabolism; in WAKIL Lipid metabolism (Academic Press, New York 1970).

314 WALKER, P. R. and BAILEY, E.: Role of 'malic enzyme' in lipogenesis. Biochim. biophys. Acta *187:* 591–593 (1969).

315 WATSON, J. A. and LOWENSTEIN, J. M.: Citrate and the conversion of carbohydrate into fat. Fatty acid synthesis by a combination of cytoplasm and mitochondria. J. biol. Chem. *245:* 5993–6002 (1970).

316 WEINHOLD, P. A.: Phospholipid metabolism in the developing liver. I. The biosynthesis of choline glycerophosphatides by liver slices from fetal, newborn and adult rats. Biochim. biophys. Acta *187:* 85–93 (1969).

317 WEISS, S. B.; SMITH, S. W., and KENNEDY, E. P.: Net synthesis of lecithin in an isolated enzyme system. Nature, Lond. *178:* 594–595 (1956).

318 WEISS, S. B. and KENNEDY, E. P.: The enzymatic synthesis of triglycerides. J. amer. chem. Soc. *78:* 3550 (1956).

319 WHEATLEY, V. R.; HODGINS, L. T.; COON, W. M.; KUMARASIRI, M.; BERENZWEIG, H., and FEINSTEIN, J. M.: Cutaneous lipogenesis. Precursors utilized by guinea pig skin for lipid synthesis. J. Lipid Res. *12:* 347–360 (1971).

320 WHEREAT, A. F.: Fatty acid synthesis in a cell-free system from rabbit aorta. J. Lipid Res. *7:* 671–677 (1966).

321 WHEREAT, A. F.: Is atherosclerosis a disorder of intramitochondrial respiration? Ann. intern. Med. *73:* 125–127 (1970).

322 WHEREAT, A. F.: Fatty acid biosynthesis in aorta and heart. Adv. Lipid Res. *9:* 119–159 (1971).

323 WIEGANDT, A.: Glycolypids. Adv. Lipid Res. *9:* 249–289 (1971).

324 WILGRAM, G. F. and KENNEDY, E. P.: Intracellular distribution of some enzymes catalyzing reactions in the biosynthesis of complex lipids. J. biol. Chem. *238:* 2615–2619 (1963).

325 WILLIAMS, M. L. and BYGRAVE, F. L.: Incorporation of inorganic phosphate into phospholipids by the homogenate and by sub-cellular fractions of rat liver. Europ. J. Biochem. *17:* 32–38 (1970).

326 WILSON, J. D.; GIBSON, K. D., and UDENFRIEND, S.: Studies on the conversion *in vitro* of serine to ethanolamine by rat liver and brain. J. biol. Chem. *235:* 3539–3543 (1960).

327 WIRTZ, K. W. A. and ZILVERSMIT, D. B.: Exchange of phospholipids between liver mitochondria and microsomes *in vitro*. J. biol. Chem. *243:* 3596–3602 (1968).

328 WRIGHT, L. D.; CLELAND, M.; DUTTA, B. N., and NORTON, J. S.: Mevaldic acid in the biogenesis of mevalonic acid. J. amer. chem. Soc. *79:* 6572 (1957).

329 WYKLE, R. L.; BLANK, M. L., and SNYDER, F.: The biosynthesis of plasmalogens in a cell-free system. FEBS Letters *12:* 57–60 (1970).

330 YOSHIDA, H. and NUKADA, T.: Increase in metabolic turnover of phosphatidic acid in brain slices caused by potassium. Biochim. biophys. Acta *46:* 408–410 (1961).

331 YOUNG, D. L.; POWELL, G., and MCMILLAN, W. O.: Phenobarbital-induced alterations in phosphatidylcholine and triglyceride synthesis in hepatic endoplasmic reticulum. J. Lipid Res. *12:* 1–8 (1971).

332 ZAHLER, W. L.; BARDEN, R. E., and CLELAND, W. W.: Some physical properties of palmityl-coenzyme A micelles. Biochim. biophys. Acta *164:* 1–11 (1968).

333 ZAKIM, D.; PARDINI, R. S.; HERMAN, R. H., and SAUBERLICH, H. E.: Mechanism for the differential effects of high carbohydrate diets on lipogenesis in rat liver. Biochim. biophys. Acta *144:* 242–251 (1967).

334 ZAKIM, D. and HERMAN, R. H.: The effect of intravenous fructose and glucose on the hepatic α-glycerophosphate concentration in the rat. Biochim. biophys. Acta *165:* 374–379 (1968).

335 ZBOROWSKI, J. and WOJTCZAK, L.: Phospholipid synthesis in rat liver mitochondria. Biochim. biophys. Acta *187:* 73–84 (1969).

336 ZIBOH, V. A.; DREIZE, M. A., and HSIA, S. L.: Inhibition of lipid synthesis and glucose-6-phosphate dehydrogenase in rat skin by dehydroepiandrosterone. J. Lipid Res. *11:* 346–354 (1970).

337 ZILVERSMIT, D. B. and MCCANDLESS, E. L.: Independence of arterial phospholipid synthesis from alterations in blood lipids. J. Lipid Res. *1:* 118–124 (1959).

338 ZILVERSMIT, D. B.; MCCANDLESS, E. L.; JORDAN, P. H., jr.; HENLY, W. S., and ACKERMAN, R. F.: The synthesis of phospholipids in human atheromatous lesions. Circulation *23:* 370–375 (1961).

Author's address: PD Dr. I. R. KUPKE, Medizinische Hochschule Hannover, Institut für Klinische Biochemie und Physiologische Chemie, Karl-Wiechert-Allee 9, *D-3 Hannover-Kleefeld* (FRG)

Progr. biochem. Pharmacol., vol. 8, pp. 125–160 (Karger, Basel 1973)

Triglyceride Turnover in Man

Effects of Dietary Carbohydrate

P. J. NESTEL

Department of Clinical Science, The John Curtin School of Medical Research, Australian National University, Canberra

Contents

I. Introduction

An increase in the proportion of dietary calories derived from carbohydrate generally leads to a rise in plasma triglyceride concentration. The magnitude of this rise is determined by many factors that are discussed in this and in the other chapters. Since the rise in the plasma triglyceride concentration reflects changes in inflow or in outflow, one could establish the relative importance of overproduction and deficient removal if the turnover of plasma triglyceride could be measured. Both overproduction and inefficient removal might operate simultaneously especially since changes in inflow and outflow probably vary throughout the day, even with fat-free diets. As will be emphasized later, the triglyceride concentration fluctuates appreciably during the day and night, when carbohydrate-rich, fat-free diets are eaten. This suggests that inflow and outflow of plasma triglyceride are not in balance throughout the 24 h and a measure of triglyceride turnover taken during only one phase of the diurnal flux will only partly describe the basis for the hypertriglyceridaemia. Most studies have hitherto been carried out during the fasting state, generally some 16 h after the last meal of a carbohydrate-rich diet; findings from such studies may give incomplete information about the events that initiate the rise in the plasma triglycerides. However, as will be discussed below, the techniques that are currently available for measuring triglyceride turnover require conditions that are more likely to be present in the fasting state than during the less steady state of carbohydrate feeding. Moreover, at least one of the more reliable methods demands that the plasma triglyceride fatty acids (TGFA) are derived from a single precursor (in this instance the plasma free fatty acids [FFA]), circumstances that do not exist during or soon after prolonged carbohydrate ingestion, when glucose itself becomes a major precursor. It will, therefore, be necessary to describe the techniques that have been used in attempts to measure triglyceride turnover and the physiological assumptions and observations upon which the various models have been based. The effects of carbohydrate-rich diets have been studied with some of these techniques which, in retrospect, have been based on partly inaccurate assumptions.

Many aspects of triglyceride metabolism are dealt with elsewhere in this book. This chapter will be confined in the main to questions of triglyceride and fatty acid kinetics and will review recent data that has been derived predominantly in man.

II. Measurement of the Turnover of Plasma Triglyceride

The major parameters that can be obtained from models that describe triglyceride transport include: (1) inflow and outflow which is the mass of material entering or leaving a compartment per unit time; in the steady state, inflow and outflow are equal and comprise the turnover which denotes transport or production, and (2) the fractional removal or fractional turnover rate (more recently referred to as turnover rate) which is the fraction of material leaving a compartment per unit time and is the reciprocal of the turnover time; it reflects removal capacity.

Measurement of triglyceride turnover by tracer techniques assumes steady state conditions, that is, balanced inflow and outflow from the plasma resulting in a constant concentration. Any change in concentration must then result from a peroid of imbalance between inflow and outflow. When comparing two steady state conditions at different triglyceride concentrations the question arises whether the higher concentration has been brought about by an increase in inflow or a decrease in outflow, or possibly both. If an increase in inflow had been responsible, the pool size would have increased until a new steady state had been reached when outflow would also have risen, once more to equal inflow; the fraction of the pool removed per unit time would be similar during both steady state periods, but turnover would be higher during the second period. If, however, the primary event had been a decrease in outflow, the pool size would also have expanded for a time, until one of two situations had occurred to prevent further expansion: (1) outflow might have risen to equal inflow once more; the turnover would then have remained constant but the fraction of the pool removed per unit time would have fallen, or (2) inflow may have fallen to equal outflow, resulting in a lower turnover but similar fractional removal rate. When comparing two distinct steady state conditions, a rise in turnover implies a probable primary increase in inflow while a fall in the fractional turnover rate implies a probable primary reduction in removal. It should be pointed out, however, that during the transition from one steady state to another there may have been a series of changes which are not reflected in the parameters of the new steady state.

The methods can be divided into those that (1) are based on the kinetics of reinjected triglyceride-containing lipoproteins, and (2) formulate models from precursor and product data derived from the rates of appearance of injected radioactively labelled precursors, usually FFA or glycerol, in plasma triglyceride.

A. Reinjection of Triglyceride-Containing Lipoproteins

Provided the reinjected material is a tracer and does not expand the pool size and provided it is metabolized physiologically, the rate of removal from the plasma should, in the steady state, reflect the endogenous turnover. This approach is conceptually straightforward but interpretations of results obtained in this way have been suspect because the injected material has probably not been handled in a manner identical to that of the native lipoproteins.

FRIEDBERG *et al.* [1], FARQUHAR *et al.* [2] and EATON *et al.* [3] reinjected human plasma or very low density lipoproteins (VLDL) after the triglyceride had been labelled beforehand. The half-times of removal were of the order of one-half to a few hours and were similar to those obtained when the triglyceride had been labelled endogenously by injecting a radiolabelled precursor. It was reasoned that the half-time of removal of the reinjected material provided a means of calculating the fractional removal rate of endogenously produced triglyceride and further, that if this rate was determined solely by the outflow from the plasma triglyceride pool, then the product of the fractional removal rate times the mass of triglyceride within the pool of VLDL was equal to triglyceride turnover. This provided values which now appear to be unduly low.

However, EATON *et al.* [3] have shown that human lipoproteins in which triglyceride is labelled *in vitro* undergo some denaturation, while HAVEL [4] has briefly reported studies in which the fractional turnover rate of reinjected labelled VLDL was very much faster than in those referred to above. Thus, this kinetically acceptable approach awaits improved methods for harvesting, labelling and reinjecting lipoproteins that are handled physiologically when injected into the recipient.

Other lipoproteins that have been used in human reinjection experiments are lymph chylomicrons. NESTEL [5] labelled the glycerides by giving radiopalmitic acid in cream to a donor with a thoracic duct fistula. The washed chylomicrons, upon reinjection into recipients, were removed at half-times of only 5–20 min in contrast to the slower rates reported by FARQUHAR *et al.* for VLDL [2]. The rates of removal were inversely related to the pools of plasma triglyceride suggesting that in humans, fasting overnight after habitual diets, an increased triglyceride concentration reflects a relative deficiency in removal mechanisms. The faster removal rate of the chylomicrons compared with that of VLDL may also indicate a difference in the handling of lipoproteins of different sizes [6].

B. Injection of Artificial Fat Emulsions

Artificial fat emulsions have also been injected to calculate rate constants for the removal of triglyceride. Unlike the studies with radiolabelled lipoproteins, the injection of the emulsions leads to an appreciable increase in the size of the triglyceride pool. Care is taken not to saturate the removal mechanisms and to maintain first-order kinetics. The maximal removal capacity in man was found [7] to vary from 42 to 88 μM triglyceride/l/min. This figure greatly exceeds the maximum turnover of endogenous triglyceride reported in fasting man by BOBERG *et al.* [8] but it does represent a measure of the removal capacity of the system which appears adequate for the amounts that may be encountered after a very fatty meal. However, in studies with lean healthy subjects receiving constant infusions of small amounts of Intralipid®, a much lower clearance capacity has been found [NESTEL and BARTER, unpublished information].

Of the various lipid emulsions Intralipid® appears to resemble most closely native chylomicrons in the characteristics of its removal. It has been used extensively by HALLBERG [9] and BOBERG *et al.* [7], who showed that its removal from plasma followed a biexponential curve. The first rate constant (K_1) indicated the maximal rate at which the removal sites were apparently being saturated, whereas the second rate constant (K_2) followed first-order kinetics and reflected the fractional removal rate of the triglyceride. BOBERG *et al.* [7] concluded from their Intralipid® studies that, in fasting subjects with endogenous hypertriglyceridaemia, the fractional removal rate of triglyceride was significantly less than in normal subjects and that removal capacity decreased with age.

The maximal removal capacity for Intralipid® appears to be increased during infusions of glucose or insulin in dogs [10].

NESTEL and BARTER [unpublished information] have infused Intralipid® at constant rates into subjects deriving their calories solely from glucose. The fractional removal was raised in studies carried out shortly after stopping the glucose drinks but decreased in studies repeated after a 16-hour fast. The results emphasize the importance of defining the time at which studies are carried out and the error of ascribing carbohydrate-induced triglyceridaemia to decreased removal alone. A preliminary report by LEWIS *et al.* [11] has also shown reduced clearance of Intralipid® in carbohydrate-fed subjects fasted overnight.

C. Isotopic Precursor-Product Models

1. Pulse Injections of Radiolabelled Precursors

FRIEDBERG *et al.* [1] were the first to utilize the precursor-product relationship between plasma FFA and TGFA which had also been reported in man by CARLSON [12] and HAVEL [13]. They observed that following the injection of radiopalmitic acid, the initial portion of the specific radioactivity-time curve of plasma TGFA was exponential; from the area below the TGFA specific activity curve they calculated the fractional conversion of plasma FFA to TGFA. This was extended by FARQUHAR *et al.* [2] and REAVEN *et al.* [14] with 3 modifications: the substitution of radiolabelled glycerol for palmitic acid as precursor, the measurement of triglyceride kinetics in the specific lipoprotein, VLDL, rather than in whole plasma, and the calculation of fractional turnover rate from the half-time of removal of radioactivity. Consequently, the specific radioactivity curve of the product remained exponential for a longer time, allowing more precise calculation of the fractional turnover rate of the triglyceride.

The crucial assumption in FARQUHAR's model [2] is that the labelled precursor, glycerol, becomes incorporated into hepatic triglyceride and discharged into the plasma at a rate that is very much faster than the rate of removal of triglyceride from the plasma. Since, in man, the production of endogenous triglyceride had been shown to occur predominantly in the liver by CARLSON and EKELUND [15] they obtained single liver biopsy specimens at the estimated times of peak plasma triglyceride specific radioactivity in order to establish their point. However, this does not now appear to be true from the later studies of QUARFORDT *et al.* [16] which suggest that the turnover characteristics of the VLDL triglyceride pool are very similar to those of the immediate precursor pool in the liver. Significant amounts of labelled triglyceride would therefore continue to be discharged from the liver into the plasma VLDL during the period when measurements of the disappearance rate of VLDL radioactivity are being carried out. The specific radioactivity curve of the VLDL triglyceride would therefore represent both continuing inflow of label from the liver and outflow of label from the plasma whereas the assumption of FARQUHAR *et al.* [2] was that it reflected outflow alone. Values for triglyceride turnover obtained by this method are appreciably less than the probable true values as discussed by SHAMES *et al.* [17].

The effects of high-carbohydrate diets have been studied with this technique. NESTEL and HIRSCH [18] and NESTEL [19] using radiopalmitate found, in normolipidaemic men, that modest increments in the triglyceride concentration brought about by 1 or 2 weeks of carbohydrate-enriched diets resulted in higher turnover of TGFA. Similar conclusions were reached in a larger study of both normolipidaemic and hyperlipidaemic subjects consuming high-carbohydrate, fat-free diets [14]. However, in all these studies, the fractional turnover rate of VLDL triglyceride was lower during carbohydrate consumption, indicating that removal was less efficient in relation to the apparently increased production. The interpretation of these earlier studies, including that by WATERHOUSE *et al.* [20], who used a somewhat different model, might be summed up as follows: (1) the models used were not consistent with current physiological information in that the slope of the VLDL triglyceride specific activity curve does not reflect the removal rate alone. Further, with high-carbohydrate diets, especially those used by REAVEN *et al.* [14] the labelled precursor would not constitute the sole precursor, and (2) both deficiencies would underestimate the true turnover, yet the overall conclusion does not appear invalid: the inflow of triglyceride is raised with carbohydrate-rich diets, though the slower fractional turnover rate denotes less than adequate outflow. This does not imply a reduced removal rate in absolute terms; on the contrary, a high inflow at steady though elevated triglyceride levels indicates a high outflow. These studies imply that the initial event is increased inflow and that outflow rises to a corresponding degree after the VLDL pool had expanded.

This approach has been extended by EATON *et al.* [3] and QUARFORDT *et al.* [16], who used the method of multicompartmental analysis developed by BERMAN. An injection of radiolabelled palmitic acid is followed by prolonged observations of the plasma FFA and VLDL-TGFA specific radioactivity curves. Several models, that appeared to be consistent with possible physiological pathways in the metabolism of FFA and TGFA, were constructed from the correspondence of the observed data with calculated data obtained with a digital computer. The studies of QUARFORDT *et al.* [16] included the effects of high-carbohydrate diets in 4 subjects.

These prolonged studies took into account the late curvilinear portion of VLDL-TGFA specific activity curves that could be explained only on the basis of a continued slow input of VLDL-TGFA tracer, i.e. the presence of several pools of fatty acids along the pathway of TGFA formation or some recycling of label. The model required a postulated slow pathway such as a slow hepatic FFA-TGFA conversion system[1]. The inflow transport of TGFA was consistently higher with the high-carbohydrate diets which appeared to occur by way of the postulated slow conversion pathway. These studies also suggested that this increased conversion was brought about by greater diversion of FFA from pathways of oxidation and ketone-formation to formation of TGFA. (A similar mechanism has been put forward to account for the increased production of VLDL-TGFA in response to ethanol by NESTEL and HIRSCH [23] and by WOLFE *et al.* [24].)

QUARFORDT *et al.* [16] also observed a reduction in the fractional removal rate of VLDL-TGFA although they concluded that an increase in glyceride production from plasma FFA was the likeliest single prime perturbation brought about by high carbohydrate consumption. Whereas both these events (increased conversion of FFA and diminished removal of TGFA participate in carbohydrate-induced hypertriglyceridaemia, it is clear from the findings of BARTER *et al.* [25] and BARTER and NESTEL [26] that a significant fraction of VLDL-TGFA is derived from the hepatic conversion of glucose to fatty acid.

Another important aspect of the studies of QUARFORDT *et al.* [16] was the finding that subjects with normotriglyceridaemia and those with endogenous hypertriglyceridaemia responded to carbohydrate diets with similar increments in triglyceride inflow. When studied in the fasting state after habitual diets both groups produced similar amounts of triglyceride but the subjects with endogenous hypertriglyceridaemia had significantly lower fractional removal rates. Although REAVEN *et al.* [14] reported that most of their subjects with endogenous hypertriglyceridaemia showed increased turnover with normal removal of triglyceride, the study of additional subjects revealed some who also showed impaired removal [27]. The discrepancy between the findings of QUARFORDT *et al.* [16] and REAVEN *et al.* [14] relates to the false assumption of the Farquhar-Reaven model.

NIKKILA and KEKKI [28] who have used the Farquhar model ex-

1 The studies of BAKER and SCHOTZ [21, 22] in rats had suggested several years previously that the conversion of plasma FFA to VLDL-TGFA might involve a series of transfers through pools of hepatic TGFA. The multicompartmental analysis of the incorporation rates of radiopalmitate into hepatic and extrahepatic lipids and into CO_2 led them to conclude that, in the fasting rat, the major portion of net hepatic TGFA formation appeared to be by way of a fatty acid pool which turned over relatively slowly compared to plasma FFA.

tensively have also recognized the problems of deviation from the single exponential in the specific radioactivity curve of plasma triglyceride after injecting radioglycerol (though they did not analyze VLDL separately). They believed this to reflect the presence of a slow hepatic 'storage or non-secreting' pool with delayed secretion of labelled triglyceride, and to early recycling of label, especially when clearance of triglyceride was rapid. In contrast to the conclusions of REAVEN *et al.* [14] and FARQUHAR *et al.* [29], who concluded that in a normolipidaemic population the plasma triglyceride concentration was determined only by triglyceride inflow and not by clearance mechanisms, NIKKILA and KEKKI [28] have observed variability in both inflow and clearance. In particular they noted more efficient removal of triglyceride in women than men at similar inflow rates. While this possibility may have special relevance to the resistance of premenopausal women to carbohydrate-inducible triglyceridaemia [30], conclusions based on this particular model must be interpreted cautiously. NIKKILA and KEKKI hoped to overcome the problem of a non-exponential specific radioactivity curve by resolving their curves with a digital computer and analyzing only those curves in which a single exponential function accounted for at least 95% of the curve. However, even an apparently single exponential curve does not rule out continuing inflow of label from the liver and underestimation of turnover. The advantages and limitations of multicompartmental analyses in relation to lipid kinetics have been comprehensively reviewed by BAKER [31] in a critical evaluation of studies of FFA and TGFA metabolism published at that time.

2. Constant Infusion of Precursor

Radiolabelled palmitic acid has been the commonest precursor used in the constant infusion studies, although SANDHOFER *et al.* [32] have reported a study with ^{14}C-glucose. RYAN and SCHWARTZ [33] proposed a model in which the rate of labelling of plasma TGFA was measured during constant infusions of radiopalmitate. Labelled triglyceride appears in plasma after a delay of about 30 min and labelling increases at a linear rate usually between the first and third hours of the infusion. After high carbohydrate diets, linearity in the rate of labelling does not occur until later, indicating an enlarged hepatic precursor pool or the operation of the slow FFA-TGFA pathway in the liver as proposed by QUARFORDT *et al.* [16]. In normolipidaemic subjects TGFA labelling reached a plateau after about 4 h when outflow of label equals inflow of label [26, 34]. This plateau is not reached until very much later in hypertriglyceridaemic subjects [25, 26].

Table I. Fractional conversions of Plasma FFA to triglyceride in response to carbohydrates and fats

Comparison, number	Mean triglyceride concentration, mg/100 ml	Mean 'TGFA inflow' μM/min
Sugar vs.	264	7.8
starch, 6	200	5.5
Fructose vs.	113	3.3
glucose, 4	115	6.6
Saturated fat vs.	100	4.2
polyunsaturated fat, 6	69	4.6

Summarized from NESTEL *et al.* [37].

The rate at which labelled TGFA appears in plasma is divided by 3 times the constant specific radioactivity of plasma FFA (since 3 moles of FFA comprise 1 mole of triglyceride). RYAN and SCHWARTZ [33] assumed that the entry of labelled TGFA would initially greatly exceed the rate of removal of label from plasma: the rate of appearance of label would, therefore, closely approximate inflow transport. Subsequent studies by HAVEL *et al.* [34] and BOBERG *et al.* [8, 35] have clearly shown that this assumption may apply when the plasma triglyceride pool is expanded and fractional removal slow, but not in normotriglyceridaemic subjects in whom the turnover would, therefore, be appreciably underestimated. Additional problems inherent in this method include the variable delay in the entry of label (referred to above) and the fact that the specific radioactivity of the true precursor FFA in the liver is lower than that in the peripheral circulation due to the addition of unlabelled splanchnic FFA [34].

Applying this model to the problem of endogenous hypertriglyceridaemia, RYAN and SCHWARTZ [33] concluded that inflow transport was not increased and hence clearance was defective, findings that were confirmed with this technique by SAILER *et al.* [36].

NESTEL *et al.* [37] have used this technique to compare the effects of different carbohydrates and fats (table I). Although quantitative data are underestimated by this method, qualitative differences can be detected, especially if the comparisons are made in the same individual and the pools of plasma triglyceride are not greatly different.

A diet very rich in sucrose led to higher plasma triglyceride levels than a diet containing an equicaloric amount of starch in each of 6 comparisons. The inflow of TGFA was also always higher, indicating greater total conversion of FFA to TGFA with sucrose than with starch. The absolute values for inflow are low for the reasons discussed above; furthermore, any contribution from lipogenesis would be missed.

In the 4 comparisons of fructose *versus* glucose diets in 4 healthy young men, the apparent conversion of FFA to TGFA with glucose was twice that with fructose at similar plasma triglyceride concentrations. Although the contribution from hepatic lipogenesis was not known, the results strongly suggest a slower turnover of TGFA with fructose. Studies with rats are conflicting; BAR-ON and STEIN [38] reported greater hepatic lipogenesis with fructose than glucose whereas FALLON and KEMP [39] and WEBB *et al.* [40] found similar increments with both sugars although all 3 studies are in accord with the above finding that fructose leads to less efficient removal of triglyceride.

Consumption of polyunsaturated fatty acids, in contrast to saturated fatty acids, led to lower plasma triglyceride levels yet slightly higher inflows in 5 of the 6 comparisons. Problems of precursors other than plasma FFA do not apply with these diets and any underestimate of TGFA inflow would have been greater with the polyunsaturated fat studies because the triglyceride pools were smaller. Polyunsaturated fatty acids lowered the plasma triglyceride levels despite an increasing turnover, presumably because of an even greater enhancement of removal mechanisms. At least one other mechanism has been observed in man by NESTEL and BARTER [41] during comparisons of palmitate and linoleate turnover and with perfused rat-livers by KOHUT *et al.* [42] that may account for the triglyceride-lowering effect of polyunsaturated fats. Compared to polyunsaturated fatty acids, saturated fatty acids appear to be preferentially incorporated into hepatic and plasma triglyceride. In recent studies with constant infusions of Intralipid®, NESTEL and BARTER [unpublished information] found greater removal rates with dietary polyunsaturated fatty acids than saturated fatty acids.

The bulk of evidence in relation to the triglyceridaemic effects of sucrose and starch suggests that the response after sucrose is greater, especially when the overall contribution from carbohydrates is high [43] and when the dietary fat is saturated [44, 45]. This subject is fully reviewed elsewhere in the book, but the influence that the kind of dietary fatty acid might have on triglyceride kinetics is apparent from the above studies.

3. Hepatic Secretion of Radiolabelled Triglyceride

Without doubt, the most direct and accurate quantification of hepatic triglyceride production by isotopic means has been provided by the studies of HAVEL *et al.* [34] and BOBERG *et al.* [8, 35]. Initially, BASSO and HAVEL [46] evaluated the model in dogs, fitted with catheters in the portal and hepatic veins and an artery. By infusing radiopalmitate at a constant rate and measuring hepatic blood flow and the appearance of label in products of FFA metabolism, it was possible to calculate the hepatic uptake of plasma FFA, the release and uptake of FFA in extrahepatic splanchnic beds and the conversion of FFA to VLDL-TGFA, CO_2 and ketones. These studies established that in normal dogs the hepatic venous FFA specific radioactivity could be equated with that of the hepatic precursor which was shown to be virtually the sole precursor of secreted TGFA. Calculations of TGFA production could, therefore, be made from the hepatic vein data alone. Studies could then be validly carried out at least in normotriglyceridaemic subjects in the postabsorptive state omitting catherization of the portal vein.

The rate of secretion of labelled TGFA was obtained from the difference between the hepatic venous and arterial values and TGFA production was then calculated from this value and the specific activity of hepatic venous FFA. As discussed later, the turnover of TGFA in a small group of hypertriglyceridaemic subjects was about 30% higher than among normotriglyceridaemic subjects, due to a greater hepatic uptake of FFA among the former [34]. In this small group no correlation was found between TGFA production and triglyceride concentration. The studies were carried out for only 4 h, which was not sufficiently long to establish clear isotopic equilibration between VLDL-TGFA and plasma FFA among the hypertriglyceridaemic subjects, and so a possible underestimate of TGFA inflow was not excluded. At least 3 of their subjects were clearly overweight and, as discussed later, hepatic fatty acids may be an additional source of VLDL-TGFA in obese subjects [26].

The studies of BOBERG *et al.* [8, 35] also suggest that this technique may underestimate TGFA inflow especially in hypertriglyceridaemia. They base this on a comparison with two other methods that they had carried out in a large number of normal and in 14 hypertriglyceridaemic subjects. The first of these additional measurements is the chemical determination of the venous-arterial difference in the triglyceride concentration across the liver: the 'chemical triglyceride secretion', for which an analytical error of less than 2% is being claimed. The second technique is the plasma 'triglyceride clearance' method that calculates the removal of triglyceride radioactivity in subjects in whom triglyceride secretion is being measured simultaneously. The difference between the amount of triglyceride secreted from the liver and that remaining in VLDL represents the amount cleared.

The correlations between the triglyceride clearance and the chemical secretion values were of a very high order, whereas the isotopic secretion method appeared to give similar results only at lower TGFA turnovers. In the studies by BOBERG *et al.* [35], several of the hypertriglyceridaemic subjects clearly secreted more triglyceride as calculated by the direct chemical and the clearance methods. Overall, there was a significant correlation between the concentration and inflow transport of triglyceride among the hypertriglyceridaemic but not the normotriglyceridaemic subjects. Thus, the hypertriglyceridaemic subjects, although few in number, appeared to fall into a group with normal and another with raised triglyceride production. Since, in the majority of subjects, a very high correlation was found between TGFA inflow and the splanchnic extraction of FFA, it seemed likely that among that small group of hypertriglyceridaemic 'overproducers' in whom this correlation did not hold, the additional TGFA was derived from a hepatic source of fatty acids. Whether this alone implies a heterogeneous aetiology for endogenous hypertriglyceridaemia, a very likely possibility for a variety of other reasons, is uncertain. The preceding dietary intake of carbohydrate and the amount of stored hepatic fatty acid might account for such findings. The hypertriglyceridaemic group as a whole was homogeneous in one respect: the fractional turnover rate was low in all, indicating some degree of impairment of removal mechanisms.

To summarize the data derived from the studies of HAVEL *et al.* [34] and BOBERG *et al.* [35] insofar as they impinge on the question of hypertriglyceridaemia: the removal rate of plasma triglyceride after an overnight fast is clearly less efficient in dealing with the existing inflow, even though many such subjects show no evidence of increased inflow. Some subjects do, however, clearly show increased production and possibly others might do so under different circumstances, e.g. in the non-fasting state. Amongst the 'overproducers' the additional source of triglyceride is probably not plasma FFA but rather hepatic fatty acids possibly derived from glucose lipogenesis. By contrast, the 'non-overproducers' derive their VLDL-TGFA from FFA extracted in the splanchnic circulation; since plasma FFA provide only a portion of TGFA under non-fasting conditions, the question of triglyceride production in the absorptive state remains to be resolved.

D. Non-Isotopic Methods

The direct chemical measurement of triglyceride production by the liver [8] has already been referred to.

PORTE and BIERMAN [47] have measured the presumed maximal rate of removal of triglyceride by stimulating removal with constant infusions of heparin. This resulted in a lowering of the plasma triglyceride concentration to a constant level and an increase in the lipolytic activity of plasma (lipoprotein lipase) to a constant rate. This lipolytic activity, as measured *in vitro,* was equated with the removal rate of triglyceride *in vivo,* an unproved assumption. The acceleration in removal might conceivably stimulate production, especially as the plasma FFA concentration would also rise considerably. It is, therefore, not surprising that their values for triglyceride turnover are higher than reported by others.

Nevertheless, the studies of PORTE and BIERMAN [47] in a large heterogeneous group with hypertriglyceridaemia, revealed some subjects in whom the removal mechanism (and presumed triglyceride turnover) was considerably less than in the remainder. Overall, a correlation between triglyceride concentration and 'turnover' was also demonstrable as with some of the other techniques. This relationship reached a very high level of significance when their subjects consumed carbohydrate-rich fat-free diets [48]. Among 47 subjects with both 'exogenous' and 'endogenous' lipaemia (defined according to whether the triglyceride concentration fell or rose respectively on fat-free diet), the relationship between concentration and turnover held with fat-free diets. This relationship has, therefore, been observed by several workers using different techniques [19, 29, 35, 48].

E. Comparison of Triglyceride Turnover with Different Methods

It is clear from the preceding discussion that the various models have made assumptions that have later been shown to be partly inaccurate or that have not been entirely proven. The direct chemical estimation of hepatic secretion must currently stand as the value most likely to be correct. It requires, however, catheterization of the hepatic veins and an extremely sensitive and accurate analytical method.

SHAMES *et al.* [17] have calculated values for triglyceride transport from the data reported by FRIEDBERG *et al.* [1], NESTEL [49], RYAN and

SCHWARTZ [33], EATON *et al.* [3], HAVEL *et al.* [34] and QUARFORDT *et al.* [16]. The values range from 15 μM TGFA/min for the RYAN and SCHWARTZ model [33] to 80 μM/min for one of 3 models proposed by QUARFORDT *et al.* [16]. The two nonisotopic methods provide higher values: BIERMAN *et al.* [48] reported from 100 to 800 μg triglyceride/min/kg and BOBERG *et al.* [8, 35] found an average value of about 110 μM/min for a normotriglyceridaemic subject of average build. The values obtained by FARQUHAR *et al.* [29] fall within these limits.

III. Precursors of Plasma Triglyceride Fatty Acids

A. Plasma Free Fatty Acids

Since many of the models that describe triglyceride transport are based on the relationship between plasma FFA and TGFA, changes in FFA transport must alter TGFA kinetics.

The transport of FFA through plasma has generally been measured by the method of ARMSTRONG *et al.* [50]. Radiolabelled fatty acid, most commonly palmitic acid, is infused intravenously at a constant rate until a constant specific radioactivity is reached in the plasma. The infusion rate per minute divided by the specific radioactivity equals the turnover or inflow from adipose tissue in μM/min. Palmitic acid has been assumed to be representative of most plasma FFA, although its kinetics clearly differ from that of linoleic acid in man [41]. The turnover can also be measured after a pulse injection of radiopalmitate but the short half-life presents problems of mixing and sampling. There is a close relationship between plasma FFA concentration and turnover [35, 51, 52]. The plasma concentration is, therefore, governed by events within adipose tissue, including the entry of glucose into adipocytes. Plasma FFA are taken up in the splanchnic region including the liver in direct proportion to the circulating concentration. Whereas the information has been available for some time in animals [53, 54], the recent technique of hepatic vein catheterization during constant infusions of radiopalmitate has provided similar data in man. HAVEL *et al.* [34] found that 34–74% of FFA turnover was taken up in the splanchnic region and the corresponding average value obtained by BOBERG *et al.* [35] was 66%; splanchnic uptake was highly correlated with FFA turnover, especially that fraction of the turnover that originated in splanchnic adipose tissue.

The magnitude of FFA transport might, therefore, be expected to determine the amount of FFA incorporated into plasma triglyceride and possibly the triglyceride concentration. Studies by both HAVEL *et al.* [34] and BOBERG *et al.* [35] do show significant correlations between the plasma FFA turnover and the secretion of TGFA from the liver as calculated from the appearance of radioactivity in the hepatic vein, during constant infusions of radiopalmitate. In other studies by SAILER *et al.* [36] and by NESTEL [55] in which the appearance of radioactivity in TGFA has been measured in peripheral and not hepatic veins during radiopalmitate infusions, similar correlations with FFA turnover have also been shown. Moreover, when the FFA turnover was suppressed with infusions of physiological amounts of insulin, there was a proportional reduction in the incorporation of radiopalmitate into plasma TGFA [55].

A correlation between plasma FFA turnover and the plasma triglyceride concentration has been reported by NESTEL *et al.* [37]: among subjects with endogenous hypertriglyceridaemia on carbohydrate-rich diets. Recently, BOBERG *et al.* [35] have also observed a significant correlation between plasma FFA turnover and the plasma triglyceride concentration among hypertriglyceridaemic but not among normotriglyceridaemic subjects. SAILER *et al.* [36] had previously reported significantly elevated concentrations of plasma FFA in patients with essential hyperlipaemia. Others have found no such correlation [16, 33, 34].

Although a correlation between FFA turnover and hypertriglyceridaemia has been a variable finding, the splanchnic uptake of FFA is elevated in patients with endogenous hypertriglyceridaemia [34]. Furthermore the fraction of FFA taken up by the liver that is subsequently incorporated into plasma TGFA is increased in some hypertriglyceridaemic states. QUARFORDT *et al.* [16], who used multicompartmental analysis to calculate the kinetic relationships between plasma FFA and VLDL-TGFA, reported an increased diversion of hepatic FFA from pathways of oxidation and ketone formation to TGFA production in subjects consuming high-carbohydrate diets. A similar mechanism has been described for the hypertriglyceridaemia of ethanol [23, 24] and of glycogenosis [56]. With perfused rat-livers, the fractional conversion of perfusate FFA to TGFA is greater with fed than fasted rats [57] and even greater with fructose-fed rats [58]. The hypertriglyceridaemia of obesity may also derive in part from the high turnover of plasma FFA which has been studied in humans by MILLER *et al.* [59] and by NESTEL and WHYTE [52].

Increased turnover of plasma FFA need not necessarily lead to hypertriglyceridaemia. With prolonged fasting, FFA is preferentially utilized for energy [60, 61] while in insulin-deficient dogs, the conversion of FFA to ketones greatly exceeds the incorporation into TGFA [46]. The distribution of FFA among the various metabolic pathways in the liver will, therefore, partly determine the availability of substrate for triglyceride formation. In fasting, healthy men, 17% of the splanchnic uptake of FFA becomes incorporated into VLDL-TGFA, compared to 29% into ketones and 34% into CO_2 [34] [2]. Thus, about 7–10% of total FFA turnover is normally destined for resecretion as TGFA. As discussed previously, these values assume that palmitic acid is a representative tracer for all plasma FFA. That this is not so, is evident from the greater utilization of palmitic than of linoleic acid (after correcting for differing turnover rates) for VLDL-TGFA formation [41].

The relative importance of plasma FFA, in contradistinction to glucose, as the substrate for plasma TGFA during high-carbohydrate diets, has recently been studied in man by BARTER and NESTEL [26]. The proportion of VLDL-TGFA derived from circulating FFA was obtained from the ratio of specific radioactivities of VLDL-TGFA: FFA during prolonged constant infusions of radiopalmitic acid. This can be calculated as soon as the product (VLDL-TGFA) has reached isotopic equilibrium with its precursors, when its specific radioactivity also becomes constant. Following 3 days of sustained carbohydrate feeding and while glucose was being consumed, plasma FFA accounted for as little as 10% of VLDL-TGFA. This contribution increased in the postabsorptive state, but even 20 h later reached only 70%. Although these studies were carried out with abnormally high intakes of glucose, they demonstrate the regulation of the substrate mixture available for TGFA formation. The reciprocal relationship between the availability of glucose and FFA inflow had been fully documented in man by FREDRICKSON and GORDON [62] and WATERHOUSE *et al.* [63]. The suppression of the plasma FFA concentration can be achieved by inhibiting inflow from adipose tissue presumably by suppressing lipolysis [63] although FREDRICKSON and GORDON reported that clearance due

2 In a series of elegant studies carried out in dogs by SPITZER *et al.* [84], of the total plasma flux of palmitic acid, 37% was removed in the liver, 22% in the extrahepatic splanchnic circulation, 6% across the heart, 5% across the kidneys with the remaining 30% being probably removed by skeletal muscle. Of the fatty acid taken up in the liver, 13% was oxidized, 13% resecreted as TGFA and 7% converted to ketones.

to re-esterification may also be enhanced. These conclusions were based on the analysis of the plasma FFA specific radioactivity curves after injections of radiopalmitate. WATERHOUSE *et al.* used multicompartmental analysis to determine possible metabolic pathways and concluded that, in man, glucose administration led to a 65-percent reduction in FFA oxidation. A similar approach to FFA kinetics in glucose-fed rats has been reported by BAKER and ROSTAMI [64].

BARTER and NESTEL [26] infused labelled FFA at constant rates for up to 36 h during the last 8 h of sustained glucose consumption and during the succeeding period of fasting. Initially, inflow was suppressed and clearance enhanced but, as fasting progressed, clearance (as measured by the fractional turnover rate) fell quickly, whereas inflow remained suppressed for up to 20–24 h, suggesting that glucose ingestion maintains its effect on lipolysis for much longer than on re-esterification.

The interrelationship of plasma FFA and triglyceride has also been studied by BARTER *et al.* [65] during alternating periods of sucrose consumption during the day and fasting by night. During the course of these pure sucrose diets, both the plasma triglyceride and FFA concentrations fell during the day and rose markedly only during the night; these diurnal fluctuations in triglyceride levels have also been reported by SCHLIERF *et al.* [66]. The significant correlations between the changes in FFA and triglyceride concentration suggest that part of the nocturnal rise in triglycerides is mediated by the flux of FFA. The nocturnal increments of both lipids could be suppressed by infusions of insulin [65], although these findings may be equally related to enhanced clearance of triglyceride (see below). In comparable experiments in rats, BAKER *et al.* [67] also demonstrated the sensitive regulation of the plasma triglyceride concentration by the flux of FFA.

B. Glucose

Carbohydrate feeding and possibly insulin administration stimulate hepatic lipogenesis from carbohydrate; the evidence for this is very extensive in experimental animals and has recently been reviewed by NIKKILÄ [68]. In human liver, ZAKIM *et al.* [69] have demonstrated *in vitro* lipogenesis from both glucose and fructose. The rôle of insulin is less certain since lipogenesis does not appear to be enhanced in subjects with endogenous hyperinsulinism [70].

However, the relative importance of increased lipogenesis and decreased removal in the genesis of hypertriglyceridaemia has not been resolved, although it is worth noting that the rate of release of triglyceride from perfused rat livers is directly correlated with fatty acid synthesis after carbohydrate feeding [40, 71]. Moreover, insulin enhances the output of triglyceride from livers of fructose-fed rats [72].

The evidence for lipogenesis from carbohydrate is more tenuous in man. GOLDRICK *et al.* [73] have shown that human adipocytes can convert glucose to triglyceride fatty acids in the presence of insulin although not to the same extent as rat adipocytes. Lipogenesis in human fat tissue may however not be quantitatively significant, since certain regulating enzymes, especially the citrate cleavage enzyme, have a much higher activity in the liver [74]. The relative contribution of liver and adipose tissue to overall lipogenesis appears to be species-related, occurring predominantly in the liver in birds [75] but in adipose tissue in pigs [76]. In rodents [77] and in ruminants [78] both organs contribute significantly.

Although *in vitro* studies with human liver have also demonstrated lipogenesis [69] it has generally been difficult to demonstrate significant incorporation of ^{14}C-glucose into plasma TGFA in contrast to labelling of the glycerol moiety, even when the label was given with a 'cold' glucose load [32]. MACDONALD [79] has also found that both radiolabelled glucose and fructose appeared in plasma triglyceride though he did not establish whether the fatty acid moiety was labelled in addition to the glycerol. His observation that the plasma triglyceride specific radioactivity was higher after fructose than glucose is impossible to evaluate without knowledge of the size and labelling of precursor pools.

Recently, BARTER *et al.* [25] have demonstrated unequivocal incorporation of ^{14}C-glucose into VLDL-TGFA in subjects consuming high-carbohydrate diets. As much as 80% of labelling was found in the fatty acid moiety (TGFA) and this incorporation persisted for up to 16 h after the last glucose meal. Since the labelling of TGFA preceded the appearance of label in plasma FFA, it was concluded that the liver was the site of lipogenesis. These studies complement those of BARTER and NESTEL [26] discussed above in demonstrating that plasma FFA are not invariably the sole precursors of VLDL-TGFA. It is worth noting in this context that TGFA are not secreted alone in increased amounts from the liver in response to dietary carbohydrate. Other constituents of VLDL are also affected: lipoprotein-protein formation is enhanced in rats [80] and cholesterol ester turnover is increased in man [81].

C. Hepatic Fatty Acids

There is abundant evidence in both animals and man that the liver is the chief source of endogenously synthetized TGFA [15, 34, 35, 82–84]. In addition to plasma FFA and fatty acids derived from carbohydrate, a further potential source of plasma TGFA is stored fatty acids in the liver. The possible addition of unlabelled fatty acids from 'stored' hepatic triglyceride to the pool of fatty acids destined for the formation of VLDL-TGFA would lead to an underestimate of TGFA transport in those isotopic studies that have assumed total derivation of TGFA from plasma FFA. The multicompartmental analysis of VLDL-TGFA specific radioactivity curves after the injection of radiopalmitate has also pointed to the existence of more than one source of hepatic fatty acids for TGFA synthesis, especially after carbohydrate-rich diets [16, 17]. The techniques used by HAVEL *et al.* [34] and by BOBERG *et al.* [8] assume complete equilibration between VLDL-TGFA and all sources of precursor hepatic fatty acids. The measurement of TGFA inflow transport from the rate of appearance of labelled TGFA in the hepatic vein during the infusion of ^{14}C-palmitic acid, provided values that were lower than those obtained by the direct chemical measurement of triglyceride secretion (8, 35]. This was especially true among the subjects with endogenous hypertriglyceridaemia; possibly these subjects have increased stores of hepatic triglyceride. HAVEL *et al.* [34] observed a greater incorporation of FFA into hepatic lipids than in normotriglyceridaemic subjects and SHAMES *et al.* [17] have suggested models of TGFA formation that include pools of slowly turning-over hepatic fatty acids that may not lie directly along the pathway of FFA to TGFA conversion.

Experimental evidence for the existence of such precursor pools in man has been obtained by comparing the maximum equilibrated specific radioactivity of VLDL-TGFA with that of plasma FFA during constant infusions of radiopalmitic acid. In addition to the high-glucose feeding regimes, already discussed, a source of fatty acid other than plasma FFA was found to provide the substrate for VLDL-TGFA under several additional sets of circumstances. Firstly, in patients with alcoholic fatty liver, at least 50% of circulating TGFA was derived from stored hepatic fatty acids and this was confirmed by the low specific radioactivity of hepatic TGFA [25]. BEZMAN-TARCHER *et al.* [85] have reported similar findings in ethanol-fed rabbits. Secondly, some obese subjects appear to derive fatty acids for VLDL-TGFA formation from sources other than

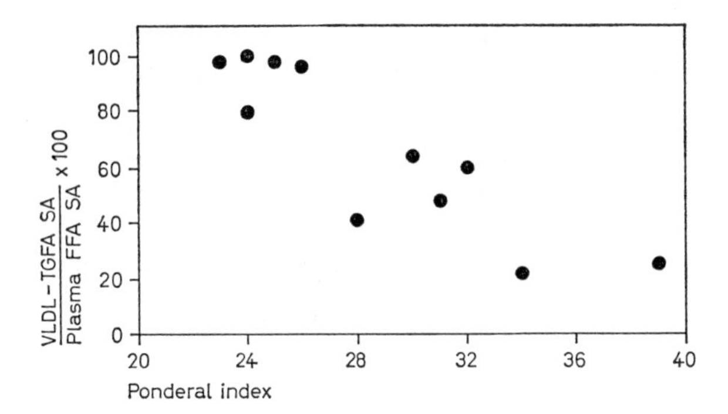

Fig. 1. Relationship between obesity (ponderal index) and the fraction of triglyceride fatty acids (TGFA) derived from plasma free fatty acids (FFA). With increasing fatness, the contribution from FFA declines and TGFA are derived increasingly from other sources of fatty acids [BARTER and NESTEL, unpublished information]. SA = specific radioactivity.

plasma FFA. HAVEL *et al.* [34] had suggested that the large amount of splanchnic adipose tissue might provide unlabelled FFA, but BARTER and NESTEL [26], using prolonged infusions of radiolabelled palmitic acid have concluded that these were derived from hepatic fat stores which are known to be increased in obesity [86]. The intervention of obesity will, therefore, influence the interpretation of triglyceride kinetics in at least two ways: (1) plasma FFA turnover is raised [52, 59] and the fractional conversion of FFA to TGFA is increased [52], and (2) the contribution of fatty acids other than plasma FFA to TGFA formation is considerable (fig. 1).

Treatment with Clofibrate appears to alter the normal FFA-TGFA kinetics. Both WOLFE *et al.* [87] and BARTER *et al.* [25] have shown that a substantial proportion of VLDL-TGFA had originated in pools of hepatic fatty acids that were not in rapid equilibrium with plasma FFA, in subjects treated for as little as 2 weeks and as long as 3 years. Clofibrate is known to increase the size and fat content of rat-livers [88]. The fatty liver of diabetes is a further instance where stored fatty acids contribute very significantly to VLDL-TGFA [46]. It seems certain that hepatic fatty acids will be found to be an important source of plasma TGFA in other disorders. It presents a problem in the measurement of triglyceride turnover by isotopic techniques.

One other highly relevant consideration of the precursors of TGFA is the contribution from intestinal sources. In the fasting rat, some 10% of circulating VLDL-TGFA is derived from reabsorbed biliary phospholipid [89, 90] and this contribution may be much greater in ethanol-fed, hypertriglyceridaemic rats [91]. What proportion, if any, of human plasma VLDL originates in the small intestine is unknown, since that found normally in chyle may have been transferred from plasma; however, VLDL-like particles have been described in small gut mucosa [92]. It is conceivable that in the studies, in which VLDL-TGFA were found to have been only partly derived from plasma FFA, some of the remaining fatty acids had originated in the intestine rather than the liver.

IV. Triglyceride Removal

In the context of carbohydrate diets, factors influencing removal mechanisms may include the secretion of insulin, the stimulation of lipoprotein lipase and the enhancement of esterification within adipose tissue. Changes in the presentation of TGFA to removal sites may also be important, since VLDL increase in size and probably in number during carbohydrate diets as the ratio of triglyceride to other lipids and to protein rises [93–95].

It is virtually certain that the removal of triglyceride is associated with the conversion of VLDL to lipoproteins of higher density [13, 16, 96]. A precursor-product relationship between the pools of TGFA transported in VLDL and low density lipoproteins has been demonstrated on the basis of specific activity-time curves after an injection of radiopalmitic acid. VLDL comprise a wide morphological spectrum, and since smaller species, especially (Sf) 20–60 VLDL accumulate in at least one form of hyperlipoproteinaemia (type III) it has been suggested that a delay in the postulated step-wise interconversion of VLDL might account for this and possibly other disorders [97, 98]. However, in a comparison with one subject with type III and one with type IV hyperlipoproteinaemia. BARTER and NESTEL [unpublished information] did not find such a delay in the conversion of large VLDL to smaller species (fig. 2). Triglyceride concentrations in the various subfractions were similar in both subjects.

The kinetics of VLDL-TGFA metabolism associated with the removal of triglyceride has been studied by BARTER and NESTEL [99]. The VLDL-TGFA specific radioactivity curves were analyzed to determine whether the pleomorphism of VLDL represented a series of precursor-product interrelationships or the existence of several independently metabolized pools of VLDL-TGFA. The data favoured the former and reduced the likelihood of several species of VLDL being secreted from the liver and then

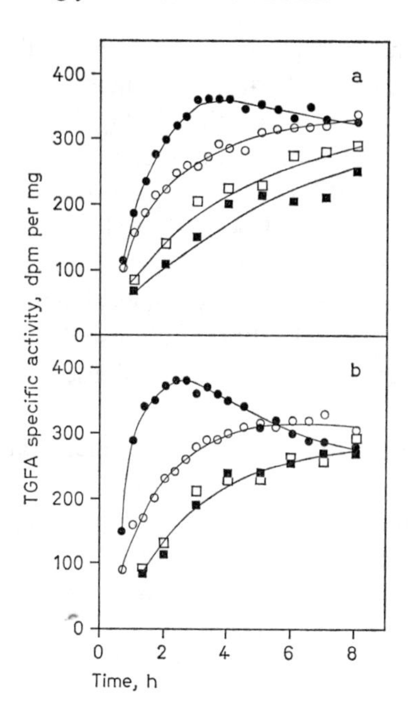

Fig. 2. Specific activity-time curves for TGFA in plasma (Sf) 100–400, 20–100, 12–20 and 0–12 lipoproteins after a pulse intravenous injection of palmitic acid-1-^{14}C. *a* Subject with type IV hyperlipoproteinaemia. *b* Subject with type III hyperlipoproteinaemia [BARTER and NESTEL, unpublished information]. dpm = Disintegrations per minute. ● = Sf 100–400; ○ = Sf 20–100; ■ = Sf 12–20; □ = Sf 0–12.

metabolized at different kinetic rates. Whether the predominance of large VLDL with high-carbohydrate diets reflects solely the hepatic secretion of these VLDL or whether the sequential conversion to smaller VLDL is diminished, requires investigation. When triglyceride production is stimulated in the rat-liver by perfusing with higher concentrations of FFA, this is achieved by the secretion of more VLDL particles which are probably also a little larger [100].

Lipoprotein lipase, the adipose tissue enzyme that mediates the hydrolysis of plasma triglyceride, is readily influenced by dietary changes. The evidence for this is persuasive in animals but is less certain in man. Fasting reduces both the post-heparin lipolytic activity of plasma [101] and the activity of lipoprotein lipase in human adipose tissue [102]. Conversely, the ingestion of glucose stimulates the release of lipoprotein lipase into plasma in normal but not in diabetic subjects [103]. There is no available data in man to show whether prolonged carbohydrate-rich diets influence lipoprotein lipase activity. According to SHIGETA *et al.* [103], increased plasma lipolytic activity is demonstrable as soon as 1 h after the ingestion of glucose: this may be a factor in the rapid fall in the plasma triglyceride concentration after an acute load of glucose [104, 105, 65]. This also coincides with peak insulin secretion and, although insulin restores the post-heparin lipolytic activity in diabetics [106], physiological amounts of insulin infused into the brachial arteries of healthy, fasting subjects did not increase the heparin-induced release of lipoprotein lipase in the forearm [107].

Insulin infusions can lower the plasma triglyceride concentration [108, 109] and reverse the hypertriglyceridaemia of uncontrolled diabetes mellitus [106]. Apart from its possible stimulation of lipoprotein lipase activity which has only been clearly shown in animal tissues [110, 111], insulin might influence the triglyceride concentration by directly enhancing removal or by suppressing mobilization of FFA. The fractional incorporation of plasma FFA into TGFA is rapidly reduced when FFA turnover is suppressed by insulin [55].

The additional possibility that insulin may lower the plasma triglyceride concentration by reducing the inflow from the liver is without support. On the contrary, insulin stimulates lipogenesis in the livers of rats pretreated with insulin [112] and the incorporation of glucose and acetate into hepatic fatty acids [113]. Furthermore, the secretion of triglycerides from perfused rat-livers is enhanced by insulin [72] and is restored to normal in livers of insulin-treated diabetic rats [114].

At present, there is no clear evidence in man that allows a distinction to be made between direct enhancement of triglyceride removal and suppression of FFA outflow as the major mechanism for insulin-induced lowering of triglyceride. Some of the data is indirect, such as the triglyceride lowering effects brought about by tolbutamide [115] and phenformin [116]. Furthermore, in the studies of JONES and ARKY [108], the triglyceride fell several hours after the insulin infusion had been stopped and when FFA levels were again elevated. The studies of BARTER *et al.* [65] carried out during the course of 100% sucrose diets consumed during the day, showed a very striking inverse correlation between the 24-hour changes in insulin and triglyceride levels. Moreover, the overnight increments in the plasma triglyceride could be significantly reduced by infusing small amounts of insulin during the night though the plasma FFA concentrations were also lowered [65]. The removal rate of intravenously administered fat emulsions can be accelerated by glucose in man [117] and by insulin in dogs [10].

At first glance, the conclusion that insulin lowers the plasma triglyceride concentration directly or indirectly, appears to be at variance with the frequently found association between hypertriglyceridaemia and hyperinsulinaemia [29, 118–120]. This correlation has led to the suggestion that endogenous hypertriglyceridaemia is due to hyperinsulinaemia. The studies of BARTER *et al.* [65], however, showed that the magnitude of the fall in the plasma triglyceride concentration during the day with acute sucrose consumption was correlated not only with the size of the insulin response, but also with the initial (fasting) insulin and triglyceride levels. Thus, the diurnal fluctuations in insulin and triglyceride were related to the fasting, overnight levels, but this does not necessarily mean that the fasting triglyceride and insulin levels were causally linked.

Furthermore, as pointed out by NIKKILÄ and TASKINEN [121] and by BIERMAN and PORTE [120] hyperinsulinaemia, hypertriglyceridaemia and glucose intolerance may be related through their common association with obesity. However, many obese subjects are only minimally hypertriglyceridaemic whereas they mostly have raised insulin levels. NIKKILÄ and TASKINEN [121] have also shown that hypertriglyceridaemic subjects may have either raised or normal insulin secretion rates and NESTEL [115] found that glucose intolerance was abolished when subjects with endogenous hypertriglyceridaemia reduced body weight to ideal levels. GLUECK *et al.* [122] have also failed to find a consistent intercorrelation between endogenous hypertriglyceridaemia, glucose intolerance and hyperinsulinaemia. However, FARQUHAR *et al.* [29] demonstrated that the triglyceride response to a high-carbohydrate diet was significantly dependent on the degree of glucose intolerance before the diet.

To some extent the intercorrelations between hypertriglyceridaemia, hyperinsulinism and glucose intolerance may be manifestations of diminished clearance of both triglyceride and glucose. This may be analogous to the situation in obesity where enlarged fat cells respond less sensitively to the actions of insulin [123]. In rats, the capacity of adipocytes to remove TGFA of chylomicrons also becomes impaired with increasing size [124].

There is certainly considerable evidence for diminished clearance of triglyceride in non-obese subjects with endogenous hypertriglyceridaemia derived from isotopic studies [16, 33–36] as well as from infusions of fat emulsions [7]. The reason for this is uncertain, although PERSSON *et al.* [125] have reported a significant inverse correlation between the plasma triglyceride concentration and the activity of lipoprotein lipase in adipose tissue. However, several recent studies of triglyceride turnover in subjects with endogenous hypertriglyceridaemia suggest that clearance is impaired only in some and adequate or even increased (together with production) in others [28, 35, 48]. It should, therefore, be possible to determine whether these groups also differ in glucose tolerance, in lipoprotein lipase activity and in their insulin response to glucose. A preliminary report by BRUNZELL *et al.* [126] does suggest that plasma post-heparin lipolytic activity may become more readily depleted in those subjects with endogenous hyperlipaemia in whom deficient removal of triglyceride can also be demonstrated by turnover techniques. Using rather large doses of heparin that may activate less readily releasable pools of lipoprotein lipase, BOBERG [127] has reported an inverse relationship between the plasma activity of this enzyme and the plasma triglyceride concentration.

Further direct evidence for deficient removal mechanisms in subjects with hypertriglyceridaemia has already been discussed and includes the reduced fractional removal of reinjected human chylomicrons [5] and Intralipid® [7]. As has already been pointed out, however, these conclusions may apply only in the fasting state.

V. Conclusions

1. The various models and techniques that have been used to measure plasma triglyceride kinetics have been reviewed. In fasting man, the mass of TGFA that is transported in plasma VLDL has been variously estimated at between 15 and over 100 μM/min. Even higher values have been reported with non-tracer techniques in which the size of the triglyceride pool has been altered. The deficiencies of some of these models have been discussed.

2. The effect of dietary carbohydrate on triglyceride turnover has been even more difficult to evaluate for the following reasons: (a) most models assume a single precursor, usually plasma FFA whereas, with carbohydrate-rich diets, additional precursors might contribute significantly; (b) hepatic lipogenesis from glucose can provide newly synthetized fatty acids for TGFA formation even in the postabsorptive state, and (c) stored hepatic fatty acids may be an additional source of TGFA.

3. Whereas TGFA transport may, therefore, be underestimated with current techniques, there is agreement that turnover is increased with carbohydrate-rich diets. The conversion of FFA to TGFA may be greater with sucrose than with starch and with glucose than with fructose. Subjects with endogenous hypertriglyceridaemia respond to carbohydrate-rich diets with an increase in triglyceride production, that is of the same order as in normotriglyceridaemic subjects.

4. The expansion in the triglyceride pool of VLDL is associated with a reduced fractional removal rate, i.e. the fraction of the pool being removed per unit time is diminished during consumption of carbohydrate. Since production of triglyceride rises with carbohydrate-rich diets, the reduced fractional removal rate implies a relative inefficiency in removal mechanisms. The relative imbalance between production and removal of triglyceride probably fluctuates throughout the 24 h. The plasma triglyceride concentration falls during the actual period of carbohydrate consumption by day and rises during the postabsorptive period at night. This rise may be related to several factors (fig. 3): (a) removal may diminish during the night as the plasma insulin level falls; (b) production from plasma FFA may rise during the night when FFA turnover rises, and (c) production from lipogenesis continues through the night with very high carbohydrate diets, though probably at a slower rate than during the day.

5. Since the plasma FFA turnover is sensitively regulated by carbohydrate, the proportion of plasma TGFA derived from FFA is similarly influenced. During glucose consumption the outflow of FFA from adipose tissue is suppressed and the removal rate of FFA accelerated. Thus, plasma FFA may provide as little as 10% of VLDL-TGFA when calories are derived exclusively from glucose, whereas virtually all TGFA is derived from FFA in the postabsorptive phase after a normal diet. However, the fraction of FFA taken up in the liver, that is converted to TGFA, is increased with carbohydrate.

6. Insulin lowers plasma triglyceride levels by suppressing FFA outflow and by accelerating removal. The nocturnal rise in triglyceride levels during carbohydrate-rich diets can be modified with insulin and the fractional incorporation of FFA into TGFA reduced.

7. In terms of triglyceride kinetics, endogenous hypertriglyceridaemia represents a heterogeneous disorder. Whereas the fractional removal of the expanded triglyceride pool is reduced in most hypertriglyceridaemic subjects in the fasting state, increased production of triglyceride has also

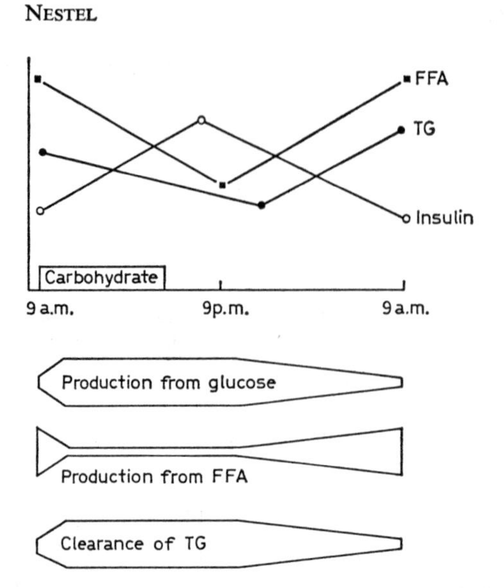

Fig. 3. Factors influencing the plasma triglyceride concentration during periods of 100 % carbohydrate diets consumed during the day. The concentrations of plasma free fatty acids (FFA) and triglyceride (TG) fall during the day and rise overnight; reciprocal changes occur with insulin [65]. The major source of triglyceride fatty acids is glucose by day and FFA by night [25, 26]. Clearance of triglyceride measured with constant infusions of Intralipid® are greater during the day than in the early morning [NESTEL and BARTER, unpublished information].

been reported in a proportion of these subjects. This results from an increased hepatic uptake of FFA and from an additional source of hepatic fatty acids that may reflect increased lipogenesis from carbohydrate. Current techniques have been unsatisfactory in clearly establishing the nature of triglyceride kinetics in this disorder: with some methods, triglyceride turnover is underestimated in normotriglyceridaemic subjects whereas, in other studies, the reverse might be true. Consequently, a correlation between triglyceride concentration and turnover has been demonstrated in some but not all investigations.

8. In obese subjects, FFA turnover and the fractional incorporation of FFA into TGFA are increased. Furthermore, even in the postabsorptive state, an appreciable fraction of circulating TGFA is derived from a source other than FFA, presumably hepatic fatty acids. The frequent association between obesity and endogenous hypertriglyceridaemia will influence the complex interrelationship of hypertriglyceridaemia, hyperinsulinaemia and

glucose intolerance which are variably present in both conditions. This may reflect inefficient removal of both glucose and triglyceride.

VI. References

1 Friedberg, S. J.; Klein, R. F.; Trout, D. L.; Bogdonoff, M. D., and Estes, E. H., jr.: The incorporation of plasma free fatty acids into plasma triglycerides in man. J. clin. Invest. *40:* 1846–1855 (1961).

2 Farquhar, J. W.; Gross, R. C.; Wagner, R. M., and Reaven, G. M.: Validation of an incompletely coupled two-compartment nonrecycling catenary model for turnover of liver and plasma triglyceride in man. J. Lipid Res. *6:* 119–134 (1965).

3 Eaton, R. P.; Berman, M., and Steinberg, D.: Kinetic studies of plasma free fatty acid and triglyceride metabolism in man. J. clin. Invest. *48:* 1560–1579 (1969).

4 Havel, R. J.: Triglyceride and very low density lipoprotein turnover; in Cowgil, Estrich and Wood Proc. of the 1968 Deuel Conference of lipids, pp. 115–130 (US Government Printing Office, Washington 1968).

5 Nestel, P. J.: Relationship between plasma triglycerides and removal of chylomicrons. J. clin. Invest. *43:* 943–949 (1968).

6 Quarfordt, S. H. and Goodman, D. S.: Heterogeneity in rate of plasma clearance of chylomicrons of different sizes. Biochim. biophys. Acta *116:* 382–385 (1966).

7 Boberg, J.; Carlson, L. A., and Hallberg, D.: Application of a new intravenous fat tolerance test in the study of hypertriglyceridaemia in man. J. Atheroscler. Res. *9:* 159–169 (1969).

8 Boberg, J.; Carlson, L. A., and Freyschuss, U.: Determination of splanchnic secretion rate of plasma triglycerides and of plasma free fatty acid total and splanchnic turnover. Europ. J. clin. Invest. *2:* 123–132 (1972).

9 Hallberg, D.: Studies on the elimination of exogenous lipids from the blood stream. Determination and separation of the plasma triglycerides after single injection of a fat emulsion in man. Acta physiol. scand. *62:* 407–421 (1964).

10 Hallberg, D.: Insulin and glucagon in the regulation of removal rate of exogenous lipids from the blood in dogs. Acta chir. scand. *136:* 291–297 (1970).

11 Lewis, B.; Mancini, M.; Ishiwata, J. I., and Mattock, M.: Dietary influences on plasma triglyceride metabolism. Abstr. Europ. Soc. clin. Invest. *1:* 380 (1971).

12 Carlson, L. A.: Studies on the incorporation of injected palmitic acid-1-C^{14} into liver and plasma lipids in man. Acta Soc. Med., Uppsala *65:* 85–90 (1960).

13 Havel, R. J.: Conversion of plasma free fatty acids into triglycerides of plasma lipoprotein fractions in man. Metabolism *10:* 1031–1034 (1961).

14 Reaven, G. M.; Hill, D. B.; Gross, R. C., and Farquhar, J. W.: Kinetics of triglyceride turnover of very low density lipoproteins of human plasma. J. clin. Invest. *44:* 1826–1833 (1965).

15 Carlson, L. A. and Ekelund, L. G.: Splanchnic production and uptake of endogenous triglycerides in the fasting state in man. J. clin. Invest. *42:* 714–720 (1963).

16 QUARFORDT, S.; SHAMES, D. M.; FRANK, A.; BERMAN, M., and STEINBERG, D.: Very low density lipoprotein triglyceride transport in type IV hyperlipoproteinemia and the effects of carbohydrate-rich diets. J. clin. Invest. *49:* 2281–2297 (1970).

17 SHAMES, D. M.; FRANK, A.; STEINBERG, D., and BERMAN, M.: Transport of plasma free fatty acids and triglycerides in man. A theoretical analysis. J. clin. Invest. 49: 2298–2314 (1970).

18 NESTEL, P. J. and HIRSH, E. Z.: Triglyceride turnover after diets rich in carbohydrate or animal fat. Austr. Ann. Med. *14:* 265–269 (1965).

19 NESTEL, P. J.: Triglyceride turnover in coronary heart disease and the effect of dietary carbohydrate. Clin. Sci. *31:* 31–38 (1966).

20 WATERHOUSE, C.; KEMPERMAN, J. H., and STORMONT, J. M.: Alterations in triglyceride metabolism as produced by dietary change. J. Lab. clin. Med. *63:* 605–620 (1964).

21 BAKER, N. and SCHOTZ, M. C.: Use of multicompartmental models to measure rates of triglyceride metabolism in rats. J. Lipid Res. *5:* 188–197 (1964).

22 BAKER, N. and SCHOTZ, M. C.: Quantitative aspects of free fatty acid metabolism in the fasted rat. J. Lipid. Res. *8:* 646–660 (1967).

23 NESTEL, P. J. and HIRSCH, E. Z.: Mechanism of alcohol-induced hypertriglyceridemia. J. Lab. clin. Med. *66:* 357–365 (1965).

24 WOLFE, B. M.; HAVEL, R. J.; MARLISS, E. B.; KANE, J. P., and SEYMOUR, J.: Effects of ethanol on splanchnic metabolism in healthy men. J. clin. Invest. *49:* 104a (1969).

25 BARTER, P. J.; NESTEL, P. J., and CARROLL, K. F.: Percursors of plasma triglyceride fatty acid in humans. The effects of glucose consumption, clofibrate administration and alcoholic fatty liver. Metabolism *21:* 117–124 (1972).

26 BARTER, P. J. and Nestel, P. J.: Plasma free fatty acid transport during prolonged glucose consumption and its relationship to plasma triglyceride fatty acids in man. J. Lipid. Res. *13:* 483–490 (1972).

27 GROSS, R. C.; FARQUHAR, J. W.; SHEN, S. W., and REAVEN, G. M.: Triglyceride production and removal in two fractions of plasma very low density lipoprotein (VLDL). Clin. Res. *18:* 140 (1970).

28 NIKKILÄ, E. A. and KEKKI, M.: Measurement of plasma triglyceride turnover in the study of hyperglyceridemia. Scand. J. clin. Lab. Invest. *27:* 97–104 (1971).

29 FARQUHAR, J. W.; FRANK, A.; GROSS, R. C., and REAVEN, G. M.: Glucose, insulin and triglyceride responses to high and low carbohydrate diets in man. J. clin. Invest. *45:* 1648–1656 (1966).

30 MACDONALD, I.: The lipid response of young women to dietary carbohydrates. Amer. J. clin. Nutr. *16:* 458–463 (1965).

31 BAKER, N.: The use of computers to study rates of lipid metabolism. J. Lipid. Res. *10:* 1–24 (1969).

32 SANDHOFER, F.; BOLZANO, K.; SAILER, S., und BRAUNSTEINER, H.: Quantitative Untersuchungen über den Einbau von Plasmaglucosekohlenstoff in Plasmatriglyceride und die Veresterungsrate von freien Fettsäuren des Plasmas zu Plasmatriglyceriden während oraler Zufuhr von Glucose bei primärer kohlenhydratinduzierter Hypertriglyceridämie. Klin. Wschr. *47:* 1087–1089 (1969).

33 RYAN, W. G. and SCHWARTZ, T. B.: Dynamics of plasma triglyceride turnover in man. Metabolism *14:* 1243–1254 (1965).

34 HAVEL, R. J.; KANE, J. P.; BALASSE, E. O.; SEGEL, N., and BASSO, L. V.: Splanchnic metabolism of free fatty acids and production of triglycerides of very low density lipoproteins in normotriglyceridemic and hypertriglyceridemic humans. J. clin. Invest. *49:* 2017–2035 (1970).

35 BOBERG, J.; CARLSON, L. A.; FREYSCHUSS, U.; LASSERS, B. W., and WAHLQVIST, M. L.: Splanchnic secretion rate of plasma triglycerides and plasma free fatty acid total and splanchnic turnover in patients with hypertriglyceridaemia. Europ. J. clin. Invest. (in press).

36 SAILER, S.; SANDHOFER, F. und BRAUNSTEINER, H.: Umsatzraten für freie Fettsäuren und Triglyceride im Plasma bei essentieller Hyperlipämie. Klin. Wschr. *44:* 1032–1036 (1966).

37 NESTEL, P. J.; CARROLL, K. F., and HAVENSTEIN, N.: Plasma triglyceride response to carbohydrates, fats and caloric intake. Metabolism *19:* 1–19 (1970).

38 BAR-ON, H. and STEIN, Y.: Effect of glucose and fructose on lipid metabolism in the rat. J. Nutr. *94:* 95–105 (1968).

39 FALLON, H. and KEMP, E. L.: Effects of diet on hepatic triglyceride synthesis. J. clin. Invest. *47:* 712–719 (1968).

40 WEBB, W.; Nestel, P. J.; Foxman, C., and LYNCH, A.: Hepatic lipogenesis, adipose lipoprotein lipase and triglyceride removal in normotriglyceridemic hexose-fed rats. Nutr. Rep. int. *1:* 189–195 (1970).

41 NESTEL, P. J. and BARTER, P.: Metabolism of palmitic and linoleic acids in man. Differences in turnover and conversion to glycerides. Clin. Sci. *40:* 345–350 (1971).

42 KOHOUT, M.; KOHOUTOVA, B., and HEIMBERG, M.: The regulation of hepatic triglyceride metabolism by free fatty acids. J. biol. Chem. *246:* 5067–5074 (1971).

43 HERMAN, R. H.; ZAKIM, D., and STIFEL, F. B.: Effect of diet on lipid metabolism in experimental animals and man. Fed. Proc. *29:* 1302–1307 (1970).

44 MACDONALD, I.: Interrelationship between the influences of dietary carbohydrates and fats on fasting serum lipids. Amer. J. clin. Nutr. *20:* 345–351 (1967).

45 ANTAR, M. A.; LITTLE, J. A.; LUCAS, C.; BUCKLEY, G. C., and CSIMA, A.: Interrelationship between the kinds of dietary carbohydrate and fat in hyperlipoproteinemic patients. Atherosclerosis *11:* 191–201 (1970).

46 BASSO, L. V. and HAVEL, R. J.: Hepatic metabolism of free fatty acids in normal and diabetic dogs. J. clin. Invest. *49:* 537–547 (1970).

47 PORTE, D., jr. and BIERMAN, E. L.: The effect of heparin infusion on plasma triglyceride *in vivo* and *in vitro* with a method for calculating triglyceride turnover. J. Lab. clin. Med. *73:* 631–648 (1969).

48 BIERMAN, E. L.; BRUNZELL, J. D.; BAGDADE, J. D.; LERNER, R. L.; HAZZARD, W. R., and PORTE, D., jr.: On the mechanism of action of Atromid-S on triglyceride transport in man. Trans. Ass. amer. Physicians *83:* 211–224 (1970).

49 NESTEL, P. J.: Metabolism of linoleate and palmitate in patients with hypertriglyceridemia and heart disease. Metabolism *14:* 1–9 (1965).

50 ARMSTRONG, D. T.; STEELE, R.; ALTSZULER, N.; DUNN, A.; BISHOP, J. S., and BODO, R. C. DE: Regulation of plasma free fatty acid turnover. Amer. J. Physiol. *201:* 9–15 (1961).

51 ISSEKUTZ, B., jr.; BORTZ, W. M.; MILLER, H. I., and PAUL, P.: Turnover rate of plasma FFA in humans and in dogs. Metabolism *16:* 1001–1009 (1967).
52 NESTEL, P. J. and WHYTE, H. M.: Plasma free fatty acid and triglyceride turnover in obesity. Metabolism *17:* 1112–1128 (1968).
53 FINE, M. B. and WILLIAMS, R. H.: Effect of fasting, epinephrine and glucose and insulin on hepatic uptake of nonesterified fatty acids. Amer. J. Physiol. *199:* 403–406 (1960).
54 MCELROY, W. T.; SIEFERT, W. L., and SPITZER, J. D.: Relationship of hepatic uptake of free fatty acids to plasma concentration. Proc. Soc. exp. Biol. Med. *104:* 20–23 (1960).
55 NESTEL, P. J.: Relationship between FFA flux and TGFA influx in plasma before and during the infusion of insulin. Metabolism *16:* 1123–1132 (1967).
56 HAVEL, R. J.; BALASSE, E. O.; WILLIAMS, H. E.; KANE, J. P., and SEGAL, N.: Splanchnic metabolism in von Gierke's disease (glycogenosis type I). Trans. Ass. amer. Physicians *82:* 305–323 (1969).
57 MAYES, P. A. and FELTS, J. M.: Regulation of fat metabolism in the liver. Nature, Lond. *215:* 716–718 (1967).
58 SCHONFELD, G. and PFLEGER, B.: Utilization of exogenous free fatty acids for the production of very low density lipoprotein triglyceride by livers of carbohydrate-fed rats. J. Lipid Res. *12:* 614–621 (1971).
59 MILLER, H. I.; BORTZ, W. M., and DURHAM, B. C.: The rate of appearance of plasma FFA in plasma triglyceride of normal and obese subjects. Metabolism *17:* 515–521 (1968).
60 FRITZ, I. B.: Factors influencing the rates of long-chain fatty acid oxidation and synthesis in mammalian systems. Physiol. Rev. *41:* 52–129 (1961).
61 OWEN, O. E. and REICHARD, G. A., jr.: Human forearm metabolism during progressive starvation. J. clin. Invest. *50:* 1536–1545 (1971).
62 FREDRICKSON, D. S. and GORDON, R. S.: The metabolism of albumin-bound ^{14}C-labelled unesterified fatty acids in normal human subjects. J. clin. Invest. *37:* 1504–1515 (1958).
63 WATERHOUSE, C.; BAKER, N., and ROSTAMI, H.: Effect of glucose ingestion on the metabolism of free fatty acids in human subjects. J. Lipid. Res. *10:* 487–494 (1969).
64 BAKER, N. and ROSTAMI, H.: Effect of glucose feeding on net transport of plasma free fatty acids. J. Lipid Res. *10:* 83–90 (1969).
65 BARTER, P. J.; CARROLL, K. F., and NESTEL, P. J.: Diurnal fluctuations in triglyceride, free fatty acids and insulin during sucrose consumption and insulin infusion in man. J. clin. Invest. *50:* 583–591 (1971).
66 SCHLIERF, G.; REINHEIMER, W., and STOSSBERG, V.: Diurnal patterns of plasma triglycerides and free fatty acids in normal subjects and in patients with endogenous (type IV) hyperlipoproteinemia. Nutr. Metabol. *13:* 80–91 (1971).
67 BAKER, N.; GARFINKEL, A. S., and SCHOTZ, M. C.: Hepatic triglyceride secretion in relation to lipogenesis and free fatty acid mobilization in fasted and glucose-refed rats. J. Lipid. Res. *9:* 1–7 (1968).
68 NIKKILÄ, E. A.: Control of plasma and liver triglyceride kinetics by carbohydrate metabolism and insulin. Adv. Lipid Res. *7:* 63–134 (1969).

69 Zakim, D.; Herman, R. H., and Gordon, W. C., jr.: The conversion of glucose and fructose to fatty acids in the human liver. Biochem. Med. *2:* 427–435 (1968).
70 Schersten, T.; Nilsson, S., and Jönsson, J.: Hepatic lipogenesis in two cases with insulin-producing tumor of the pancreas. Acta med. scand. *190:* 353–357 (1971).
71 Windmueller, H. G. and Spaeth, A. E.: *De novo* synthesis of fatty acid in perfused rat liver as a determinant of plasma lipoprotein production. Arch. Biochem. Biophys. *122:* 362–369 (1967).
72 Topping, D. L. and Mayes, P. A.: The immediate effects of insulin and fructose on the metabolism of the perfused liver. Changes in lipoprotein secretion, fatty acid oxidation and esterification, lipogenesis and carbohydrate metabolism. Biochem. J. *126:* 295–311 (1972).
73 Goldrick, R. B.; Ashley, B. C. E., and Lloyd, J. L.: Effects of prolonged incubation and cell concentration on lipogenesis from glucose in isolated human omental fat cells. J. Lipid Res. *10:* 253–259 (1969).
74 Shrago, E.; Glennon, J. A., and Gordon, E. S.: Comparative aspects of lipogenesis in mammalian tissues. Metabolism *20:* 54–62 (1971).
75 Goodridge, A. G. and Ball, E. G.: Lipogenesis in the pigeon. *In vivo* studies. Amer. J. Physiol. *213:* 215–219 (1967).
76 O'Hea, E. K. and Leveille, G. A.: Significance of adipose tissue and liver as sites of fatty acid synthesis in the pig and the efficiency of utilization of various substrates for lipogenesis. J. Nutr. *99:* 338–344 (1969).
77 Jansen, G. R.; Hutchison, D. F., and Zanetti, M. E.: Studies of lipogenesis *in vivo*. Effect of dietary fat or starvation on conversion of ^{14}C glucose into fat and turnover of newly synthesized fat. Biochem. J. *99:* 333–340 (1966).
78 Ballard, F. J.; Hanson, R. W., and Kronfeld, D. S.: Gluconeogenesis and lipogenesis in tissue from ruminant and non-ruminant animals. Fed. Proc. *28:* 218 (1969).
79 Macdonald, I.: Ingested glucose and fructose in serum lipids in healthy men after myocardial infarction. Amer. J. clin. Nutr. *21:* 1366–1373 (1968).
80 Eaton, R. P. and Kipnis, D. M.: Effect of glucose feeding on lipoprotein synthesis in the rat. Amer. J. Physiol. *217:* 1153–1168 (1969).
81 Nestel, P. J.: Turnover of plasma esterificed cholesterol. Influence of dietary fat and carbohydrate and relation to plasma lipids and body weight. Clin. Sci. *38:* 593–600 (1970).
82 Stein, Y. and Shapiro, B.: Uptake and metabolism of triglycerides by the rat liver. J. Lipid Res. *1:* 326–331 (1960).
83 Havel, R. J.; Felts, J. M., and Duyne, C. M. van: Formation and fate of endogenous triglycerides in blood plasma of rabbits. J. Lipid Res. *3:* 297–308 (1962).
84 Spitzer, J. J.; Nakamura, H.; Hori, S., and Gold, M.: Hepatic and splanchnic uptake and oxidation of free fatty acids. Proc. Soc. exp. Biol. Med. *132:* 281–286 (1969).
85 Bezman-Tarcher, A.; Nestel, P. J.; Felts, J. M., and Havel, R. J.: Metabolism of hepatic and plasma triglycerides in rabbits given ethanol or ethionine. J. Lipid Res. *7:* 248–257 (1966).

86 ZELMAN, S.: The liver in obesity. Arch. intern. Med. *90:* 141–156 (1952).

87 WOLFE, B.; KANE, J.; HAVEL, R. J., and BREWSTER, H.: Splanchnic metabolism in healthy young men given clofibrate. Circulation *41:* suppl. 3, p. 2 (1970).

88 GOULD, R. G.; SWYRYD, E. A.; COAN, B. J., and AVOY, D. R.: Effects of chlorophenoxyisobutyrate (CPIB) on liver composition and triglyceride synthesis in rats. J. Atheroscler. Res. *6:* 555–564 (1966).

89 BAXTER, J. H.: Origin and characteristics of endogenous lipid in thoracic duct lymph in rats. J. Lipid Res. *7:* 158–166 (1966).

90 OCKNER, R. F.; HUGHES, F. B., and ISSELBACHER, K. J.: Very low density lipoproteins in intestinal lymph. Origin, composition and role in lipid transport in the fasting state. J. clin. Invest. *48:* 2079–2088 (1969).

91 MISTILIS, S. P. and OCKNER, R. K.: Alcohol-induced fatty liver. Importance of endogenous intestinal lipoproteins. J. clin. Invest. *49:* 66a (1970).

92 JONES, A. L. and OCKNER, R. K.: An electron microscopic study of endogenous very low density lipoprotein production in the intestine of rat and man. J. Lipid Res. *12:* 580–589 (1971).

93 SCHONFELD, G.: Changes in the composition of very low density lipoproteins during carbohydrate induction in man. J. Lab. clin. Med. *75:* 206–211 (1970).

94 BARTER, P. J. and NESTEL, P. J.: The distribution of triglyceride in subclasses of very low density plasma lipoproteins. J. Lab. clin. Med. *76:* 925–932 (1970).

95 RUDERMAN, N. B.; JONES, A. L.; KRAUSS, R. M., and SHAFRIR, E.: A biochemical and morphologic study of very low density lipoproteins in carbohydrate-induced hypertriglyceridemia. J. clin. Invest. *50:* 1355–1368 (1971).

96 LAROSA, J. C.; LEVY, R. I.; VIRGIL BROWN, W., and FREDRICKSON, D. S.: Changes in high density lipoprotein protein composition after heparin-induced lipolysis. Amer. J. Physiol. *220:* 785–791 (1971).

97 HAZZARD, W. R.; PORTE, D., jr., and BIERMAN, E. L.: Heterogeneity of very low density lipoproteins in man. Evidence for a functional role of a beta migrating fraction in triglyceride transport and its relation to broad-beta disease (type III hyperlipoproteinemia). J. clin. Invest. *49:* 40a (1970).

98 QUARFORDT, S.; LEVY, R. I., and FREDRICKSON, D. S.: On the lipoprotein abnormality in type III hyperlipoproteinemia. J. clin. Invest. *50:* 754–761 (1971).

99 BARTER, P. J. and NESTEL, P. J.: Precursor-product relationship between pools of VLDL triglyceride. J. clin. Invest. *51:* 174–180 (1972).

100 RUDERMAN, N. B.; RICHARDS, K. C.; VALLES, V., and JONES, A. I.: Regulation of production and release of lipoprotein by the perfused rat liver. J. Lipid Res. *9:* 613 619 (1968).

101 ARONS, D. L.; SCHREIBMAN, P. H., and ARKY, R. A.: Post heparin lipolytic and monoglyceridase activities in fasted man. Proc. Soc. exp. Biol. Med. *137:* 780–782 (1971).

102 PERSSON, B.; HOOD, B., and ANGERVALL, G.: Effects of prolonged fast on lipoprotein lipase activity eluted from human adipose tissue. Acta med. scand. *188:* 225–229 (1970).

103 SHIGETA, Y.; KIM, M.; HOSHI, M., and ABE, H.: Effect of glucose and fat loading on lipoprotein lipase activity in plasma and tissues of diabetes. Endocrin. jap. *16:* 541–546 (1969).

104 HAVEL, R. J.: Early effects of fasting and of carbohydrate ingestion on lipid and lipoproteins of serum in man. J. clin. Invest. *36:* 855–859 (1957).

105 PERRY, W. F. and CORBETT, B. N.: Changes in plasma triglyceride concentration following the intravenous administration of glucose. Canad. J. Physiol. Pharmacol. *42:* 353–356 (1964).

106 BAGDADE, J. D.; PORTE, D., jr., and BIERMAN, E. L.: Diabetic lipemia. A form of acquired fat-induced lipemia. New Engl. J. Med. *276:* 427–433 (1967).

107 NESTEL, P. J.: The depletion and restoration of post-heparin lipolytic activity in the human forearm. Proc. Soc. exp. Biol. Med. *134:* 896–899 (1970).

108 JONES, D. P. and ARKY, R. A.: Effects of insulin on triglyceride and free fatty acid metabolism in man. Metabolism *14:* 1287–1293 (1965).

109 SCHLIERF, G. and KINSELL, L. W.: Effect of insulin in hypertriglyceridemia. Proc. Soc. exp. Biol. Med. *120:* 272–274 (1965).

110 EAGLE, C. R. and ROBINSON, D. S.: The ability of actinomycin D to increase the clearing-factor lipase activity of rat adipose tissue. Biochim. J. *93:* 10c (1964).

111 AUSTIN, W. and NESTEL, P. J.: The effect of glucose and insulin *in vitro* on the uptake of triglyceride and on lipoprotein lipase activity in fat pads from normal, fed rats. Biochim. biophys. Acta *164:* 59–63 (1968).

112 LETARTE, J. and RUSSELL FRASER, T.: Stimulation by insulin of the incorporation of ^{14}C-U-glucose into lipids released by the liver. Diabetologia *5:* 358–359 (1969).

113 HAFT, D. E.: Effects of insulin on glucose metabolism by the perfused normal rat liver. Amer. J. Physiol. *213:* 219–230 (1967).

114 HEIMBERG, M.; HARKEN, D. R. VAN, and BROWN, T. O.: Hepatic lipid metabolism in experimental diabetes. II. Incorporation of (1-^{14}C) palmitate into lipids of the liver and of the $d < 1.020$ perfusate lipoproteins. Biochim. biophys. Acta *137:* 435–445 (1967).

115 NESTEL, P. J.: Carbohydrate-induced hypertriglyceridemia and glucose utilization in ischemic heart disease. Metabolism *15:* 787–795 (1966).

166 CRAWFORD OWEN, W.; KREISBERG, R. A., and SIEGAL, A. M.: Carbohydrate-induced hyper-riglyceridemia. Inhibition by phenformin. Diabetes, N.Y. *20:* 739–744 (1971).

117 PELKONEN, R.; NIKKILÄ, E. A., and TASKINEN, M.-R.: Effect of glucose on the disposal of intravenous fat emulsions from circulation. Scand. J. clin. Lab. Invest. *19:* suppl. 95, p. 93 (1967).

118 REAVEN, G. M.; LERNER, R. L.; STERN, M. P., and FARQUHAR, J. W.: Role of insulin in endogenous hypertriglyceridemia. J. clin. Invest. *46:* 1756–1767 (1967).

119 FORD, S., jr.; BOZIAN, R. C., and KNOWLES, H. C., jr.: Interactions of obesity and glucose and insulin levels in hypertriglyceridemia. Amer. J. clin. Nutr. *21:* 904–910 (1968).

120 BIERMAN, E. L. and PORTE, D. P., jr.: Carbohydrate intolerance and lipemia. Ann. intern. med. *68:* 926–933 (1968).

121 NIKKILÄ, E. A. and TASKINEN, J. R.: Hypertriglyceridemia and insulin secretion, a complex causal relationship; in JONES Proc. 2nd Int. Symp. Atherosclerosis, pp. 220–230 (Springer, New York 1970).

122 GLUECK, C. J.; LEVY, R. I., and FREDRICKSON, D. S.: Immunoreactive insulin, glucose tolerance, and carbohydrate inducibility in types II, III, IV and V hyperlipoproteinemia. Diabetes, N.Y. *18:* 737–747 (1969).

123 SALANS, L. B.; KNITTLE, J. L., and HIRSCH, J.: The role of adipose cell size and adipose tissue insulin sensitivity in the carbohydrate intolerance of human obesity. J. clin. Invest. *47:* 153–165 (1968).

124 NESTEL, P. J.; AUSTIN, W., and FOXMAN, C.: Lipoprotein lipase content and triglyceride fatty acid uptake in adipose tissue of rats of differing body weights. J. Lipid Res. *10:* 383–387 (1969).

125 PERSSON, B.; BJÖRNTORP, P., and HOOD, B.: Lipoprotein lipase activity in human adipose tissue. Metabolism *15:* 730–741 (1966).

126 BRUNZELL, D.; PORTE, D., jr., and BIERMAN, E. L.: Evidence for a common saturable removal system for dietary and endogenous triglyceride in man. J. clin. Invest. *50:* 15a (1971).

127 BOBERG, J.: Heparin-released blood plasma lipoprotein lipase activity in patients with hyperlipoproteinemia. Acta med. scand. *191:* 97–102 (1972).

Author's address: Dr. PAUL J. NESTEL, Department of Clinical Science, The John Curtin School of Medical Research, The Australian National University, *Canberra, A.C.T.* (Australia)

Progr. biochem. Pharmacol., vol. 8, pp. 161–188 (Karger, Basel 1973)

Influence of Fructose on Hepatic Synthesis of Lipids

D. ZAKIM

Department of Medicine, University of California Medical Center, and Division of Molecular Biology and Department of Medicine, Veterans Administration Hospital, San Francisco, Calif.

Contents

I. Introduction

Interest in the effect of specific carbohydrates on the intermediary metabolism of lipids has been stimulated in the last several years by observations in animals and man that variations in the type and amount of dietary carbohydrate influence the concentration of serum lipids [4, 35, 51, 59]. Several studies have shown also that the content of liver lipids is altered by variations in the source and amount of dietary carbohydrate [13, 93]. The biochemical basis for the differential effects of various sugars on lipid metabolism in whole animals are not completely clarified; but a large body of indirect knowledge suggests that the effects of different sugars on metabolism of lipids within the liver must have an important bearing on the overall problem in the intact animal. The reason for this is the central rôle of the liver for fatty acid metabolism in the whole animal. For example, the liver not only removes fatty acids from the blood, but resecretes them in the form of lipoprotein triglyceride and phospholipid, and in some species of adult animals including man [68] the liver is the major site for *de novo* synthesis of fatty acids from carbohydrate precursors. Another important consideration is that in the liver fructose and glucose are metabolized, in part, in different glycolytic pathways which have differing control parameters and capacities [95].

On the basis of *in vitro* studies the synthesis of fatty acids appears to be controlled in a rigorous manner by inter-relations between reactions for the synthesis and degradation of fatty acids and complex lipids, and possibly also by variations in the concentrations of intermediates in the glycolytic and tricarboxylic acid pathways. In this review we consider the data relating to the possible ways in which the intrahepatic metabolism of fructose could alter the rate of fatty acid synthesis. For this reason the pathway for fatty acid synthesis and its regulatory parameters *in vitro* and *in vivo* are discussed along with the direct effects of fructose on the rate of hepatic lipogenesis and esterification of fatty acids. Major emphasis in this review is placed on the evidence that fructose has effects on hepatic lipogenesis which are different from those due to other sugars, and that the biochemical basis for the unique action of fructose on hepatic synthesis of fatty acids is related to the rapid rate of glycolysis of fructose. We would also point out that although fatty acids are synthetized by mitochondria and microsomes in addition to the particle-free fraction of the cell [31, 38, 60], synthesis in the former two compartments represents chain elongation. In this review *de novo* synthesis of the entire fatty acid by the cytoplasmic system only is discussed.

II. The Synthesis of Fatty Acids

A. Pathway for Fatty Acid Synthesis

Prior to 1953, synthesis of fatty acids was thought to proceed via reversal of the β-oxidation cycle [96]. Studies initiated by WAKIL [79–82] and LYNEN [48–50], however, eventually led to the complete elucidation of the correct pathway for *de novo* synthesis of fatty acids. It was found that pigeon-liver supernatant could be fractioned into 3 fractions which were needed for the synthesis of fatty acids from acetate [66], and that these fractions had no activity for β-oxidation cycle enzymes, indicating that the pathway of fatty acid synthesis was separate from that for their catabolism. It was observed also that HCO_3^- was required for fatty acid synthesis, but was not consumed in the course of the reaction [25]. The overall reaction for synthesis of palmitate appeared to be

$$\text{8-acetyl-CoA} + 10\ \text{ATP} + 16\ \text{NADPH} + 16\ \text{H} + \xrightarrow[HCO_3-]{Mn^{2+}} \text{palmitic acid} + 16\ \text{ADP} + 16\ P_i + 16\ \text{NADP} + 8\ \text{CoA}. \quad (1)$$

The mechanism of this overall reaction was clarified further by the demonstration that one of the purified fractions contained biotin and converted acetyl-CoA into malonyl-CiA, which was shown to be a precursor of fatty acids [21, 50, 80]. Thus, the synthesis of malonyl-CoA was established as the first step unique to the synthesis of fatty acids. It is now known that the synthesis of palmitate proceeds as follows:

$$\text{Acetyl-CoA} + HCO_3^- \xrightarrow[\text{carboxylase}]{\text{acetyl-CoA}} \text{malonyl-CoA}. \quad (2)$$

ATP, Mn_2+

$$\text{Acetyl-CoA} + 7\ \text{malonyl-CoA} + 14\ \text{NADPH} + 14\ \text{H} + \xrightarrow[\text{synthetase}]{\text{fatty acid}} \text{palmitic acid} + 14\ \text{NADP}^+ + 7\ Co_2 + 8\ \text{CoA} + 6\ H_2O. \quad (3)$$

The fatty acid synthetase portion of the pathway (equation 3) has been studied in great detail. In bacteria the synthesis of palmitate from malonyl-CoA and acetyl-CoA consists of a series of reactions each catalyzed by a separate enzyme [50]. On the other hand, in higher systems the fatty acid synthetase, which is a macromolecular complex, cannot be broken down into functioning component parts, though the general properties of this sequence in the synthesis of fatty acids are the same with *Escherichia coli* or mammalian systems.

In nonruminants acetate is not an important precursor of fatty acids [6, 32]; and although glycolysis of fructose is reviewed elsewhere in this volume, it is essential to point out the pathway for the supply of precursors for the synthesis of fatty acids, especially the production of cytoplasmic acetyl-CoA. This proceeds via the oxidation of pyruvate in the mitochondria to acetyl-

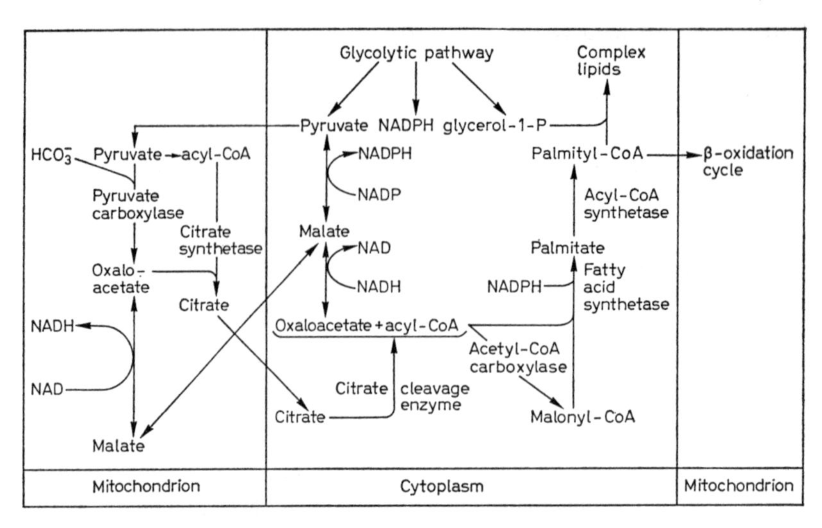

Fig. 1. Schematic representation of interrelations between glycolytic and lipogenic pathway. It should be noted that the synthesis of complex lipids occurs in the microsome, though it is not shown this way in the figure.

CoA, condensation of oxaloacetate and acetyl-CoA to form mitochondrial citrate, passage of citrate out of the mitochondria to the cytosol, and subsequent cleavage of cytoplasmic citrate to oxaloacetate and acetyl-CoA [41, 42]. The pathway from pyruvate to fatty acids and the inter-relations between fatty acid oxidation, esterification and NADPH generating reactions are illustrated in figure 1.

B. Regulation of Fatty Acid Synthesis

In a general way, the rate of flux of substrate through a pathway depends on the amount of enzyme in the pathway, the functional status of the enzyme, and the availability of substrate. All three of these basic regulatory parameters are probably important for the effects of fructose on the synthesis of fatty acids; and in order to gain some insight into the potential complexity of the regulation of fatty acid metabolism within the liver, and of the interdependence of various portions of the fatty acid metabolizing pathways, it is essential to consider the dynamic regulation of acetyl-CoA carboxylase. Acetyl-CoA carboxylase is the first enzyme unique to fatty acid synthetizing pathway and is subject to regulation by small molecules *in vitro*. It is, therefore, in a position to modulate the rate of fatty acid synthesis, though as will be evident below it is not truly the 'rate-limiting' reaction with intact tissues.

Purified acetyl-CoA carboxylase from animal tissues is a polymeric enzyme consisting of identical protomers of approximately 410,000 molecular weight [see reference 49 for extensive review of this subject]. The protomers are catalytically inactive whereas the polymeric form is the enzymatically active species. Enzyme activity depends therefore on the protomer-polymer equilibrium. A variety of physiological intermediates influence this equilibrium *in vitro* [57]. For example, citrate activates acetyl-CoA carboxylase by inducing a conformational change which favors polymerization as evidenced by an increase in the sedimentation velocity of the enzyme [49, 57]; and citrate must be added, in fact, to *in vitro* assay systems in order to measure the activity of acetyl-CoA carboxylase. Long-chain acyl-CoA compounds shift the equilibrium toward the protomer form, tending to inactivate the enzyme [49, 57]. The inhibition of activity by long-chain acyl-CoA is competitive with respect to citrate and K_I for acyl-CoA is approximately 10^{-7} M the order of magnitude of the *in vivo* concentration of these compounds [11, 90]. Fatty acids have an effect which is similar to that for acyl-CoA derivatives. The effect of these hydrophobic compounds is countered by (+)-palmitoyl carnitine which displaces them from the enzyme [22].

It has been considered that the activity of acetyl-CoA carboxylase is subject to feedback regulation by the end-products of fatty acid synthesis or to modulation of activity by alterations in the concentration of citrate within the cell. This possibility has not been established in an independent way as yet. In fact, despite the attractiveness of end-product regulation by acyl-CoA, the concentration of which increases in conditions associated with falling rates of fatty acid synthesis [11, 61, 90], there are several problems in postulating that the intracellular concentrations of long-chain acyl-CoA compounds or citrate have any influence on the *in vivo* activity of acetyl-CoA carboxylase.

With regard to acyl-CoA compounds, inhibition of acetyl-CoA carboxylase is not stoichiometric, but depends on the ratio of protein to acyl-CoA concentration, and can be blocked completely by concentrations of protein below those in the supernatant of the intact cell [see reference 96 for review of this problem]. Also, inhibition of acetyl-CoA carboxylase by long-chain acyl-CoA compounds is not a specific effect since these compounds inhibit citrate synthetase, glucose-6-phosphate dehydrogenase, fatty acid synthetase, glutamic dehydrogenase, and several other enzymes [77]. The difficulty in postulating that citrate is a regulatory molecule for fatty acid synthesis is that its concentration *in vivo* is far below that needed to activate acetyl-CoA carboxylase *in vitro* [30, 54]. The observation that citrate activation of acetyl-CoA carboxylase is time-dependent and interacts with stimulation by Mg^{2+} may be important in this regard [30], but fatty acid synthesis in homogenates of rat-liver is stimulated only slightly by addition of citrate [39]. Moreover, it is possible to extract active forms of acetyl-CoA carboxylase from 100,000 *g* liver supernatants of carbohydrate-fed rats which do not require citrate for activity [73, 83]. Thus, the effect of citrate on the conformation of acetyl-CoA carboxylase may be important only *in vitro*.

Although the physiologic importance of long-chain acyl-CoA compounds and citrate for the regulation of fatty acid synthesis is open to question, the rate of fatty acid synthesis is not related solely to the total amount of enzyme in the fatty acid synthetizing pathway. For example, after a 24-hour fast the activity of acetyl-CoA carboxylase, determined directly in cell-free supernatants from rat-liver, is approximately one-half of that in fed controls, but the rate of synthesis of fatty acids in liver slices from fasted animals falls to 1% of that in fed rats [40]. Similar discrepancies between the functional status of acetyl-CoA carboxylase and the activity of the enzyme measured directly, in the presence of citrate, have been observed in alloxan-diabetic or fat-fed rats [10, 61, 96]. The nature of the substances responsible for constraining the activity of acetyl-CoA carboxylase in fat-fed, diabetic or fasted rats is unclear. The data of FRITZ and HSU [22] suggest, however, that it might be a hydrophobic compound since treatment of fractions of liver supernatant from fasted rats with (+)-palmitoylcarnitine increases 10-fold the incorporation of acetate into fatty acids [22, 57]. It appears that the activity of acetyl-CoA carboxylase is subject to dynamic regulation *in vivo*, and this regulation is related almost certainly to the kinds of conformational changes which alter the protomer polymer equilibrium. On the other hand, the effector for these effects *in vivo* has not been identified with any degree of certainty.

As shown in figure 1, the supply of NADPH could limit the synthesis of fatty acids. In alloxan-diabetic or fasted rats, the activities of glucose-6-P and 6-phosphogluconate dehydrogenases fall [27] in association with diminished rates of hepatic synthesis of fatty acids. It was postulated, in fact, that the availability of NADPH was the controlling factor in lipogenesis [45, 70], but this work was completed prior to the elucidation of the importance of the acetyl-CoA carboxylase step and there is little to support the notion that production of NADPH regulates fatty acid synthesis. For example, the capacity for NADPH production in liver homogenates from alloxan-diabetic, pancreatectomized or fat-fed rats is not decreased in comparison to control animals [1, 10, 63]. Further, although the synthesis of fatty acids from acetate may require exogenous NADPH (and/or NADH), metabolism in the pentose-phosphate pathway, and oxidation of pyruvate and citrate in the tricarboxylic acid pathway should insure adequate reducing power for the synthesis of fatty acids when a sugar is used as precursor.

It is interesting that all studies of the dynamic regulation of fatty acid synthesis have concentrated on fat-fed, fasted or diabetic animals, situations in which fatty acid synthesis is switched-off. There is little data relating to the problem of moment-to-moment fluctuation of rates of hepatic fatty acid synthesis in carbohydrate-fed animals. This is not a trivial point since it bears directly on the question of end-product inhibition and the importance of NADPH for regulating fatty acid synthesis. The synthesis of fatty acids allows for the storage of excess calories and feedback inhibition of fatty acid synthesis in fed animals would decrease storage of excess calories. In terms of body economy there does not appear to be a need to modulate the rate of fatty acid synthesis, except to spare carbon for hepatic production of glucose. In a setting of excess endogenous carbohydrate there is no useful purpose in limiting the rate of fatty acid synthesis. Also, in a fasted or alloxan-diabetic animal the synthesis of fatty acids from acetate could be limited by the supply of NADPH and NADH, but this metabolic setting is completely different from that in a carbohydrate-fed animal using a physiologic substrate for the synthesis of fatty acids.

III. The Effect of Fructose on the Amounts of Fatty Acid Synthetizing Enzymes

Short periods of feeding diets containing large amounts of fructose increase the amount of the hepatic acetyl-CoA carboxylase as compared to animals fed a chow diet [93, 98]. The amount of fatty acid synthetase is increased also [11a, 98]. However, the effect of feeding fructose for 2 days on the specific activities of hepatic acetyl-CoA carboxylase and fatty acid synthetase was no greater than that due to an isocaloric amount of glucose [93, 98]. On the other hand, in one study the activity of fatty acid synthetase was greater in liver from rats fed fructose or sucrose for 30 days than from rats fed an isocaloric amount of glucose [11a]. In pair-feeding experiments with rats, SZEPESI and FREEDLAND [74, 75] observed that a high-fructose diet as compared to glucose increased significantly the weight of the liver and the amount of soluble liver protein per 100 mg body weight. Similar observations were made by ADELMAN *et al.* [2] and CARROLL [13]. Thus, in the rat the total capacity for hepatic fatty acid synthesis is increased by fructose as compared to glucose feeding, despite nearly identical specific activities for acetyl-CoA carboxylase and fatty acid synthetase in both diet groups.

The amount of acetyl-CoA carboxylase in liver depends on the balance between rates of synthesis and degradation which appear to be independent and subject to change during alteration of the diet [53, 58]. Although diets containing large amounts of carbohydrate increase the half-time for the degradation of this enzyme, and its rate of synthesis [53], the specific effects of fructose have not been studied.

In experiments of somewhat different design from those above HICKS *et al.* examined the effect of refeeding diets high in sucrose or glucose to rats previously fasted for 40 h [36]. In these experiments acetyl-CoA carboxylase activity was assayed by measuring the incorporation of ^{14}C-acetyl-CoA into fatty acids in 100,000 *g* liver supernatant rather than by fixation of $H^{14}CO_3^-$. The activities of acetyl-CoA carboxylase and fatty acid synthetase were higher in rats refed glucose as compared to those refed sucrose, despite the large amount of fructose contained in the sucrose moiety. Unfortunately, it is uncertain whether this interesting difference is peculiar to the comparison of sucrose and glucose diets or whether fructose and sucrose would have similar actions in rats.

In addition to acetyl-CoA carboxylase and fatty acid synthetase several other enzymes in the fatty acid synthetizing pathway are adaptive and increase in response to fructose feeding as compared to standard chow diets [10, 20, 84]. The rôle of these other enzymes in fatty acid synthesis is to provide substrate and cofactor as indicated in figure 1, but it has never been shown that they are rate-limiting at any time. FITCH and CHAIKOFF reported that the hepatic activities of malic enzyme, glucose-6-phosphate dehydrogenase, 6-phosphogluconate dehydrogenase, and citrate cleavage enzyme are greater in animals fed fructose than those fed a hexose-free diet or isocaloric amounts of glucose [20]. This has not been a consistent finding, however, in that levels of NADP-malate dehydrogenase (malic enzyme) and citrate cleavage enzyme have also been found to be identical in rats fed isocaloric amounts of fructose or glucose, specific activities in both instances being expressed per milligram supernatant protein [93]. Thus, in normal rats, the effects of fructose on the activities of adaptive enzymes of the fatty acid synthetizing pathway seem to reflect the general response to feeding a large amount of carbohydrate rather than a specific effect of fructose.

As noted above, alloxan-diabetes decreases the capacity of the liver to synthetize fatty acids, and there is an associated decline in the activities of lipogenic enzymes. Fructose feeding, but not glucose, can repair the defect in hepatic fatty acid synthesis in diabetic rats as measured by the incorporation of ^{14}C-acetate into fatty acids [5]. A similar difference between the actions of fructose and glucose is seen for the activity of hepatic citrate cleavage enzyme in alloxan-diabetic rats in that a fructose diet, but not glucose, raises the activity of this enzyme to normal in these animals [41]. The depressed hepatic activity of glucose-6-P dehydrogenase in diabetic rats is increased also by fructose feeding, but not by glucose [19]. The effect of fructose feeding on the activity of acetyl-CoA carboxylase in livers from alloxan-diabetic animals has not been examined directly; but TAKEDA *et al.* [76] have found that hepatic acetyl-CoA carboxylase activity is normal in alloxan-diabetic rats fed glycerol.

The chemical basis for the effects of fructose on the level of activities of fatty acid synthetizing enzymes is not completely clarified; but it seems reasonably certain that there is increased synthesis and decreased catabolism of these enzymes during ingestion of fructose as compared to a chow diet. The problem is to identify the factor primarily responsible for these events. In this regard, the studies in alloxan-diabetic animals are especially important since they suggest that the essential feature of enzyme induction of the fatty acid synthetizing enzymes is the rate of metabolism of sugars in the glycolytic pathway. For example, the activity of glucokinase, which is essential for hepatic glycolysis of glucose, is dependent on the presence of glucose and insulin; and in the absence of insulin there is no glycolysis of glucose. In contrast, fructose metabolism proceeds normally in livers from alloxan-diabetic rats or even diabetic man since fructokinase is a constitutive enzyme [2, 56, 69, 71], there being relatively small variability in the activity of fructokinase as a result of changes in the fructose content of the diet [2, 93]. Even in the absence of insulin, fructose is glycolyzed at a rapid rate and can provide substrate for the fatty acid synthetizing enzymes. One cannot exclude the possibility that maintenance of high activities of fatty acid synthetizing enzymes is attributable to effects of fructose on the concentration of an intermediate which is an inducer of these enzymes, but it is of importance that fructose feeding allows for simultaneous high rates of gluconeogenesis and lipogenesis in normal and diabetic animals. That feeding glycerol mimics most of the effects of fructose in alloxan-diabetic animals, including an increase in the activity of acetyl-CoA carboxylase [76], strengthens the notion that the differential effects of fructose and glucose on the fatty acid synthetizing capacity of livers from diabetic rats is related to respective controls on their rates of metabolism in the Embden-Meyerhoff pathway.

IV. The Effect of Fructose on the Dynamic Regulation of Fatty Acid Synthesis

Modulation of the rates of fatty acid synthesis might be achieved not only by variation in the amounts of fatty acid synthetizing enzymes but also by alterations of the functional status of acetyl-CoA carboxylase, the 'rate-limiting' enzyme in the pathway. Changes in the concentrations of metabolic intermediates which are effectors (negative and positive) of the acetyl-CoA carboxylase step or changes in the rate of esterification of fatty acids independently of alterations in the concentrations of intermediates could influence the activity of this enzyme. Alternatively, fructose simply could provide more fatty acid precursor (cytoplasmic acetyl-CoA) than other sugars. Each of these possibilities has been examined and the data are reviewed below.

A. The Effects of Fructose on the Concentrations of Metabolic Intermediates

Because the catalytic properties of acetyl-CoA carboxylase can be influenced, at least *in vitro*, by metabolic intermediates (see section II B), fructose metabolism could alter the rates of fatty acid synthesis by changing the concentration of these intermediates *in vivo*. For example, increases in the concentration of citrate, if they occurred, might activate acetyl-CoA carboxylase. Similarly the metabolism of fructose could lead to reduced intracellular levels of long-chain acyl-CoA compounds.

Data are available for the effects of fructose administered *in vivo* on the concentration of a variety of glycolytic and Krebs cycle intermediates, though not for all known *in vitro* effectors of acetyl-CoA carboxylase. In the studies of BURCH *et al.* [12] rats, fasted for 24 h, were anesthetized and given 20 mmoles of fructose/kg intraperitoneally; 60 min later there were marked increases in the hepatic concentrations of glucose-6-P (900%), glycerol-1-P (400%), pyruvate and lactate. Comparison of the effects of glucose and fructose revealed that the administration of fructose led to greater increases in the concentrations of glycerol-1-P, lactate and pyruvate than glucose. Of interest in these experiments were the comparative effects of dihydroxyacetone, glycerol and fructose on the concentrations of metabolic intermediates. Equal amounts of fructose and glycerol, on a molar basis, had similar effects on all intermediates except for glycerol-1-P which was 20 times higher in glycerol- than fructose-treated animals, and lactate and pyruvate which were 3 times greater in the fructose-treated rats. The effects of dihydroxyacetone were closer to those of fructose than glycerol.

The results of more acute studies [94] differ from those reported above in that fructose did not produce sustained increases in the concentrations of metabolic intermediates. Thus, 5 min after intravenous injection of 200 mg fructose/400 g rat, the hepatic concentration of glycerol-1-P increased 5-fold, nearly the same as observed by BURCH *et al.* [12]; but 20 min after the administration of fructose the concentration of glycerol-1-P was equal to that at zero time [94]. In fact, the hepatic concentration of glycerol-1-P was greater 20 min after the administration of glucose than after fructose. The concentration of fructose in the blood of these animals declined between 5 and 20 min, but was still 24.3 mg/100 ml 20 min after the injection of fructose, which is sufficient to saturate fructokinase [95]. A similar result was found for the effect of fructose injection on the concentration of pyruvate – an initial 4-fold increase at 5 min followed by a decline to control levels by 20 min. In acute experiments, KUPKE and LAMPRECHT also failed to find an increased hepatic concentration of glycerol-1-P, or pyruvate, after the intraportal injection of fructose [44], despite concentrations of fructose in the portal blood great enough to saturate fructokinase. We would point out that the disagreement between the results from acute and longer-term experiments *in vivo* could be related to experimental design in that the former studies utilized fed rats whereas in the latter experiments animals had been fasted approximately 24 h prior to the administration of the sugars. The effects of long-term feeding of fructose on the concentration of metabolic intermediates have been studied in rat-livers from animals fed diets of variable carbohydrate composition [92, 93]. Diets high in fructose as compared to glucose or standard chow diets increased significantly the hepatic levels of pyruvate, malate and acetyl-CoA. There was also a small increase in the concentration of citrate, from approximately 100 to 150 nmoles/g wet weight of liver. Based on the concentrations of citrate required for activation of acetyl-CoA carboxylase *in vitro* [30], this change is too small to lead to a significant activation of acetyl-CoA carboxylase if the enzyme is partly in an inactive form *in vivo*. On the other hand, the data suggest that fructose-feeding increases the steady state level of Krebs' cycle intermediates, and as a result, could provide greater amounts of substrate for fatty acid synthesis than glucose or starch (chow diet).

Using perfused livers from fasted rats, EXTON and PARK [16] demonstrated that fructose (20 mM) increased the concentrations of lactate, pyruvate, glucose-6-P and glycerol-1-P after 60 min of perfusion (only time reported).

The concentration of malate was increased also, but no direct comparisons were made using different sugars. WOODS *et al.* [87] perfused livers from fed rats with 10 mM fructose and observed qualitatively similar changes. Serial assays were done in these experiments; and, of interest, the concentration of glycerol-1-P was greatest at the first sample time (10 min) and declined progressively during the remainder of the experiment, although the fructose concentration was still high at the end of the perfusion. The concentrations of pyruvate and lactate remained elevated for the entire experiment. WIELAND and MATSCHINSKY also found that the addition of fructose to the perfusate increased the concentration of glycerol-1-P in perfused livers, but the increase was only twofold and was not sustained [84].

The exact significance of the effects of fructose on the concentration of glycolytic intermediates is unclear except that the data indicate that fructose is metabolized at a greater rate than glucose in mammalian liver. On the other hand, by increasing the concentration of glycerol-1-P, even transiently, fructose might enhance fatty acid synthesis by stimulating the rate of esterification of fatty acids, lowering the concentration of long-chain acyl-CoA compounds, and thereby relieving constraint on the activity of acetyl-CoA carboxylase. There is, in fact, an inverse relationship in the liver between the concentrations of these two intermediates in fasted or fat-fed rats [11, 90]; but there is no data for this tissue on the response of the concentration of long-chain acyl-CoA to administration of different sugars to carbohydrate-fed animals. WILLIAMSON *et al.* [86], however, have measured the effect of glycerol and dihydroxyacetone on the concentration of these intermediates. After intramuscular injection of glycerol in fasted rats, the hepatic concentration of glycerol-1-P increased 40-fold, but the level of long-chain acyl-CoA compounds was unchanged. Dihydroxyacetone had essentially no effect on the level of either of these intermediates. No associated measurement of fatty acid synthetizing activity was made. The question of the importance of the concentration of glycerol-1-P and long-chain acyl-CoA compounds as they relate to control of fatty acid synthesis is discussed more fully in the section on esterification.

B. The Effect of Fructose on the Esterification of Fatty Acids

In an *in vitro* system the rate of esterification of fatty acids with glycerol-1-P is dependent on the concentration of this compound and that of the long-chain acyl-CoA compound [42]. Since the administration of fructose leads in some experimental settings to increases in the hepatic concentration of glycerol-1-P, it has been postulated that fructose, as a result of its effect on the concentration of glycerol-1-P, also should enhance the rate of esterification of fatty acids. Results of studies of the relationship between rates of esterification of fatty acid and the concentration of glycerol-1-P in liver and other tissue are surprising, however, in that there appears to be no correlation between rates of esterification and the concentration of glycerol-1-P [14, 67]. In extensive studies examining directly the effects of variation in the concentration of glycerol-1-P on rates of triglyceride synthesis in adipose tissue, DENTON and HALPERIN [14] demonstrated that large decreases in the concentration of glycerol-1-P after administration of insulin and adrenaline were associated with 5-fold increases in the rate of triglyceride synthesis. In contrast, insulin alone increased the concentration of glycerol-1-P 8.5-fold, but was associated with only a doubling in the rate of triglyceride synthesis. In these studies it was shown also that increasing rates of esterification were not associated with decreasing concentrations of long-chain acyl-CoA compounds.

The problem of the effect of regulation of fatty acid esterification has not been examined as carefully or directly in liver as in adipose tissue, but a number of investigators have been interested in the antiketogenic effects of fructose and other sugars and their data reflect on the problem of how hepatic fatty acid esterification is modulated [15, 17, 43, 55, 62].

In perfused livers from fasted rats, MCGARRY and FOSTER found that addition of fructose to the perfusate increased the recovery of [1-^{14}C-]oleate twofold in the lipid fraction at the end of the experiment, and the concentration of glycerol-1-P rose 4-fold in comparison to the controls [55]. On the other hand, after addition of lactate to the perfusion, the concentration of glycerol-1-P increased only slightly; but the recovery of [1-^{14}C]oleate as liver lipids increased to the same extent as in perfusions with fructose. Addition of ethanol or fructose to the perfusate increased the concentration of glycerol-1-P to the same extent, but the amount of [1-^{14}C]oleate recovered in liver lipids was nearly twofold greater after perfusion with ethanol than with fructose.

Direct comparison of fructose and glucose indicated that the effect of fructose on esterification was greater than that for glucose; and in association with increased recovery in the lipid fraction, fructose also inhibited production of ketone bodies to a greater extent than glucose. Of importance, fructose had no effect on the total amount of fatty acids metabolized during the course of the experiment. This parameter was independent also of the nutritional status of the donor animal or addition of other sugars, lactate or ethanol to the perfusate.

Krebs and Hems [43], also studying ketone body synthesis by perfused liver from fasted rats, found that fructose decreased the endogenous production of ketone bodies, but in contrast to the data of McGarry and Foster [55] there was essentially no difference between the antiketogenic actions of glucose and fructose. As noted in section II A, Williamson *et al.* [86] studying ketogenesis in intact rats found no correlation in the hepatic concentration of glycerol-1-P and the production of ketone bodies, and presumably entry of fatty acids into oxidative pathways. In fasted, intact rats, glycerol and dihydroxyacetone were equally antiketogenic but glycerol raised the intrahepatic concentration of glycerol-1-P 40-fold whereas dihydroxyacetone had almost no effect.

Topping and Mayes [78] studied the effects of fructose on the esterification and oxidation of [14C]-oleate in perfused livers from fed rats. As in fasted animals, addition of fructose to the perfusate increased the recovery of oleate as esterified products and decreased oxidation to CO_2 and ketone bodies.

Using suspensions of liver cells, Ontko [62] obtained results identical to those of McGarry and Foster [55]. Thus, addition of fructose to cell suspensions prepared from fasted rats increased the incorporation of [1-^{14}C] palmitate into phospholipids, and di- and triglycerides. Glucose had a qualitatively similar effect, but in some experiments the stimulation of the synthesis of phospholipids and glycerides was greater after the addition of fructose than glucose. Fructose also was a more potent inhibitor of the oxidation of fatty acids to CO_2 and ketone bodies than was glucose, an effect which was amplified in the presence of relatively high concentrations of free fatty acids. Glycerol had a greater effect on the synthesis of glycerides and reduction of ketone body production than fructose; but unfortunately the concentrations of glycerol-1-P were not measured in these experiments. On the other hand, in separate experiments with an identical system Berry [personal communication] has found that the addition of fructose to cell suspensions does not increase the concentration of glycerol-1-P.

Fallon and Kemp fed rats diets containing large amounts of different sugars for 7 days [17]. In comparison to control animals fed a chow diet, fructose feeding increased the activity of glycerol-P-acyl-transferase in liver microsomes and also increased the synthesis of triglycerides by this system. Interestingly, fructose-feeding had a greater effect on the synthesis of triglycerides by microsomes than isocaloric amounts of starch; but the effects of each sugar on the activity of glycerol-P-acyltransferase were identical, indicating again the complex nature of the control of esterification reactions.

The effects of glucose and fructose feeding were the same either on enzyme activity or the overall synthesis of triglycerides.

In all experiments in which it has been measured, an interesting result is the constancy of fatty acid uptake by the liver. Thus, in cell suspensions [62] or perfused livers [55] from fed or fasted rats supplemented with a variety of sugars in the perfusion media, the uptake of fatty acids per unit time is the same. Since fructose and other sugars influence only the disposition of compounds into oxidation or esterification pathways but do not increase the total uptake of fatty acids by the liver, the activities of acyl-CoA synthetases may be important in regulating the overall metabolism of fatty acids in liver. There are no data available on the specific effects of fructose in acute or chronic experiments on the activity of this enzyme, but it has been shown that feeding large amounts of sucrose for 36 h depresses slightly the palmityl-CoA synthetase activity in whole homogenates of rat liver, as compared to rats fed a stock diet [18]. Activity of this enzyme is increased in fasted rats [18]. LIPPEL [47] has confirmed the data of FARSTAD [18] for the sucrose-induced decline in palmityl CoA synthetase in whole homogenates of liver, but the latter author found that the activity of the microsomal enzyme was unchanged by 72 h of sucrose feeding.

It is clear that fructose increases the rate of esterification of fatty acids in liver. However, this effect is not mediated via an action of fructose on the concentrations of glycerol-1-P or acyl-CoA compounds, nor on the level of activity of the esterifying enzymes. Possibly, the primary effect of fructose is to inhibit the oxidation of fatty acids, enhanced esterification reflecting a rerouting of substrate to an alternative pathway. There are no data which bear on this point in a direct manner, but fructose does not alter the production of ketone bodies from short-chain fatty acids which are extremely poor substrates for acyltransferases in mammalian tissues [62]. Another observation pertinent to this question is that with perfused liver fructose is not antiketogenic at very high concentrations of perfused oleate (>2 mM) [43]. Although the mechanism of this effect of high concentrations of fatty acids is unclear, these data indicate that fructose does not directly block oxidation of fatty acids. It seems, therefore, that the metabolism of fructose increases directly the rate of esterification of fatty acids in the liver. Although this effect is not unique to fructose in most experiments, this sugar has a quantitatively greater effect on esterification than glucose. We would stress that the concentration of substrates must influence the rate of esterification, but primary control over this reaction is exerted by factors other than the availability of substrates. The nature of the factors controlling esterification are unclear.

Because of its effect on esterification, fructose will increase the triglyceride content of the liver by sparing fatty acids from oxidation. Whether this effect is important also for the *de novo* synthesis of fatty acids is difficult to assess. As mentioned above, increased rates of esterification resulting from the metabolism of fructose do not decrease the concentration of long-chain acyl-CoA compounds and therefore cannot be expected to alter the functional status of acetyl-CoA carboxylase on this basis. It is possible, however, that the action of fructose on the rate of esterifiaction in liver microsomes leads to an increase in the rate of fatty acid synthesis by the liver.

In bacteria the growing fatty acid chain is synthetized attached to a protein designated 'acyl-carrier protein' [52, 75], and the fatty acid-acyl carrier protein complex is a substrate for esterification in these systems [3, 28]. In bacteria, therefore, the rate of esterification can affect the rate of fatty acid synthesis by increasing the availability of free acyl-carrier protein. There is no acyl-carrier protein, as such, in higher organisms; but in the fatty acid synthetase complex from yeast there is an analogous protein which functions as an integral component of the fatty acid synthetase [85]. Whether the fatty acid bound to fatty acid synthetase is a substrate for esterification in yeast or higher systems is unknown. If it were, however, fructose could stimulate fatty acid synthesis by enhancing the rate of removal of products from the synthetase complex thereby allowing for initiation of a new process of chain elongation. Alternatively, increased rates of esterification could prevent competition between product and substrate for binding to the synthetase. In either case, the effect of fructose on esterification of fatty acids could contribute to its overall effect on the rate of fatty acid synthesis. Since malonyl-CoA is an inhibitor of acetyl-CoA carboxylase [57, 96], the acetyl-CoA carboxylase – fatty acid synthetase – and acyltransferases could function as a tightly coupled regulatory system.

C. The Conversion of ^{14}C-Fructose to Fatty Acids

In rats fed a stock chow diet ^{14}C-fructose is converted to fatty acids in liver slices at a rate 4-fold greater than that for ^{14}-C-glucose [93]. A similar result is obtained using liver slices obtained during elective surgery in humans with no known liver disease [97]. The difference between recovery of ^{14}C-fructose and ^{14}C-glucose as fatty acids depends, however, on the concentration of glucose used. Thus, as the concentration of glucose is increased from 5 to 30 mM the recovery of ^{14}C-glucose as fatty acids approaches that of

5 mM ^{14}C-fructose. The previous dietary history of rats also influences the extent of conversion of ^{14}C-fructose and ^{14}C-glucose to fatty acids by liver slices. A diet containing a large amount of glucose or fructose enhances the synthesis of fatty acids from either sugar; but irrespective of the diet more ^{14}C-fructose is recovered as fatty acids than ^{14}C-glucose. On the other hand, the glucose diet enhances selectively the conversion of ^{14}C-glucose to fatty acids, as compared to ^{14}C-fructose; and the fructose diet has a similar selective effect on the conversion of fructose to fatty acids. The general effect of increased synthesis of fatty acids with either of the carbohydrate-enriched diets correlates best with increases in the activity of fatty acid synthesizing enzymes.

The relatively selective effect of diet in enhancing the conversion of fructose or glucose to fatty acids is mediated via increases in the activity of enzymes responsible for the glycolysis of fructose or glucose [93]. In glucose fed rats the activity of glucokinase increased relative to that in chow-fed or fructose-fed animals. In the latter group of rats, glucokinase activity decreased. It is not possible to examine in detail the effects of preceding diet on the conversion of fructose to fatty acids in human liver, but SCHERSTEN *et al.* studied the effect of fructose on the synthesis of lipids in human liver slices by administering fructose intravenously the day prior to abdominal surgery [66]. Control patients were fed a normal hospital diet until the evening before surgery. The patients in the experimental group were given 400 g of fructose intravenously the day prior to surgery and their caloric intake from the diet was diminished accordingly. Preoperative loading with fructose increased nearly 4-fold the conversion of ^{14}C-fructose to the fatty acid moiety of triglycerides in liver slices as compared to control patients. Fructose supplementation also increased 3-fold the conversion of ^{14}C-glycerol to triglyceride fatty acids. On the other hand, intravenous fructose did not alter the synthesis of glyceride-glycerol, or the total incorporation of label into phospholipids with either ^{14}C-glycerol or ^{14}C-fructose as substrates. No comparisons were made between fructose and glucose as substrates for fatty acid synthesis.

D. Acute Effects of Fructose on the Synthesis of Fatty Acids from Acetate

In several studies the effect of *in vivo* or *in vitro* fructose on the conversion of ^{14}C-acetate to fatty acids has been examined. With newly-hatched chicks, GOODRIDGE [29] found that addition of fructose to the incubation media of liver slices from unfed animals stimulated the incorporation of ^{14}C-acetate and $^{3}H_2O$ into fatty acids. In comparison to studies without added substrate, 11 mM fructose increased the rate of fatty acid synthesis nearly 30-fold from 20 mM ^{14}C-acetate. The same concentration of glucose increased the rate of fatty acid synthesis by only 50%; 20 mM glycerol increased the incorporation of ^{14}C-acetate into fatty acids 15-fold, as compared to the experiment with no added sugar. The effects of fructose on recovery of $^{3}H_2O$ in fatty acids synthetized from 20 mM acetate were less dramatic; the recovery of ^{3}H in fatty acids increased only 6-fold in comparison to studies with no added substrate. In identical experiments glucose had no discernible effect on the rate of fatty acid synthesis. Fatty acids were not synthetized in these experiments from glucose, fructose or glycerol, since in the absence of acetate, fructose, glucose or glycerol had no effect on the incorporation of $^{3}H_2O$ into fatty acids. The effect of fructose on the synthesis of fatty acids from acetate was not inhibited by the administration of cycloheximide. The oxidation of acetate to CO_2 was unaffected by fructose, glucose or glycerol.

Although fructose *in vitro* stimulated fatty acid synthesis to a greater extent than glucose, administration of glucose *in vivo* increased the subsequent conversion of ^{14}C-acetate to fatty acids in liver slices 100-fold, whereas *in vivo* fructose had only a small stimulating effect. In fact, fatty acid synthesis in fructose-supplemented liver slices from fasted chicks was 7 times greater than the level of synthesis in slices from fructose-fed animals. With liver slices from glucose-fed chicks, addition of fructose or glucose to the media did not alter the rate of conversion of ^{14}C-acetate or $^{3}H_2O$ to fatty acids. Again, there was little synthesis of fatty acids in the absence of acetate.

YEH and LEVEILLE [89], also studying chicks, obtained results in many ways similar to those of GOODRIDGE [29], though the experimental designs were not exactly the same. After a 2-hour fast, the synthesis of fatty acids from ^{14}C-acetate by liver slices decreased 95% as compared to the controls. Addition of fructose (final concentration 5 mM) to liver slices increased the rate of fatty acid synthesis approximately 3.5-fold. In contrast, addition of either 5 mM glucose of 10 mM glycerol to the liver slices had very little effect on the rate of lipogenesis. Even in the presence of fructose, however, lipogenesis was reduced in fasted as compared to fed chicks. In chicks fasted 6 h, fructose stimulated fatty acid synthesis in liver slices 8-fold as compared to no addition of substrate. Glucose and glycerol had less effect on lipogenesis than fructose, though again it should be emphasized that the rate of fatty acid synthesis in the presence of 5 mM fructose was, at most, 10% of the control rate for fed animals. In contrast to the results of GOODRIGDE, intravenous administration of fructose to fasted chicks increased the subsequent incorporation of acetate into hepatic fatty acids in slice experiments [89]; glucose was less effective than fructose in this regard. Since the time between injection of the sugar and the preparation of liver slices was only 45 min, the effects of fructose on the subsequent synthesis of fatty acids was almost certainly not related to enzyme induction but to alteration of the dynamic properties of the fatty acid synthetizing pathway. Also, in intact chicks the recovery of intravenously administered ^{14}C-fructose in liver lipids was 6-fold greater than that for ^{14}C-glucose [89].

The reason for the discrepancies in the data of GOODRIDGE [29] and YEH and LEVEILLE [89] are not immediately apparent. One possibility, however, is that the age of experimental animals may influence the results of studies of different sugars on hepatic lipogenesis. There is no direct data on this point, but HILL [37] has found that the age of rats influences the response of the serum triglyceride concentration to the ingestion of sucrose. Also of interest are observations by BENDER *et al.* [8] that both age and strain of rats are important for the effects of sucrose-feeding on the synthesis of fatty acids from ^{14}C-glucose.

In experiments with perfused livers from fed rats, addition of fructose to the perfusate had no effect on the rate of lipogenesis from [Me-^{3}H] acetate [71]. Rat-insulin had a stimulating effect on lipogenesis and fructose plus bovine insulin increased lipogenesis from [Me-^{3}H] acetate still further; but unfortunately the control experiment with bovine insulin alone was not reported.

E. The Conversion of ^{14}C-Fructose to Glyceride-Glycerol

A number of laboratories have documented that fructose is converted rapidly to glyceride-glycerol in liver both after *in vivo* administration of the sugar or in *in vitro* experiments with liver slices [7, 23, 24, 35, 37, 44, 88]. In all instances in which glucose and fructose have been compared, the recovery of ^{14}C-glucose as glyceride-glycerol is substantially less than that for ^{14}C-fructose [35, 97] and, also, the amount of labeled carbon as glyceride-glycerol is always greater than that found in the fatty acid moiety of the complex lipids. This has led KUPKE and LAMPRECHT [44] to conclude that fructose is converted predominantly to the glyceride-glycerol portion of the triglyceride molecule and that the synthesis of fatty acids from fructose is quantitatively less important than its contribution as a precursor for glycerol-1-P. The problem in resolving this question is that the exact rate of fatty acid synthesis is not estimated by isotopic techniques. The data reported in the literature do not include corrections for dilution of label along the fatty acid synthetizing pathway. In considering the pathway from fructose to fatty acids it is obvious, however, that the label incorporated into the fatty acid moiety will be diluted by endogenous intermediates far more than that incorporated into the glyceride-glycerol moiety of triglycerides or phospholipids. In fact, in the metabolism of fructose there should be almost no dilution of label to the glycerol-1-P level. For this reason, we feel that it is unwarranted to draw conclusions about the relative contributions of fructose, or other precursors of fatty acids, to the synthesis of the fatty acids or glycerol moiety of triglycerides. What are required, but are not available, are chemical estimates of the rates of *de novo* synthesis of fatty acids by perfused liver cell suspensions, or liver *in situ*. The only data relating to the possible quantitative significance of hepatic synthesis of fatty acids from carbohydrate precursors are values for the activity of enzymes in the fatty acid synthetizing pathway. These data suggest, for both man and rat, that the rates of hepatic synthesis of fatty acids can be high in comparison to the turnover rates of triglyceride fatty acids in the serum [91, 93, 98].

V. Mechanism of the Differential Effects of Fructose and Glucose on Hepatic Lipogenesis

In the experiments examining the conversion of ^{14}C-precursors to fatty acids the only manner in which fructose could have altered the capacity for fatty acid synthesis is via a dynamic change, since the duration of the experiments precluded inductive effects of fructose. Further, comparison of the inductive effects of fructose and glucose on the activities of fatty acid synthetizing enzymes indicate identical effects with either sugar; yet fructose is a better precursor of fatty acids than glucose, and in most reported experiments also enhances conversion of acetate to fatty acids to a greater extent than glucose. As reviewed in section III, fructose does not lower the concentration of acyl-CoA compounds even though it enhances esterification. Moreover, regulation of acetyl-CoA carboxylase by the concentration of acyl-CoA has not been shown to occur in an *in vivo* system, and the small changes in the concentration of citrate in fructose-fed rats are unlikely to have much effect on the conformational state of acetyl-CoA carboxylase considering the concentration of citrate required for an effect *in vitro* [30, 54]. Thus, although there is considerable evidence to indicate that the activity of acetyl-CoA carboxylase could be modulated *in vivo* via interconversions between constrained and active forms of the enzyme, there are no data which indicate in a clear manner how the metabolism of fructose can affect the equilibrium between the different forms of the enzyme. An alternative to regulation by altering the functional status of fatty acid synthetizing pathways is regulation of fatty acid synthesis in fed animals by the availability of substrate since, for example, at less than saturating concentrations of acetyl-CoA the rate of fatty acid synthesis will depend on the flux of substrate through the fatty acid synthetizing pathway independent of the functional capacity of the system. The evidence that this pathway functions at less than saturating concentrations of substrate, and that fructose can influence the flux of substrate through the pathway is abundant.

As for most reactions *in vivo*, the K_m of the enzymes in the fatty acid synthetizing pathway are close to the tissue concentrations of their substrates [93]. The pathway between citrate cleavage enzyme and acetyl-CoA carboxylase is not saturated. There is also not a uniform distribution of these substrates within the various intracellular compartments, and citrate seems to be concentrated preferentially in the mitochondria. It is important, further, that the concentration of enzymes in this pathway may be approximately the same as the concentrations of their substrates [72]. As a result K_m values, as determined *in vitro* with small quantities of enzyme, do not reflect the true K_m for the *in vivo* situation, increasing the importance of flux rates for determining the rate of fatty acid synthesis. In this setting changes in the supply of precursors will increase the rate of fatty acid synthesis.

In the liver, fructose is metabolized by a pathway separate from that for glucose, up to the triose level [95]. It is phosphorylated by fructokinase which has a low $K_{fructose}$ and which has a greater activity at V_{max} than glucose phosphorylating enzymes in liver. That glucokinase, the principle enzyme for the phosphorylation of glucose, has a high $K_{glucose}$ amplifies the differences between the rates of glycolysis for fructose and glucose. The metabolism of fructose also avoids any potential regulation at the phosphofructokinase level. As a result of these factors the rate of glycolysis of fructose is greater than that for glucose, and this is reflected by the effects of fructose administration, as compared to glucose, on the levels of several metabolic intermediates within the liver, the accumulation of lactate in the plasma of both man and rats after injection of fructose, and the greater oxidation of fructose to CO_2. Since fructose is metabolized in the glycolytic pathway at a far greater rate than glucose, it should provide more substrate for fatty acid synthesis and be a better precursor of fatty acids than glucose, a prediction which fits with all data collected to date. An important observation in this regard is that the rate of synthesis of fatty acids in liver slices is dependent on the concentration of glucose [97] and on the rate of synthesis of glucose-6-P, apparently the rate-limiting step for the hepatic glycolysis of glucose [46].

In recent experiments, WOODS and KREBS [88] came to the conclusion that the capacity of perfused liver for the synthesis of fatty acids is equivalent to rates of glycolysis. They further 'pointed out that hepatic glycolysis (producing acetyl CoA in aerobic conditions) is not primarily an energy-providing process but part of the mechanism converting carbohydrate into fat'.

To a large extent it seems that the effects of fructose on rates of hepatic fatty acid synthesis in fed animals are related to the uniqueness of the pathway for glycolysis of fructose in the liver rather than to direct effects on the enzymes of the fatty acid synthetizing pathway. On the other hand, it is not possible to ascribe all the effects of fructose to its influence on flux rates in the Embden-Meyerhof pathway, and this is especially true in fasted animals, in which case fructose seems to relieve constraint on the capacity of the fatty acid synthetizing enzymes. The data of GOODRIDGE [29] cited above indicate, for example, that fructose increases the synthesis of fatty acids from acetate, but is not itself converted to fatty acids. Possibly the metabolism of fructose in this setting is important for generating reduced pyridine nucleotides, but the mechanism of the type of fructose effect described by GOODRIDGE remains uncertain.

That fructose is a more potent agent than glucose in reversing the effects of fasting on hepatic lipogenesis also may be related to the greater glycolytic capacity for the former as compared to the latter sugar. With regard to this point, it is important to stress that the glycolysis of fructose is constitutive whereas that for glucose is adaptive. Thus, the activity of fructokinase changes little in fasted animals, but the activity of glucokinase depends on the presence of glucose and insulin. In fasted animals refed fructose or glucose, it is likely that the differences between the glycolytic rates of these two sugars will be greater than in animals fed *ad libitum*. As a result, the differences between the conversion of fructose and glucose to fatty acids in the liver may be amplified by a pattern of meal-eating.

VI. Conclusions

Fructose has both general and unique effects on hepatic lipogenesis. It shares with glucose the capacity for increasing by induction the synthesis of fatty acid synthetizing enzymes in normal animals. In alloxan-diabetic animals, however, fructose-feeding prevents the decline in the capacity for hepatic fatty acid synthesis seen in glucose-fed animals. Both sugars stimulate esterification of fatty acids by the liver, but fructose has a greater effect than glucose on this reaction. Although fructose can increase the hepatic concentration of glycerol-1-P, there is no correlation between the rate of esterification and the concentration of this intermediate; and the mechanism of action of fructose on esterification of fatty acids is unclear. It is also still uncertain whether the rate of esterification influences directly the *de novo* synthesis of fatty acids in the liver.

In fed animals fructose is converted to fatty acids by the liver at a greater rate than glucose. The available evidence indicates that the metabolism of fructose does not alter the functional status of acetyl-CoA carboxylase, potentially a rate controlling enzyme for the synthesis of fatty acids, but increases the supply of substrate available for entry into the fatty acid synthetizing pathway. It seems that to a larger extent the rapid synthesis of fatty acids form fructose is a reflection of the high hepatic capacity for the metabolism of fructose in the Embden-Meyerhof pathway.

VII. References

1 Abraham, S.; Migliorini, R. H.; Bortz, W., and Chaikoff, I. L.: The relation of lipogenesis to reduced triphosphopyridine nucleotide generation and to certain enzyme activities in the liver of the 'totally' depancreatized rat. Biochim. biophys. Acta *62:* 27–34 (1962).

2 Adelman, R. C.; Spolter, P. D., and Weinhouse, S.: Dietary and hormonal regulation of enzymes of fructose metabolism in rat liver. J. biol. Chem. *241:* 5467–5472 (1966).

3 Ailhaud, G. P.; Vagelos, P. R., and Goldfine, H.: Involvement of acyl carrier protein in acylation of glycerol-3-phosphate in *Clostridium butyricum*. I. Purification of *Clostridium butyricum* acyl carrier protein and synthesis of long chain acyl derivations of acyl carrier protein. J. biol. Chem. *242:* 4459–4465 (1967).

4 Anderson, J. T.; Grande, F.; Matsumato, Y., and Keep, A.: Glucose, sucrose and lactose in the diet and blood lipids in man. J. Nutr. *79:* 349–359 (1963).

5 Baker, N.; Chaikoff, I. L., and Schusdek, J.: Effect of fructose on lipogenesis from lactate and acetate in diabetic liver. J. biol. Chem. *194:* 435–443 (1952).

6 BALLARD, F.J.; HANSON, R.W., and KRONFELD, D.S.: Gluconeogenesis and lipogenesis in tissue from ruminant and non-ruminant animals. Fed. Proc. *28:* 218–231 (1969).

7 BAR-ON, H. and STEIN, Y.: Effect of glucose and fructose administration on lipid metabolism in the rat. J. Nutr. *94:* 95–105 (1968).

8 BENDER, A. E.; DAINJI, K. B., and YAPA, C. G. R.: Effects of dietary sucrose on the metabolism *in vitro* of liver from rats of different strains. Biochem. J. *119:* 351–352 (1970).

9 BERGSTROM, J.; HULTMAN, E., and ROCH-NORLUND, A. E.: Lactic acid accumulation in connection with fructose infusion. Acta med. scand. *184:* 359–364 (1968).

10 BORTZ, W.; ABRAHAM, S., and CHAIKOFF, I. L.: Localization of the block in lipogenesis resulting from feeding fat. J. biol. Chem. *238:* 1266–1272 (1963).

11 BORTZ, W. M. and LYNEN, F.: Elevation of long chain acyl CoA derivatives in liver of fasted rats. Biochem. Z: *339:* 77–82 (1963).

11a BRUCKDORFER, K. R.; KHAN, I. H., and YUDKIN, J.: Dietary carbohydrate and fatty acid synthetase activity in rat liver and adipose tissue. Biochem J. *123:* 7 (1971).

12 BURCH, H. B.; LOWRY, O. H.; MEINHARDT, L.; MAX, P., jr., and CHYL, K.-J.: Effect of fructose, dihydroxyacetone, glycerol and glucose on metabolites and related compounds in liver and kidney. J. biol. Chem. *245:* 2092–2102 (1970).

13 CARROLL, C.: Influences of dietary carbohydrate-fat combinations on various enzymes associated with glycolysis and lipogenesis in rats. II. Glucose vs. fructose with corn oil and two hydrogenated oils. J. Nutr. *82:* 163–172 (1964).

14 DENTON, R. M. and HALPERIN, M. L.: The control of fatty acid and triglyceride synthesis in rat epididymal adipose tissue. Roles of coenzyme A derivatives, citrate and L-glycerol-3-phosphate. Biochem. J. *110:* 27–33 (1968).

15 EXTON, J. H. and EDSON, N. L.: The antiketogenic actions of sorbitol. Biochem. J. *91:* 478–483 (1964).

16 EXTON, J. H. and PARK, C. R.: Control of gluconeogenesis in liver. III. Effects of L-lactate, pyruvate, fructose, glucagon epinephene, and adenosine 3′5′-monophosphate on gluconeogenic intermediates in perfused rat liver. J. biol. Chem. *244:* 1424–1433 (1969).

17 FALLON, H. J. and KEMP, E. L.: Effects of diet on hepatic triglyceride synthesis. J. clin. Invest. *47:* 712–719 (1968).

18 FARSTAD, M.: *In vivo* and *in vitro* studies of the regulation of palmityl-CoA synthetase activity in rat liver. Acta physiol. scand. *74:* 568–576 (1968).

19 FITCH, W. M.; HILL, R., and CHAIKOFF, I. L.: Hepatic glycolytic enzyme activities in the alloxan diabetic rat. Response to glucose and fructose feeding. J. biol. Chem. *243:* 2811–2813 (1959).

20 FITCH, W. M. and CHAIKOFF, I. L.: Extent and patterns of adaptation of enzyme activities in liver of normal rats fed diets high in glucose and fructose. J. biol. Chem. *235:* 554–557 (1960).

21 FORMICA, J. V. and BRADY, R. O.: The enzymatic carboxylation of acetyl coenzyme A.: J. amer. Chem. Soc. *81:* 752 (1959).

22 FRITZ, I. B. and HSU, M. P.: Studies on the control of fatty acid synthesis. I. Stimulation by (+)-palmitylcarnitine of fatty acid synthesis in liver preparations from fed and fasted rats. J. biol. Chem. *242:* 865–872 (1967).

23 GALE, M. M. and CRAWFORD, M. A.: Different rate of incorporation of glucose and fructose into plasma and liver lipids in guinea pig. Metabolism *18:* 1021–1025 (1969).

24 GERICKE, C.; RAUSCHENBACH, P.; KUPKE, I. und LAMPRECHT, W.: Über die Biosynthese von Lipiden aus Fructose in Leber. III. Biosynthese des Glyceridglycerins aus [6-^{14}C] Fructose. Z. Physiol. Chem. *349:* 1055–1061 (1968).
25 GIBSON, D. M.; TITCHENER, E. B., and WAKIL, S. J.: Requirement for biocarbonate in fatty acid synthesis. J. amer. Chem. Soc. *80:* 2908 (1958).
26 GIBSON, D. M.; HICKS, S. E., and ALLMANN, D. W.: Adaptive enzyme formation during hyperlipogenesis. Adv. Enzyme Reg. *4:* 239–246 (1966).
27 GLOCK, G. E. and MCLEAN, P.: Preliminary investigation of the hormonal control of the hexose monophosphate exodative pathway. Biochem. J. *61:* 390–397 (1955).
28 GOLDFINE, H.; AILHAUD, G. P., and VAGELOS, P. R.: Involvement of acyl carrier protein in acylation of glycerol-3-phosphate in *Clostridium butyricum*. II. Evidence for the participation of acyl thioester of acyl carrier protein. J. biol. Chem. *242:* 4466–4475 (1967).
29 GOODRIDGE, A. G.: Regulation of lipogenesis. Stimulation of fatty acid synthesis *in vivo* and *in vitro* in the liver of the newly hatched chick. Biochem. J. *118:* 259–263 (1970).
30 GREENSPAN, M. D. and LOWENSTEIN, J. M.: Effects of magnesium ions, adenosine triphosphate, palmitylcarnitine, and palmityl coenzyme A on acetyl coenzyme A carboxylase. J. biol. Chem. *243:* 6273–6280 (1968).
31 GUCHHAIT, R. B.; PUTZ, G. R., and PORTER, J. W.: Synthesis of long chain fatty acids by microsomes of pigeon liver. Arch. Biochem. *117:* 541–549 (1966).
32 HANSON, R. W. and BALLARD, F. J.: The relative significance of acetate and glucose as precursors for lipid synthesis in liver and adipose tissue from ruminants. Biochem. J. *105:* 529–536 (1967).
33 HARADA, T.; OKUDA, H., and FUJII, J.: Effect of citrate on acyl-CoA synthetase of rat liver. J. Biochem. *65:* 497–501 (1969).
34 HEINZ, F.; LAMPRECHT, W., and KIRSCH, J.: Enzymes of fructose metabolism in human liver. J. clin. Invest. *47:* 1826–1832 (1968).
35 HERMAN, R. H.; ZAKIM, D., and STIFEL, F. B.:Effect of diet on lipid metabolism in experimental animals and man. Fed. Proc. *29:* 1302–1307 (1970).
36 HICKS, S. E.; ALLMANN, D. W., and GIBSON, D. M.: Inhibition of hyperlipogenesis with puromycin or actinomycin D. Biochim. biophys. Acta *106:* 441–444 (1965).
37 HILL, P.: Effect of fructose on rat lipids. Lipids *5:* 621–627 (1970).
38 HOWARD, C. F.: Synthesis of fatty acids in outer and inner membranes of mitochondria. J. biol. Chem. *245:* 462–468 (1970).
39 ILIFFE, J. and MYANT, N. B.: The sensitivity of acetyl-coenzyme A carboxylase to citrate stimulation in a homogenate of rat liver containing subcellular particles. Biochem. J. *117:* 385–395 (1970).
40 KORCHAK, H. M. and MASORO, E. J.: Changes in the level of fatty acid synthesizing enzymes during starvation. Biochim. biophys. Acta *58:* 354–356 (1965).
41 KORNACKER, M. S. and LOWENSTEIN, J. M.: Citrate and the conversion of carbohydrate into fat. Activities of citrate cleavage enzyme and acetate thiokinase in liver of normal and diabetic rats. Biochem. J. *95:* 832–837 (1965).
42 KORNBERG, A. and PRICER, W. E., jr.: Enzymatic esterification of α-glycerophosphate by long chain fatty acids. J. biol. Chem. *204:* 345–357 (1953).
43 KREBS, H. A. and HEMS, R.: Fatty acid metabolism in the perfused rat liver. Biochem. J. *119:* 525–523 (1970).

44 Kupke, I. und Lamprecht, W.: Ueber die Biosynthese von Lipiden aus Fructose in Leber. I. Einbau von uniform markierter [^{14}C] Fructose in Leberlipids. Z. Physiol. Chem. *348:* 17–26 (1967).

45 Langdon, R. G.: The biosynthesis of fatty acids in rat liver. J. biol. Chem. *226:* 615–629 (1957).

46 Lea, M. A. and Walker, D. G.: Factors affecting hepatic glycolysis and some changes that occur during development. Biochem. J. *94:* 655–665 (1965).

47 Lippel, K.: Regulation of rat liver acyl-CoA synthetase activity. Biochim. biophys. Acta *239:* 384–392 (1971).

48 Lynen, F.: Participation of acyl-coenzyme A (Co-A) in carbon chain biosynthesis. J. Cell. Comp. Physiol. *54:* suppl 1, pp. 33–49 (1959).

49 Lynen, F.; Matsuhashi, M.; Numa, S., and Schweizer, E.: The cellular control of fatty acid synthesis at the enzymatic level; in Grant The control of lipid metabolism. Biochemical Society Symp. 24, p. 43 (Academic Press, New York 1963).

50 Lynen, F.: The role of biotin-dependent carboxylations in biosynthetic reactions. Biochem. J. *102:* 381–399 (1967).

51 MacDonald, I. and Braithwaite, D. M.: The influence of dietary carbohydrates on the lipid pattern in serum and in adipose tissue. Clin. Sci. *27:* 23–30 (1964).

52 Majerus, P. W.; Alberts, A. W., and Vagelos, P. R.: The acyl carrier protein of fatty acid synthesis. Purification, physical properties, and substrate binding site. Proc. nat. Acad. Sci., Wash. *51:* 1231–1238 (1964).

53 Majerus, P. W. and Kilburn, E.: Acetyl coenzyme A carboxylase. The roles of synthesis and degradation in regulation of enzyme levels in rat liver. J. biol. Chem. *244:* 6254–6262 (1969).

54 Martin, D. B. and Vagelos, P. R.: The mechanism of tricarboxylic acid cycle regulation of fatty acid synthesis. J. biol. Chem. *237:* 1787–1792 (1962).

55 McGarry, J. D. and Foster, D. W.: The regulation of ketogenesis from oleic acid and the influence of antiketogenic agents. J. biol. Chem. *246:* 6247–6253 (1971).

56 Miller, M.; Drucker, W. R.; Owens, J. E.; Craig, J. W., and Woodward, H., jr.: Metabolism of intravenous fructose and glucose in normal and diabetic subjects. J. clin. Invest. *31:* 115–125 (1953).

57 Moss, J. and Lane, M. D.: The biotin-dependent enzymes. Adv. Enzymol. *35:* 321–442 (1971).

58 Nakaniski, and Numa, S.: Rate of synthesis and degradation of acetyl CoA carboxylase. Europ. J. Biochem. *16:* 161–173 (1970).

59 Nikkila, E. A. and Ojala, K.: Induction of hyerglyceridemia by fructose in the rat. Life Sci. *4:* 937–943 (1965).

60 Nugteren, D. H.: Enzymatic chain elongation of fatty acids by rat liver microsomes. Biochim. biophys. Acta *106:* 280–290 (1965).

61 Numa, S.; Matsuhashi, M. und Lynen, F.: Zur Störung der Fettsäuresynthese bei Hunger und Alloxan-diabetes. I. Fettsäuresynthese in der Leber normaler und hungernder Ratten. Biochem. Z. *334:* 203–217 (1961).

62 Ontko, J.: Metabolism of free fatty acids in isolated liver cells. Factors affecting the partition between esterification and oxidation. J. biol. Chem. *247:* 1788–1800 (1972).

63 Patterson, D. S. P.: The formation of reduced nicotinamide-adenine dinucleotide

phosphate in relation to the impaired hepatic lipogenesis of the rat adapted to a high fat diet. Biochim. biophys. Acta *84:* 198–200 (1964).

64 PEREIRA, J. N. and JANGAARD, N. O.: Different rates of glucose and fructose metabolism in rat liver tissue *in vitro*. Metabolism *20:* 392–400 (1971).

65 PORTER, J. W.; WAKIL, S. J.; TIETZ, A.; JACOB, M. J., and GIBSON, D. M.: Studies on the mechanism of fatty acid synthesis. II. Cofactor requirements of the soluble pigeon liver system. Biochim. biophys. Acta *25:* 35–41 (1952).

66 SCHERSTEN, T.; NILSSNO, S. V.; CAHLIN, E., and JILDEROS, B.: Synthesis of phospholipids and triglycerides in human liver slices. II. Influence of preoperative fructose load. Scand. J. clin. Lab. Invest. *26:* 399–406 (1970).

67 SHAFRIR, E. and KERPEL, S.: Fatty acid esterification and release as related to the carbohydrate metabolism of adipose tissue. Effect of epinephrine, cortisol, and adrenalectomy. Arch. Biochem. *105:* 237–246 (1964).

68 SHRAGO, E.; SPENNETTA, T., and GORDON, E.: Fatty acid synthesis in human adipose tissue. J. biol. Chem. *244:* 2761–2766 (1969).

69 SILLERO, M. A. G.; SILLERO, A., and SOLS, A.: Enzymes involved in fructose metabolism in liver and the glyceraldehyde metabolic crossroads. Europ. J. Biochem. *10:* 345–350 (1969).

70 SIPERSTEIN, M. D. and FAGAN, V. M.: Role of glycolysis in fatty acid and cholesterol synthesis in normal and diabetic rats. Science *126:* 1012–1013 (1957).

71 SMITH, L. H., jr.; ETTINGER, R. H., and SELIGSON, D.: A comparison of the metabolism of fructose and glucose in hepatic disease and diabetes millitus. J. clin. Invest. *32:* 273–282 (1953).

72 SRERE, P. A.: Enzyme concentrations in tissues. Science *158:* 936–937 (1967).

73 SWANSON, R. F.; CURRY, W. M., and ANKER, H. S.: The activation of acetyl-CoA carboxylase by incubation. Biochim. biophys. Acta *159:* 390–397 (1968).

74 SZEPESI, B. and FREEDLAND, R. A.: Alterations in the activities of several rat liver enzymes at various times after initiation of a high protein regimen. J. Nutr. *93:* 301–306 (1967).

75 SZEPESI, B. and FREEDLAND, R. A.: Alterations in the activities of several rat liver enzymes at various times after the feeding of high carbohydrate diets to rats previously adapted to a high protein regimen. J. Nutr. *94:* 37–46 (1968).

76 TAKEDA, Y.; INOUE, H.; HONJO, K.; TANIOKA, H., and DAIKUHARA, Y.: Dietary response of various key enzymes related to glucose metabolism in normal and diabetic rat liver. Biochim. biophys. Acta *136:* 214–222 (1967).

77 TAKETA, K. and POGELL, B. M.: The effect of palmityl coenzyme A on glucose-6-phosphate dehydrogenase and other enzymes. J. biol. Chem. *241:* 720–726 (1966).

78 TOPPING, D. L. and MAYES, P. A.: The immediate effects of insulin and fructose on the metabolism of the perfused liver. Changes in lipporotein secretion, fatty acid oxidation and esterification, lipogenesis and carbohydrate metabolism. Biochem. J. *126:* 295–311 (1972).

79 WAKIL, S. J.; Porter, J. W., and GIBSON, D. M.: Studies on the mechanism of fatty acid synthesis. I. Preparation and purification of an enzyme system for reconstruction of fatty acid synthesis. Biochim. biophys. Acta *24:* 453–461 (1957).

80 WAKIL, S. J.: A malonic acid derivative as an intermediate in fatty acid synthesis. J. amer. chem. Soc. *80:* 6465 (1958).

81 WAKIL, S. J.; TITCHENER, E. B., and GIBSON, D. M.: Evidence for the participation of biotin in the enzymic synthesis of fatty acids. Biochim. biophys. Acta *29:* 225–226 (1958).

82 WAKIL, S. J.; PUGH, E. L., and SAUER, F.: The mechanism of fatty acid synthesis. Proc. nat. Acad. Sci., Wash. *52:* 106–114 (1964).

83 WELBOURNE, W. P.; SWANSON, R. F., and ANKER, H. S.: Active acetyl-CoA carboxylase from liver of rats fed carbohydrate. Biochem. biophys. Res. Commun. *37:* 933–937 (1969).

84 WIELAND, O. und MATSCHINKY, F.: Zur Natur der antiketegenen Wirkungen von Glycerin und Fructose. Life Sci. *2:* 49–54 (1962).

85 WILLECKE, K.; RITTER, E., and LYNEN, F.: Isolation of an acyl carrier protein component from the multienzyme complex of yeast fatty acid synthetase. Europ. J. Biochem. *8:* 503–509 (1969).

86 WILLIAMSON, D. H.; VELOSO, D.; ELLINGTON, E. V., and KREBS, H. A.: Changes in the concentrations of hepatic metabolites on administration of dihydroxyacetone or glycerol to starved rats and their relationships to the control of ketogenesis. Biochem. J. *114:* 575–584 (1969).

87 WOODS, H. F.; EGGLESTON, L. V., and KREBS, H. A.: The cause of hepatic accumulation of fructose-1-phosphate on fructose loading. Biochem. J. *119:* 501–510 (1970).

88 WOODS, H. F. and KREBS, H. A.: Lactate production in the perfused rat liver. Biochem. J. *125:* 129–139 (1971).

89 YEH, Y. Y. and LEVEILLE, G. A.: *In vitro* and *in vivo* restoration of hepatic lipogenesis in fasted chicks. J. Nutr. *101:* 803–810 (1971).

90 ZAKIM, D.: Effect of ethanol on hepatic acyl-coenzyme A metabolism. Arch. Biochem. *111:* 253–256 (1965).

91 ZAKIM, D.: The effect of fructose on the hepatic synthesis of fatty acids. Acta med. scand. (in press).

92 ZAKIM, D.; PARDINI, R.; HERMAN, R. H., and SAUBERLICH, H.: The relation of hepatic α-glycerophosphate concentration to lipogenesis in rat liver. Biochim. biophys. Acta *137:* 179–180 (1967).

93 ZAKIM, D.; PARDINI, R. S.; HERMAN, R. H., and SAUBERLICH, H. E.: Mechanism for the differential effects of high carbohydrate diets on lipogenesis in rat liver. Biochim. biophys. Acta *144:* 242–251 (1967).

94 ZAKIM, D. and HERMAN, R. H.: The effect of intravenous fructose and glucose on the hepatic α-glycerophosphate concentration in the rat. Biochim. biophys. Acta *165:* 374–379 (1968).

95 ZAKIM, D. and HERMAN, R. H.: Fructose metabolism. II. Regulatory control to the triose level. Amer. J. clin. Nutr. *21:* 315–319 (1968).

96 ZAKIM, D. and HERMAN, R. H.: Regulation of fatty acid synthesis. Amer. J. clin. Nutr. *22:* 200–213 (1969).

97 ZAKIM, D.; HERMAN, R. H., and GORDON, W. C., jr.: The conversion of glucose and fructose to fatty acids in the human liver. Biochem. Med. *2:* 427–437 (1970).

98 ZAKIM, D. and HO, W.: The acetyl-CoA carboxylase and fatty acid synthetase activities of rat intestinal mucosa. Biochim. biophys. Acta *222:* 558–559 (1970).

Author's address: Dr. DAVID ZAKIM, Department of Medicine, University of California Medical Center, *San Francisco, CA 94122* (USA)

Progr. biochem. Pharmacol., vol. 8, pp. 189–215 (Karger, Basel 1973)

Dietary Carbohydrates and Adipose Tissue Metabolism

A. VRÁNA and P. FÁBRY

Research Centre of Metabolism and Nutrition, Institute for Clinical and Experimental Medicine, Prague

Contents

I. Homeostatic Rôle of Adipose Tissue

Cells of the animal organism depend essentially on the continuous supply of energy yielding substances. The ability of some specialized tissues to form, deposit and mobilize caloric reserves, however, enables the organism not only to overcome minor deviations in the food supply but also to survive prolonged periods of food deprivation. Moreover, some tissues are able to adapt to changes in the composition of the substrate mixture supplied in the diet or from reserves in such a way that the substrate which is available in adequate amount is used preferentially for energy production. This applies in particular to the alternative utilization of glucose and free fatty acids (FFA) which can replace each other to a considerable extent [98] as sources of energy. Glucose and FFA are directly available energy sources; with regard to the rate of their utilization in tissues and the size of their pool they are, however, a very short-term reserve. Their continuous uptake in tissues is balanced under normal conditions by the supply of the required substances in the diet, usually in a more complex form, and by the supply from bodily reserves or, in the case of glucose, by gluconeogenesis from non-

Table I. Quantitative importance of different body components as sources of energy in man (70 kg). Modified and re-calculated according to CAHILL [17]

	kcal	Period for which basal metabolic need of 72 kcal/h is met
Glucose (extracellular fluid)	80	1 h
Plasma free fatty acids (FFA)	3	2.5 min
Plasma triglycerides	30	25 min
Triglycerides of adipose tissue	141,000	81 days
Protein (mainly muscle)	24,000	14 days
Glycogen (muscle)	600	8 h
Glycogen (liver)	300	4 h

carbohydrate precursors. Tissue reserves of carbohydrates which are represented mainly by liver and muscle glycogen also suffice to meet the caloric requirements for only a relatively short time. From the quantitative aspect, by far the most important calorie reserve is adipose tissue triglyceride. The relative importance of different bodily components as the energy source of man (70 kg) is apparent from table I. Similar quantitative relations of different calorie reserves were described also in the laboratory rat [3].

Physiological and biochemical mechanisms associated with the formation, liberation and utilization of lipid reserves were intensely investigated in recent decades and reviewed in considerable detail [62, 66, 102]. By hydrolysis of adipose tissue triglycerides, catalyzed by 'hormone-sensitive' lipase, FFA and glycerol are formed. FFA are transported by the blood stream bound to albumin and are taken up and utilized by the liver and other tissues, in particular by skeletal muscle and the heart muscle. Their uptake by tissues is a function of their blood concentration. Adipose tissue triglyceride hydrolysis, the release of FFA and their utilization in tissues are accentuated under conditions when the glucose utilization in tissues is reduced either as a result of its shortage (fasting) or as a result of the reduced ability of tissues to utilize glucose which is present in adequate or even excessive amounts (diabetes). These are situations when adipose tissue serves as the main donor of energy substrate. On the other hand, when the energy supply exceeds its consumption in tissues (food supply) and the need arises to accumulate energy, adipose tissue acts as an acceptor. It takes up preformed fatty acids supplied by the diet or formed in the liver, it forms fatty acids and esterifies them. The relationship between anabolic and catabolic processes in adipose tissue is markedly influenced by nutritional factors, either directly by substrates or by means of hormonal and nervous regulatory mechanisms [45, 55, 61].

Adipose tissue plays a part in the regulation of caloric homeostasis not only by the uptake, deposition and release of fatty acids but also by their formation from carbohydrates and their metabolites. The part played by adipose tissue and the liver in these processes differs in different species. In some, e.g. in birds [84], the liver is almost the sole site of *de novo* synthesis of fatty acids. In rats and mice adipose tissue is considered an important site of this process [35, 65, 82, 83], although a certain portion of fatty acids in adipose tissue triglycerides, in particular fatty acids 18:0 and 18:1, are formed in the liver [21].

As far as human adipose tissue is concerned, a low synthesis of fatty acids from carbohydrate has been repeatedly demonstrated [110]. Adipose tissue of adults has a low sensitivity to insulin [72, 48] and does not respond by marked changes in lipogenesis to nutritional stimuli such as the composition of the diet [57] or fasting and re-feeding [110]. It seems that in the anabolic stage of energy homeostasis in man, adipose tissue participates by taking up and esterifying fatty acids transported as lipoproteins rather than by *de novo* synthesis of fatty acids.

II. Carbohydrate Metabolism in Adipose Tissue

The utilization of free glycerol which is formed by the breakdown of triglycerides is, due to the lack or very low activity of glycerokinase in adipose tissue, very low. Therefore, the availability and actual utilization of carbohydrates in adipose tissue are factors of primary importance in the regulation of the triglyceride turnover in this tissue, since they are the main supplier of suitable materials from which α-glycerophosphate can be formed [112, 118]. After entering the adipocyte, glucose is phosphorylated by hexokinase and the formed glucose-6-phosphate can be metabolized either in the glycogen cycle, in the Embden-Meyerhoff pathway or in the pentose cycle. More recent investigations indicate that 1–5% of the glucose taken up are transformed into glycogen, 10–20% are metabolized in the pentose cycle and the remaining major portion enters the Embden-Meyerhoff pathway [36, 79].

This ratio is markedly influenced by hormones, e.g. by insulin and epinephrine [80]. The fate of glucose in adipose tissue incubated *in vitro* in a

Table II. Effect of insulin on U-^{14}C-fructose and U-^{14}C-glucose metabolism of rat epididymal adipose tissue. Data from FROESCH and GINSBERG [37]

	Hexose uptake, μmoles sugar/g wet tissue/3 h	μmoles sugar carbon/g wet tissue/3 h		
		CO_2	total lipids	fatty acids
U-^{14}C-glucose, 200 mg/100 ml				
No insulin added	16.86±0.20	2.20±0.14	9.60–0.16	5.91±0.03
Insulin, 1,000 μU/ml	52.95±1.80	6.80±0.48	36.23±3.42	30.31±2.95
U-^{14}C-fructose, 200 mg/ml				
No insulin added	7.79±0.57	1.43±0.06	3.79±0.03	1.92±0.14
Insulin, 1,000 μU/ml	15.11±0.38	3.01±0.17	7.10±0.64	4.65±0.27

medium without addition of hormones is apparent from the findings of CAHILL *et al.* [16]. The incorporation of universally labelled ^{14}C-glucose (5 mmol) into different products in their experiments was as follows (in per cent of the total amount): glycogen, 1; CO_2, 42; fatty acids (mainly in triglycerides), 13; glycerol (mainly in triglycerides), 44.

Adipose tissue is able to utilize, in addition to glucose, other hexoses to a different extent. Fructose utilization in rat adipose tissue is quantitatively approximately equal to glucose utilization. As compared with glucose, however, the transport and utilization of fructose in adipose tissue are more stimulated by an increase of the extracellular concentration of hexose and less by insulin [37, 38]. Mannose utilization in adipose tissue is somewhat lower than glucose utilization and is also stimulated by insulin [44, 133]. Galactose transport into adipose tissue is also stimulated by insulin, its subsequent metabolism is, when compared with glucose, approximately two orders lower [44]. The ratio of utilization of the above hexoses in adipose tissue is apparent from tables II and III.

In addition to insulin, hexose uptake and its further metabolism in adipose tissue are stimulated by a number of other hormones, e.g. by epinephrine, norepinephrine, prolactin, oxytocin [130]. Some other organic and inorganic substances are also effective [3]. Insulin, unlike the above substances, is, however, effective in physiological concentrations [6, 43].

Part of the glucose [107], fructose [38] and mannose [133] taken up by adipose tissue is transformed into glycogen. Insulin enhances the glycogen synthesis in adipose tissue by stimulating hexose transport and by activation

Table III. Effects of insulin on the metabolism of various U-^{14}C-hexoses in rat epididymal adipose tissue. Data from GOODMAN [44]

	Microgram hexose carbon/gram fresh tissue/hour		
	taken up	converted to fatty acids	oxidized to CO_2
Glucose, 1 mg/ml			
No insulin added	87.2±32	29.8 ± 5.7	35.4 ± 4.3
Insulin, 500 μU/ml	686.0±36	310.0 ±16.0	248.0 ±11.3
Mannose, 1 mg/ml			
No insulin added	34.9±21.6	10.4 ± 1.9	19.4 ± 2.6
Insulin, 500 μU/ml	281.0±23.6	138.0 ±16.0	122.0 ±11.0
Fructose, 1 mg/ml			
No insulin added	80.4±14.1	10.3 ± 3.4	11.9 ± 2.1
Insulin, 500 μU/ml	230.0±22.8	36.4 ± 6.1	51.2 ± 4.7
Galactose, 1 mg/ml			
No insulin added	18.2± 3.4	0.22± 0.02	0.61± 0.20
Insulin, 500 μU/ml	39.9± 7.3	0.19± 0.02	0.76± 0.15

of uridine diphosphoglucose (UDPG)-glycogen transglukosylase]129], an enzyme catalyzing the transfer of the glucose unit from UDPG to glycogen. Glycogen in adipose tissue can be locally used for the synthesis of fatty acids and α-glycero-phosphate for their esterification [70, 107, 118]. As glucose-6-phosphatase and fructose-1, 6-diphosphatase are absent in adipose tissue [124], glycogenolysis in adipose tissue does not lead to the formation of free glucose.

Fatty acid synthesis in adipose tissue from various precursors, e.g. from hexose, pyruvate, lactate, acetate and amino acids, has been shown. In addition to activities of enzymes involved directly in the fatty acid synthesis such as acetyl-CoA carboxylase and fatty acid synthetase, in the regulation of fatty acid synthesis other factors also play an important part, such as the formation of reducing equivalents, the concentration of α-glycerophosphate and long-chain fatty acyl-CoA and the substrate concentration [4, 42, 52, 89, 110, 134].

A reduced fatty acid synthesis in adipose tissue was repeatedly found during fasting [53, 63, 64], in diabetes [131], after the ingestion of a high-fat diet [53, 65, 78] and in old-age [41]. Re-feeding with a high-carbohydrate diet [1, 63, 64], its long-term administration [53, 63, 65], infrequent feeding of a diet adequate in carbohydrates [32, 85] or the administration of insulin

[54] markedly promote fatty acid synthesis in adipose tissue (see part III of this chapter).

Recent work indicates that insulin acts in the regulation of fatty acid synthesis in adipose tissue both by stimulating the uptake of hexoses and by a more direct mechanism – an increase of the pyruvate conversion to acetyl-CoA [51, 71]. The latter mechanism involves stimulation of the pyruvate dehydrogenase activity by insulin [25, 70, 71] as a result of the conversion of the inactive form of the enzyme into the active form [125].

Insulin and glucose enhance triglyceride synthesis from labelled fatty acid in adipose tissue [101]. In the mechanism obviously the increased production of α-glycerophosphate and increased production of adenosine triphosphate (ATP) plays a part [118]. Fasting reduces the uptake of fatty acids by adipose tissue and their esterification [27, 109]. During fasting, in addition to the reduced availability of α-glycerophosphate, obviously the increased intracellular FFA concentration plays a part which, *per se*, can enhance the esterification process [118] but can, at the same time, exert an inhibitory action by inhibiting the transfer of long-chain acyl groups to α-glycerophosphate [40].

The uptake in adipose tissue, of fatty acid from triglycerides of very low density lipoproteins and chylomicrons is controlled by lipoprotein lipase activity [8, 61]. The activity of this enzyme, which correlates well with the uptake of triglyceride fatty acids by adipose tissue [8], is reduced during fasting [59] and in diabetes [77] and is increased after ingestion of food and administration of glucose or insulin [60, 61]. Glucose and insulin are effective also *in vitro;* when added to adipose tissue of fasted rats, the lipoprotein lipase activity increases [105, 132] and the uptake of triglycerides is enhanced [2]. *In vitro* glucose can be replaced by fructose [132]. *In vivo,* however, the administration of fructose, contrary to glucose, did not lead in BAR-ON's and STEIN's experiments [5] to an increased lipoprotein lipase activity in adipose tissue. The interpretation of this phenomenon is not clear.

The transport form, in which fat is mobilized from adipose tissue, is FFA which are formed by hydrolysis of adipose tissue triglycerides; this process is influenced in a marked way by hormonal and nutritional factors [28, 46, 47, 126, 127]. Stimulation of lipolysis is the result of activation of 'hormone-sensitive' lipase by cyclic 3', 5'-adenosine monophosphate [15, 103]. In addition to lipolysis, the release of FFA from adipose tissue is influenced also by the rate of re-esterification which, in turn, depends on the availability of α-glycerophosphate formed by glycolysis [112]. Insulin reduces lipolysis in the presence and absence of glucose [20, 34, 68] or fructose [39] and reduces the concentration of cyclic nucleotide in adipose tissue [14]. Findings pertaining to the effect of insulin on enzymes involved in the formation and degradation of nucleotide in adipose tissue are not uniform [56, 86].

III. Effects of Carbohydrate Ingestion on Adipose Tissue Metabolism

Under physiological conditions, dietary carbohydrates exert a marked effect on basic metabolic processes in adipose tissue. This can be readily demonstrated on metabolic and enzymatic changes which are induced by fasting and subsequent re-feeding with a carbohydrate-containing diet or adaptation to diets with a different carbohydrate ratio.

Fasting reduces the glucose uptake by adipose tissue and its further metabolism [53, 63]. The activities of enzymes involved already in the initial steps of glucose utilization are reduced, i.e. hexokinase, phosphofructokinase and aldolase activities [108]. The blood insulin concentration is reduced during fasting [114] and the sensitivity of adipose tissue to its action is also reduced [93, 117]. By a complex mechanism in which insulin deficiency, reduced glucose utilization in adipose tissue and activation of humoral and nervous mechanisms play a part, lipolysis and FFA release from adipose tissue is activated [29, 45, 98]. At the same time synthesis, uptake and esterification of fatty acids are reduced [27, 55, 109]. As a result of these changes the flow of fatty acids is directed to other tissues which may use them readily as a source of energy. The reduced glucose utilization in adipose tissue spares glucose, the limited resources of which can be used preferentially by tissues for the metabolism of which they are indispensible [17, 98]. During fasting other synthetic processes in adipose tissue decline, such as ribonucleic acid and protein synthesis [73].

Re-feeding precipitates changes in adipose tissue which can be characterized as a general shift from catabolic to anabolic processes. Insulin secretion [114] and adipose tissue sensitivity to its action [69, 93] rise to or above the pre-fasting level. The activity of a number of adipose tissue enzymes catalyzing various steps of carbohydrate metabolism is increased (table IV). Already during the initial hours of re-feeding the activity of hexokinase and other glycolytic enzymes increases [108]. The hexokinase activity increases before the aldolase activity. This renders it possible in the initial stage of re-feeding, when fatty acid synthesis is not yet restored, to direct glucose-6-phosphate to glycogen synthesis which is, moreover, facilitated by the fact that glycogen formation is less inhibited by fasting

Table IV. Enzymes, the activities of which are increased in the adipose tissue of rats re-fed a high-carbohydrate, low-fat diet

Enzyme	Reference No.
Hexokinase	19, 108
Phosphoglucomutase	19
Phosphofructokinase	108
Aldolase	19, 104
Transketolase	49
Triose phosphate isomerase	104
Phosphoglycerate kinase	104
UDPG-glycogen transferase	50
NADP malate dehydrogenase	33, 67, 110
Citrate cleavage enzyme	67, 110

than is fatty acid synthesis. The glycogen concentration in adipose tissue reaches high levels between 24 and 48 h of re-feeding, then it declines as the glycogen is used as a precursor for fatty acid and α-glycerophosphate synthesis [70, 107]. Re-feeding a high-carbohydrate diet increases in a marked way fatty acid synthesis [1, 63–65]. It may be assumed that accentuation of these processes is associated with the *de novo* synthesis of enzymes. This is suggested by the chronological parallelism of changes in protein synthesis [73] and lipogenesis [1], as well as by the findings of JOMAIN and HANSON [67] who demonstrated that dietary protein is essential for the increase of enzyme activities in adipose tissue during re-feeding.

Accentuation of anabolic processes in adipose tissue during re-feeding is not the consequence of simple caloric repletion but is proportional to the ratio of dietary carbohydrate. A high-fat, low-carbohydrate or carbohydrate-free diet prevents the manifestation of the above-mentioned changes [33, 63]. As carbohydrates are not only an energy substrate and precursor for fatty synthesis but also the main physiological stimulus for insulin secretion [90], the blocking effect of the high-fat diet may be either the manifestation of substrate deficiency or insulin deficiency, or of both. Therefore, FÁBRY *et al.* [33] investigated the effect of the composition of the diet and of exogenous insulin on the activity of NADPH-generating enzymes in adipose tissue of rats re-fed after previous fasting (table V). They found that the activity of malic enzyme and of pentose cycle dehydrogenases declined during fasting as was expected. Re-feeding a high-carbohydrate diet increased the malic

Table V. Influence of diet and of exogenous insulin on activities of NADPH generating enzymes in rat adipose tissue. Data from FÁBRY *et al.* [33]

Group	Malic enzyme activity[1]	Pentose shunt enzymes activity
Controls	2.09±0.17	5.51±1.14
Fasted 72 h	1.47±0.15	3.32±0.56
Re-fed high-carbohydrate diet	11.59±1.03	9.16±1.14
Re-fed high-carbohydrate diet plus insulin	13.26±0.39	10.02±1.66
Re-fed high-fat diet	2.69±0.40	6.10±0.97
Re-fed high-fat diet plus insulin	3.18±0.47	5.84±0.63

1 Units/mg protein × 100. One unit of enzyme was defined as the amount of enzyme preparation which reduces 1.0 μmol of NADP/min at 25 °C.

enzyme activity 5 times and that of pentose cycle dehydrogenases to twice the value recorded in fed animals.

Exogenous insulin did not influence in any instance the activity of the investigated enzymes in a marked way and it is thus probable that the availability of carbohydrates or their metabolites is essential for the characteristic response of adipose tissue enzymes to re-feeding. In the above experiments of FÁBRY *et al.* [33], in addition to enzyme determinations, histological examinations of adipose tissue were made. Fasting resulted in a decrease in nucleolar and cytoplasmic ribonucleic acid (RNA) staining and in a reduction of the mean fat cell size. Re-feeding the high-carbohydrate diet induced a marked enlargement of nucleoli and increased cytoplasmic RNA staining with essentially no changes in mean cell size. Re-feeding the high-fat, carbohydrate-free diet produced a striking hypertrophy of fat cells, whereas the nucleolar and cytoplasmic RNA staining was comparable to the pre-fast state. Exogenous insulin did not produce a marked difference in either of the two diet groups.

Experiments of KAZDOVÁ *et al.* [75] revealed that re-feeding of rats with a standard or high-carbohydrate diet resulted in an increase of RNA and desoxyribonucleic acid (DNA) content and an enhanced ^{14}C-2-thymidine incorporatioin to adipose tissue DNA. Re-feeding with a high-fat, carbohydrate-free diet did not cause any significant change in comparison with the pre-fasting values.

Table VI. Influence of the content of fat in the diet on metabolic and enzymatic activities in rat epididymal adipose tissue *in vitro*. Data from LEVEILLE [83]

Investigated parameter	Content of fat in the diet, weight %		
	10	20	30
Conversion of U-^{14}C glucose to			
CO_2[1]	559±81	188± 7	165±24
Fatty acids[1]	455±79	75± 9	45±10
Glyceride-glycerol[1]	208±10	122± 9	109±11
Glycogen[1]	25± 3	15± 1	1± 3
Conversion of 1-^{14}C-acetate to			
CO_2[1]	245±12	232± 1	287±18
Fatty acids[1]	438±28	195±21	143±28
Glucose-6-phosphate dehydrogenase[2]	0.334±0.03	0.207±0.01	0.188±0.02
6-Phosphogluconate dehydrogenase[2]	0.175±0.08	0.139±0.08	0.132±0.06
Malic enzyme[2]	0.316±0.02	0.120±0.01	0.093±0.007

1 μmoles substrate incorporated/100 mg tissue/3 h.
2 μmoles substrate utilized/min/mg N.

The results of these studies indicate that there are essentially two different types of adipose tissue response to re-feeding. Re-feeding with a carbohydrate diet results in an elaborate re-arrangement of the biochemical cell systems to perform the function which has been designated as 'adaptive hyperlipogenesis' [116]. In addition, the adipose tissue of such fed rats becomes more cellular as the restoration of its weight is connected with adipose tissue hyperplasia. In contrast, the fat-fed animals exhibit a marked fat cell hypertrophy without the overshoot of enzymatic activities typical for the carbohydrate re-fed animals [10, 74].

Chronic ingestion of a diet with a high-carbohydrate and low-fat ratio causes similar, although less dramatic, changes of metabolic activity of adipose tissue than their ingestion during a single period of re-feeding. The ingestion of a high-carbohydrate diet increases markedly the glucose uptake by adipose tissue and its further metabolism. At the same time, also, the adipose tissue sensitivity to the effect of insulin is increased [53, 78]. A high-carbohydrate diet has a particularly marked effect on fatty acid synthesis in adipose tissue [53, 63, 64, 81–83]. Changes of metabolic activity of adipose tissue are also in keeping with changes in the activities of some enzymes [82] (table VI). As apparent from the table, the depression of metabolic activity and reduction of activities of the investigated enzymes in adipose tissue is proportional to the ratio of dietary fat. Deficiency of carbohydrates and their replacement by fat has the most profound effect on fatty acid synthesis from glucose, which declines by one order when the ratio of dietary fat is increased from 10 to 30% (and carbohydrates are correspondingly reduced).

In addition, although to a smaller extent, glucose oxidation, glycerol and glycogen formation from glucose and acetate oxidation are reduced as well as the formation of fatty acids from acetate. The activity of NADPH generating enzymes declines also, the decline of malic enzyme activity correlating best with the diminishing lipogenesis.

Diets with a low-carbohydrate and high-protein ratio lead to a less marked inhibition of lipogenesis and associated enzymatic changes than high-fat diets [82]. This can be explained by the fact that the specific inhibitory effect of fat on lipogenesis which probably is inherent to fat as such is eliminated [9, 53, 128], as well as by the fact that glucogenic amino acids are a rich source for glucose formation by gluconeogenesis.

In certain situations the administration of exogenous insulin leads to changes in adipose tissue similar to the ingestion of carbohydrates. This is obviously conditioned, in addition to specific effects of insulin on intracellular systems, by an increased transport of the particular monosaccharide into the fat cell [51, 54, 71, 76, 125].

IV. Effects of the Type of Dietary Carbohydrate on Adipose Tissue Metabolism

In other chapters of this volume ample evidence is provided that, in addition to the dietary carbohydrate ratio, the type of carbohydrate may influence in a marked way the metabolism of experimental animals and man. In recent years special attention has been paid in particular to the effects of sucrose and fructose respectively, which, as some work indicates, enhance lipid synthesis and content in the liver and the blood lipid concentration, as compared with diets containing the same caloric volume of starch and glucose, respectively [58, 87, 95].

As far as some other effects of sucrose are concerned, e.g. on glucose tolerance, the results of different laboratories are in many respects controversial. COHEN and TEITELBAUM [22] and COHEN [24] found in rats and man that the ingestion of a high-sucrose diet leads to a deteriorated glucose tolerance; other workers [30], however, did not confirm these findings in

Table VII. Effect of dietary sucrose and of the length of feeding period on the serum insulin concentrations in the rat. Data from VRÁNA *et. al.* [122]

Type of diet	Feeding period days	Serum insulin, μU/ml	
		fed rats	rats fasted 18 h
Experiment I			
Starch	50	21.0±2.0	13.5±1.0
Starch plus sucrose[1]	50	26.5±2.4	10.3±1.5
Sucrose	50	23.4±2.0	10.8±1.1
Experiment II			
Starch	60	22.1±3.0	–
Sucrose	10	23.5±3.0	–
Sucrose	30	22.7±2.0	–
Sucrose	60	24.3±2.0	–

Mean values±SEM (n=6–7). Differences between diets are not significant.
1 Diet containing a 1:1 mixture of starch and sucrose.

man. The possible rôle of insulin is not clear either; COHEN and TEITELBAUM [22] consider the low serum insulin-like activity to be the cause of the deteriorated glucose tolerance in rats fed a sucrose diet. SZANTO and YUDKIN [113] investigated insulinaemia after a glucose load in subjects previously ingesting a high-sucrose diet and recorded a higher degree of hyperinsulinaemia in one-third of the investigated subjects.

Adipose tissue is one of the important physiological consumers of glucose and could, therefore, participate in metabolic changes induced by the type of ingested carbohydrates. The authors of this chapter started, some years ago, experiments focused on this problem. At that time, in the literature, data from investigations concerned with the effect of diets with different types of dietary carbohydrates on the metabolic process in adipose tissue were lacking. Information on hexose utilization in adipose tissue in relation to the type of carbohydrate was needed, however; among other reasons for the interpretation of the impaired glucose tolerance in sucrose-fed rats which, in keeping with COHEN and TEITELBAUM [22], was found also in our laboratory. This effect of a sucrose diet is well illustrated in figure 1. It is obvious that in rats fed for 8 weeks a diet with a 70% caloric volume of sucrose, the hyperglycaemia after a glucose load is more marked and

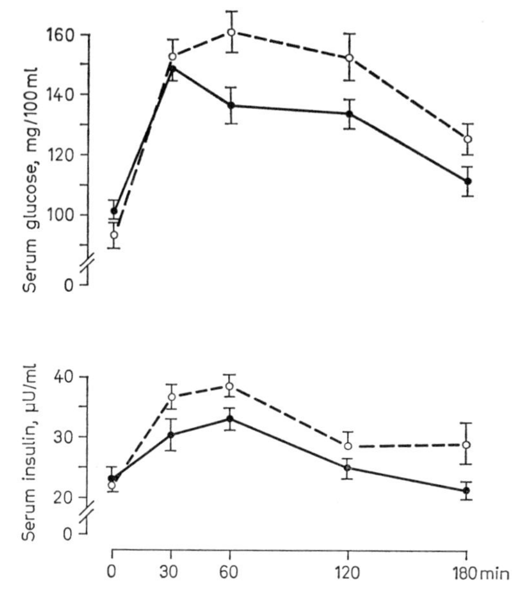

Fig. 1. Concentrations of serum glucose and immunoreactive insulin after oral glucose load in rats previously fed diets with a high content of sucrose (○ – – – ○) or starch (●——●). Glucose (3.5 g/kg) was administered by means of a gavage to animals fasting 18 h. Animals were sacrificed by decapitation at given intervals in groups comprising 6 rats each [VRÁNA, SLABOCHOVÁ and FÁBRY, unpublished data].

persists longer than in rats fed a diet with a corresponding proportion of starch[1]. Simultaneously assessed serum concentrations of immunoreactive insulin in the sucrose-fed group were raised in proportion to the higher serum sugar levels. Therefore, it is not likely that the reduced glucose tolerance is due to an inadequate insulin secretion as a manifestation of β-cell exhaustion after a previous period of sucrose-induced hyperinsulinaemia. This is suggested also by the investigation of basal values of insulinaemia at different intervals of experimental feeding (10–60 days). At none of the investigated intervals was a significant difference found between groups fed the sucrose or starch diets (table VII). It could thus be assumed that the impaired glucose tolerance induced by the sucrose diet is rather the con-

1 The composition of diets used in this and following experiments was as follows: protein, 20 cal %; fat, 10 cal %; carbohydrates (starch, sucrose, glucose, fructose, respectively), 70 cal%. Supplementation of diets by mineral and vitamin mixtures [31] was the same for all groups.

Table VIII. Insulin sensitivity of adipose tissue *in vitro*. Incorporation of ^{14}C-U glucose into total lipids of adipose tissue. Data from VRÁNA and KAZDOVÁ (121]

Type of diet	cpm/mg protein/2 h		
	0[1]	250	1,000
Fed rats			
Starch diet	4.951 ± 524[2]	7.876 ± 1.445[3]	8.625 ± 1.574[4]
Sucrose diet	3.635 ± 356	4.205 ± 239	4.216 ± 289
Rats fasted 18 h			
Starch diet	2.321 ± 187	4.101 ± 653[3]	4.863 ± 927[3]
Sucrose diet	2.025 ± 66	2.390 ± 272	2.575 ± 149

cpm = Counts per minute.
1 Concentration of insulin in incubation medium, μU/ml.
2 Mean values ± SEM (n = 8).
Symbols for statistical significance (starch vs. sucrose diet):
3 $p<0.05$.
4 $p<0.02$.

sequence of a reduced capacity of tissues to utilize glucose and/or their resistance to insulin. In subsequent series, therefore, the basal as well as insulin-stimulated glucose utilization was investigated in the isolated hemidiaphragm and parametrial adipose tissue of rats fed diets with a high proportion of sucrose or starch.

The results revealed that the type of dietary carbohydrate did not affect glucose utilization nor the stimulating effect of insulin in the diaphragm. Marked and readily reproducible changes were, however, revealed in the metabolic response of the adipose tissue of rats fed for prolonged periods (4–8 weeks) a starch or sucrose diet. Although under basal conditions of incubation (medium containing 5 mmol glucose, without insulin) there was no difference in glucose utilization between groups fed the starch or sucrose diet, the stimulating effect of insulin was substantially lower in tissues of rats fed the sucrose diet than in rats fed the starch diet. This decrease of the hormone effect was apparent regardless of whether the tissue was removed from fed rats or from rats fasted for 18 h (table VIII). Other experiments revealed that the sucrose-induced reduction of adipose tissue sensitivity to insulin is roughly proportional to the sucrose ratio in the diet [122].

The decreased sensitivity of adipose tissue to insulin in sucrose-fed rats could be caused by different factors. Among others, an altered cellularity of adipose tissue could be involved [106, 119], or the possible alteration of the cell membrane as a result of nutritional deficiency of some factors essential for the interaction between insulin and membrane receptors for insulin such as, for example, trivalent chromium [91] or choline [115]. Absolute or relative chromium or choline deficiency as a result of chronic ingestion of a high-sucrose diet was described or suggested by several authors [18, 111]. Our experiments in which we tested the participation of the above factors on the sensitivity of adipose tissue in rats fed a sucrose diet gave negative results. The cellularity of adipose tissue investigated from the DNA concentration did not differ in relation to the type of dietary carbohydrates [120, 121]. Administration of trivalent chromium in amounts which, according to data in the literature, compensate deficiency signs [91] did not improve the sucrose-induced deterioration of glucose tolerance [Vrána and Fábry, unpublished results] nor the reduced sensitivity of adipose tissue to insulin (table IX). The administration of choline also did not affect the deteriorated glucose utilization in adipose tissue of sucrose-fed rats (table IX).

Further experiments revealed that the effect of sucrose on adipose tissue is not specific for the situation when glucose utilization is stimulated by insulin but becomes apparent also under different conditions when the demands of cellular systems for substrate processing are increased. In some experimental series we increased the glucose concentration in the incubation medium from 5 to 25 mmol. Moreover, we investigated also the incorporation of labelled fructose into adipose tissue lipids. Fructose transport into adipose tissue cells is, as is known, less sensitive to insulin than glucose transport. The results thus could indicate whether the sucrose-induced changes in adipose tissue pertain primarily to the substrate transport into tissue, or whether rather an alteration of intracellular systems involved in the processing of the substrate is in play.

Data summarized in table X indicate that merely an increase of the glucose concentration in the medium was sufficient to make the inhibitory effect of sucrose on the incorporation of labelled glucose into adipose tissue lipids apparent even in the absence of exogenous insulin. Fructose utilization responded to the ingestion of the sucrose diet even more sensitively than glucose utilization. Fructose incorporation into lipids, in keeping with data in the literature [38], was in all specimens higher as compared with glucose. A marked inhibition of this process was apparent already after 1 week on the sucrose diet and also in tissue specimens incubated with the lower

Table IX. Sensitivity of adipose tissue to insulin in rats fed starch or sucrose diets after administration of trivalent chromium and choline. Incorporation of labelled glucose into total lipids of adipose tissue is expressed as nmol/mg protein/2 h [VRÁNA and FÁBRY, unpublished results]

	Insulin concentration in incubation medium	
	0	1,000 μU/ml
Experiment I		
Starch diet	60.0 ± 6.7[1]	113.1 ± 5.6[3]
Sucrose diet	56.9 ± 2.1	72.6 ± 4.1
Starch diet plus Cr^{+++}	62.3 ± 4.2	110.5 ± 9.9[2]
Sucrose diet plus Cr^{+++}	52.5 ± 4.8	68.0 ± 3.8
Experiment II		
Starch diet	54.6 ± 6.3	123.0 ± 7.4[3]
Sucrose diet	49.7 ± 4.6	68.3 ± 5.5
Starch diet plus choline	58.1 ± 5.8	117.0 ± 8.3[3]
Sucrose diet plus choline	48.1 ± 6.0	72.2 ± 3.2

1 Mean values ± SEM (n = 6–7).
Cr^{+++} was administered as chromium acetate, 25 ppm, in drinking water, for a period of 2 weeks.
Choline chloride was administered intraperitoneally, 25 mg/day for 5 days. Before administration of chromium or choline the rats were fed the respective diets for 4 weeks.
Symbols for statistical significance (starch diet vs. sucrose diet):
2 $p < 0.01$.
3 $p < 0.001$.

fructose concentration (5 mmol). Insulin further enhanced fructose incorporation, although the percentage increase was, as expected, lower than that of glucose [VRÁNA and FÁBRY, unpublished results].

From these results it could be concluded that the intake of a high-sucrose diet does not primarily affect the substrate transport into the cell but affects most probably intracellular systems involved in the fatty acid synthesis and possibly their esterification. This is also suggested by our data on the reduced incorporation of labelled acetate into adipose tissue lipids in sucrose-fed rats (table XI) [unpublished results] as well as by recent findings of workers of Queen Elizabeth College, London, who found in

Table X. Incorporation of labelled glucose and fructose into total tissue lipids *in vitro* in relation to the length of administration of sucrose diet. Data given as nmol/mg protein/h [123]

Administration of sucrose diet weeks	Type and concentration of labelled hexose in incubation medium			
	U-^{14}C-glucose		U-^{14}C-fructose	
	5 mmol	25 mmol	5 mmol	25 mmol
0[1]	94.1±17.3[2]	309.8±51.3	304.6±21.4	2311.9±163.8
1	58.8± 9.1	118.6±17.7[3]	62.6±14.9[5]	918.8±240.5[5]
2	76.8±12.9	121.7± 9.3[4]	68.8± 9.1[5]	858.2±153.2[5]
5	73.6±12.4	131.3±10.6[4]	70.5±11.5[5]	725.1±155.3[5]

1 Rats fed starch diet.
2 Mean±SE (n=5–6).
Significance (compared with group fed starch diet):
3 $p<0.02$.
4 $p<0.01$.
5 $p<0.001$.

adipose tissue of rats fed a sucrose diet, in addition to reduced lipogenesis [7], reduced activities of pyruvate kinase, glucose-6-phosphate dehydrogenase [95] and fatty acid synthetase [11].

The decision whether the different metabolic effect of sucrose and starch is caused by different rates of breakdown and absorption of a polymerized carbohydrate such as starch, and simple sugars such as sucrose, or whether mainly the effect of the ample supply of fructose formed by hydrolysis of sucrose in the gut is involved seems to be in favour of the latter possibility. This applies to the effect on blood lipids [88, 95–97] as well as to the metabolism of adipose tissue. In some recent experiments we investigated the incorporation of ^{14}C-labelled glucose, fructose and acetate into adipose tissue lipids of rats fed diets with high ratio (70 cal%) of glucose, sucrose and fructose. While the sucrose diet caused a decline of incorporation of all three substrates as compared with the glucose diet, the decline was even more marked and in all instances highly significant in the tissues of animals fed the fructose diet (table XI). Similar conclusions were reached recently also by NAISMITH [95] who did not discover any differences between the glucose and starch diets as far as the activity of pyruvate kinase and glucose-6-phosphate

Table XI. Incorporation of labelled glucose, fructose and acetate into total lipids of adipose tissue of rats fed glucose, sucrose and fructose diets. Data given as nmol/mg protein/2 h [VRÁNA and FÁBRY, unpublished results].

Type of diet	U-^{14}C-glucose 25 mmol	U-^{14}C-fructose 25 mmol	1-^{14}C-acetate 10 mmol
Glucose diet	925±272[1]	2710±397	127± 8
Sucrose diet	336± 80	1099± 87[3]	61±13[3]
Fructose diet	285± 33[2]	622±130[4]	55±18[3]

1 Mean±SEM (n=6–7).
Symbols for statistical significance (compared with group fed glucose diet):
2 $p<0.05$.
3 $p<0.01$.
4 $p<0.001$.

The tissues were incubated for 2 h in 3 ml Krebs-Ringer bicarbonate buffer ($^1/_2$ Ca^{++}), pH 7.4, containing precursors listed in table; incubation medium with labelled acetate also contained 5 mmol glucose.

dehydrogenase in adipose tissue is concerned, but found a significant drop of both enzymes on the sucrose and even more on the fructose diet.

The mechanism of the effect of dietary sucrose and fructose on the above metabolic parameters is not clear so far. From table XII, which summarizes the main effects of the sucrose and fructose diets on adipose tissue, it is apparent that the glucose conversion to CO_2 and fatty acids and triglycerides, respectively, is inhibited. The reduced activity of some enzymes involved in this process is also in keeping with the reduced lipogenesis. Recent experiments from our laboratory indicate that the effect of dietary fructose is not immediate, as the inhibition of hexose utilization develops after administration of the experimental diet for 1 week or longer. For example, 3 days re-feeding with the high-sucrose diet after previous fasting did not cause any changes in the incorporation of labelled substrates into adipose tissue lipids as compared with re-feeding of glucose. Nor did short-term oral or intra-peritoneal administration of fructose to animals before tissue specimens were taken or *in vitro* pre-incubation of the tissue in a medium containing fructose have an effect on glucose utilization [VRÁNA and FÁBRY, unpublished results].

Table XII. Effect of a high proportion of sucrose or fructose in the diet on some metabolic parameters in rat adipose tissue. If not otherwise stated, the results are from *in vitro* experiments

Investigated parameter	Change[1]	Reference No.	Remark
Conversion of glucose to			
CO_2	reduction	7	
Triglycerides	reduction	7, 123	
Conversion of fructose to triglycerides	reduction	123	
Conversion of acetate to triglycerides	0	23	*in vivo*
	reduction	see table XI	
Effect of insulin on glucose utilization	reduction	121, 123	
Cellularity of adipose tissue	0	120, 121	
FFA release and effect of epinephrine	0	121	
Activity of lipoprotein lipase	increase	94	
	reduction	13	
Activity of glucose-6-phosphate dehydrogenase	reduction	95	
Activity of pyruvate kinase	reduction	95	
Activity of fatty acid synthetase	reduction	11	68 cal% sucrose
	0	12	30 cal% sucrose

1 Compared with group fed starch of glucose diet respectively.

It would be tempting to speculate whether a series of changes is involved in the effect of dietary sucrose or fructose starting by enhanced lipid formation in the liver and ensuing hypertriglyceridaemia. The increased supply of triglycerides could then inhibit fatty acid synthesis in adipose tissue. Evidence for this possibility is so far only indirect [134]. We were able to confirm the marked hypertriglyceridaemia in fructose-fed rats of the Wistar strain used by us [Vrána and Fábry, unpublished results]. The findings of Naismith [95] who observed a dissociation of the effect of dietary glucose and fructose on the triglyceridaemia and of pyruvate kinase and glucose-6-phosphate dehydrogenase activity in adipose tissue indicate, however, that the above inter-relationships might be rather complex.

The problem of the biological importance of changes in adipose tissue induced by dietary sucrose and fructose still remains open. In view of the controversial data on the quantitative importance of lipogenesis in adipose tissue [26, 35, 92, 99], so far it cannot be said whether inhibition of this process is or is not fully responsible for the poorer glucose tolerance in sucrose-fed rats. Even more so because in our experimental set-up only the isolated hemidiaphragm was used as a representative of muscle. At present we are conducting experiments to find out whether a high sucrose or fructose intake alters the metabolism of skeletal muscle.

V. References

1 ALLMAN, D. W.; HUBBARD, D. D., and GIBSON, D. M.: Fatty acid synthesis during fat-free refeeding in starved rats. J. Lipid Res. *6:* 63–74 (1965).

2 AUSTIN, P. and NESTEL, P. J.: The effect of glucose and insulin *in vitro* on the uptake of triglyceride and on lipoprotein lipase activity in fat pads from normal fed rats. Biochim. biophys. Acta 164: 59–63 (1968).

3 BALL, E. G. and JUNGAS, R. L.: Some effects of hormones on the metabolism of adipose tissue. Recent Progr. Hormone Res. *20:* 183–214 (1964).

4 BALL, E. G.: Regulation of fatty acid synthesis in adipose tissue. Adv. Enzyme Regulat. *4:* 3–18 (1966).

5 BAR-ON, H. and STEIN, Y.: Effect of glucose and fructose administration on lipid metabolism in the rat. J. Nutr. *94:* 95–105 (1968).

6 BEIGELMAN, P. M.: Bio-assay for insulin-like activity utilizing glucose uptake by adipose tissue. Metabolism *9:* 580–586 (1960).

7 BENDER, A. E. and THADANI, P. V.: Some metabolic effects of dietary sucrose. Nutr. Metabol. *12:* 22–39 (1970).

8 BEZMAN, A.; FELTS, J. M., and HAVEL, R. J.: Relation between incorporation of triglyceride fatty acids and heparin released lipoprotein lipase from adipose tissue slices. J. Lipid Res. *3:* 427–431 (1962).

9 BORTZ, W.; ABRAHAM, S., and CHAIKOFF, I. L.: Localization of the block in lipogenesis resulting from feeding fat. J. biol. Chem. *238:* 1266–1272 (1963).

10 BRAUN, T.: KAZDOVÁ, L.; FÁBRY, P.; LOJDA, Z., and HROMÁDKOVÁ, V.: Meal eating and refeeding after a single fast as a stimulus for increasing the number of fat cells in abdominal adipose tissue of rats. Metabolism *17:* 825–832 (1968).

11 BRUCKDORFER, K. R.; KHAN, I. H., and YUDKIN, J.: Fatty acid synthetase activity in rat liver and adipose tissue of rats fed with carbohydrates. Biochem. J. *129:* 439–446 (1972).

12 BRUCKDORFER, K. R.; KANG, S. S., and YUDKIN, J.: Does dietary lactose produce hyperlipaemia in the rat? Proc. Nutr. Soc. *31:* 10–11 A (1971).

13 BRUCKDORFER, K. R., KANG, S. S. and YUDKIN, J.: The hyperlipaemic effect of sucrose in male and female rats. Proc. Nutr. Soc. *31:* 11 (1971).

14 BUTCHER, R. W.; BAIRD, C. E., and SUTHERLAND, E. W.: Effect of lipolytic and antilipolytic substances on adenosine 3'5'-monophosphate levels in isolated fat cells. J. biol. Chem. *243:* 1705–1712 (1968).

15 BUTCHER, R. W.: The role of cyclic AMP in the actions of some lipolytic and anti-lipolytic agents; in JEANRENAUD and HEPP Adipose tissue. Regulation and metabolic functions, pp. 5–10 (Thieme, Stuttgart, and Academic Press, New York 1970).
16 CAHILL, G. F., jr.; LEBOEUF, B., and RENOLD, A. E.: Studies on rat adipose tissue *in vitro*. III. Synthesis of glycogen and glyceride-glycerol. J. biol. Chem. *234:* 2540–2543 (1959).
17 CAHILL, G. F., jr.: Starvation in man. New Engl. J. Med. *282:* 668–675 (1970).
18 CHALVARDJIAN, A. and STEPHENS, S. Lipotropic effect of dextrin *versus* sucrose in choline-deficient rats. J. Nutr. *100:* 397–403 (1970).
19 CHANDLER, M. A. and MOORE, R. O.: Glycogen deposition in adipose tissue: Variations in levels of glycogen cycle enzymes during fasting and refeeding. Arch. Biochem. *108:* 183–192 (1964).
20 CHLOUVERAKIS, C.: Factors affecting the inhibitory action of insulin on lipolysis in a glucose-free medium. Endocrinology *81:* 521–526 (1967).
21 CHRISTOPHE, J.; WINAND, J., and FURNELLE, J.: Distribution of newly synthesized fatty acids and metabolic heterogeneity of NEFA and glycerides within adipose tissue; in MAŠEK, OŠANCOVÁ, CUTHBERTSON Nutrition, Proc. 8th Int. Congr., Prague 1969, pp. 254–260 (Excerpta Medica, Amsterdam 1970).
22 COHEN, A. M. and TEITELBAUM, A.: Effect of dietary sucrose and starch on oral glucose tolerance and insulin-like activity. Amer. J. Physiol. *206:* 105–109 (1964).
23 COHEN, A. M. and TEITELBAUM, A.: Effect of different levels of protein in sucrose and starch diets on lipid synthesis in the rat. Israel J. med. Sci. *2:* 727–732 (1966).
24 COHEN, A. M.: Effect of dietary carbohydrate on the glucose tolerance curve in the normal and the carbohydrate-induced hyperlipemic subjects. Amer. J. clin. Nutr. *20:* 126–130 (1967).
25 COORE, H. G.; DENTON, R. M.; MARTIN, B. R., and RANDLE, P. J.: Regulation of adipose tissue pyruvate dehydrogenase by insulin and other hormones. Biochem. J. *125:* 115–127 (1971).
26 CURTIS-PRIOR, P. B.; TRETHEWEY, J.; STEWART, G. A., and HANLEY, T.: The contribution of different organs and tissues of the rat to assimilation of glucose. Diabetologia *5:* 384–391 (1969).
27 DANIEL, S. M. and RUBINSTEIN, D.: The effect of fasting on the esterification of palmitate by rat epididymal adipose tissue *in vitro*. Canad. J. Biochem. *45:* 965–972 (1966).
28 DOLE, V. P.: A relation between nonesterified fatty acids in plasma and the metabolism of glucose. J. clin. Invest. *35:* 150–154 (1956).
29 DOLE, V. P.: Energy storage; in RENOLD and CAHILL Handbook of physiology, section 5, pp. 13–18 (American Physiological Society, Washington 1965).
30 DUNNINGAN, M. G.; FYFE, T.; MCKIDDIE, M. T., and CROSBIE, S. M.: The effect of isocaloric exchange of dietary starch and sucrose on glucose tolerance, plasma insulin and serum lipids in man. Clin. Sci. *38:* 1–9 (1970).
31 FÁBRY, P.; POLEDNE, R.; KAZDOVÁ, L., and BRAUN, T.: The effect of feeding frequency and type of dietary carbohydrate on hepatic lipogenesis in the albino rat. Nutr. Diet. *10:* 81–90 (1968).
32 FÁBRY, P.: Feeding pattern and nutritional adaptations (Academia, Prague, and Butterworth, London 1969).

33 FÁBRY, P.; KLEINFELD, R.; TEPPERMAN, H. M., and TEPPERMAN, J.: Effect of diet and insulin on the morphology and TPNH generating enzyme activities of rat adipose tissue. Proc. Soc. exp. Biol. Med. *133*: 577–581 (1970).

34 FAIN, J. N.; KOVACEV, V. P., and SCOW, R. O.: Antilipolytic effect of insulin in isolated fat cells of rat. Endocrinology *78:* 773–778 (1966).

35 FAVARGER, P.: Relative importance of different tissues in the synthesis of fatty acids; in RENOLD and CAHILL Handbook of physiology, section 5, pp. 19–23 (American Physiological Society, Washington 1965).

36 FLATT, J. P. and BALL, E. G.: Pathways of glucose metabolism, part II; in RENOLD and CAHILL Handbook of physiology, section 5, pp. 273–280 (American Physiological Society, Washington 1965).

37 FROESCH, E. R. and GINSBERG, J. L.: Fructose metabolism in adipose tissue. I. Comparison of fructose and glucose metabolism in epididymal adipose tissue of normal rats. J. biol. Chem. *237:* 3317–3324 (1962).

38 FROESCH, E. R.: Fructose metabolism in adipose tissue from normal and diabetic rats; in RENOLD and CAHILL Handbook of physiology, section 5, pp. 281–294 (American Physiological Society, Washington 1965).

39 FROESCH, E. R.; BÜRGI, H.: BALLY, P., and LABHART A.: Insulin inhibition of spontaneous adipose tissue lipolysis and effects upon fructose and glucose metabolism. Molec. Pharmacol. *1:* 280–296 (1965).

40 GALTON, D. J. and WILSON, J. P. D.: The effects of starvation on lipogenesis in human adipose tissue. Europ. J. clin. Invest. *1:* 94–98 (1970).

41 GELLHORN, A. and BENJAMIN, W.; Effects of aging on the composition and metabolism of adipose tissue in the rat; in RENOLD and CAHILL Handbook of physiology, section 5, pp. 399–406 (American Physiological Society, Washington 1965).

42 GIBSON, D. M.: The biosynthesis of fatty acids. J. chem. Educ. *42:* 236–243 (1965).

43 GLIEMANN, J.: Studies on the action of insulin in isolated fat cells; in JEANRENAUD and HEPP Adipose tissue. Regulation and metabolic functions, pp. 116–119 (Thieme, Stuttgart, and Academic Press, New York 1970).

44 GOODMAN, H. M.: A comparative study of the effects of insulin and growth hormone on hexose metabolism in adipose tissue. Endocrinology 80: 45–51 (1967).

45 GOODNER, J.; TUSTISON, W. A.; DAVIDSON, M. B.; CHU, P. C., and CONWAY, M. J.: Studies of substrate regulation in fasting. I. Evidence for central regulation of lipolysis by plasma glucose mediated by the sympathetic nervous system. Diabetes, N.Y. *16:* 576–589 (1967).

46 GORDON, R. S., jr. and CHERKES, A.: Unesterified fatty acids in human blood plasma. J. clin. Invest. *35:* 206–212 (1956).

47 GORDON, R. S., jr. and CHERKES, A.: Production of unesterified fatty acids from isolated rat adipose tissue incubated *in vitro*. Proc. Soc. exp. Biol. Med. *97:* 150–151 (1958).

48 GRIES, F. A. and STEINKE, J.: Insulin and human adipose tissue *in vitro*. A brief review. Metabolism *16:* 693–696 (1967).

49 GUMAA, K. A.; NOVELLO, F., and MCLEAN, P.: The pentose phosphate pathway of glucose metabolism. Hormonal and dietary control in the oxidative and nonoxidative reactions and related enzymes of the cycle in adipose tissue. Biochem. J. *114:* 253–265 (1969).

50 GUTMAN, A. and SHAFRIR, E.: Metabolic influences on enzymes of glycogen synthesis and breakdown in adipose tissue. Amer. J. Physiol. *207:* 1215–1220 (1964).
51 HALPERIN, M. L. and ROBINSON, B. H.: Mechanism of insulin action on control of fatty acid synthesis independent of glucose transport. Metabolism *20:* 78–86 (1971).
52 HANSON, R. W.; PATEL, M. S.; JOMAIN-BAUM, M., and BALLARD, F. J.: Role of mitochondria in metabolism of pyruvate and lactate by rat adipose tissue. Metabbolism *20:* 27–42 (1971).
53 HAUSBERGER, F. X. and MILSTEIN, S. W.: Dietary effects on lipogenesis in adipose tissue. J. biol. Chem. *214:* 483–488 (1965).
54 HAUSBERGER, F. X.: Action of insulin and cortisone on adipose tissue. Diabetes, N.Y. *7:* 211–220 (1958).
55 HAVEL, R. J.: Metabolism of lipids in chylomicrons and very low-density lipoproteins; in RENOLD and CAHILL Handbook of physiology, section 5, pp. 499–507 (American Physiological Society, Washington 1965).
56 HEPP, K. D.; MENAHAN, L. A.; WIELAND, O., and WILLIAMS, R. H.: Studies on the action of insulin in isolated adipose tissue cells. II. 3′5′-nucleotide phosphodiesterase and antilipolysis. Biochim. biophys. Acta *184:* 554–565 (1969).
57 HIRSCH, J. and GOLDRICK, R. B.: Metabolism of human adipose tissue *in vitro;* in RENOLD and CAHILL Handbook of physiology, section 5, pp. 455–470 (American Physiological Society, Washington 1965).
58 HODGES, R. E. and KREHL, W. A.: The role of carbohydrates in lipid metabolism. Amer. J. clin. Nutr. *17:* 334–346 (1965).
59 HOLLENBERG, C. H.: The effect of fasting on the lipoprotein lipase activity of rat heart and diaphragm. J. clin. Invest. *39:* 1282–1287 (1960).
60 HOLLENBERG, C. H.: Adipose tissue lipases, part II; in RENOLD and CAHILL Handbook of physiology, section 5, pp. 301–307 (American Physiological Society, Washington 1965).
61 HOLLENBERG, C. H.; VOST, A., and PATTEN, R. L.: Regulation of adipose mass. Control of fat cell development and lipid content. Recent Progr. Hormone Res. *26:* 463–503 (1970).
62 HOLMES, W. L. and BORTZ, W. M. (ed.): Biochemistry and pharmacology of free fatty acids. Progr. biochem. Pharmacol., vol. 6 (Karger, Basel 1971).
63 JANSEN, G. R.; HUTCHISON, C. F., and ZANETTI, M. E.: Studies on lipogenesis *in vivo.* Effect of dietary fat and starvation on conversion of ^{14}C glucose into fat and turnover of newly synthesized fat. Biochem. J. *99:* 323–332 (1966).
64 JANSEN, G. R.; ZANETTI, M. E., and HUTCHISON, C. F.: Studies on lipogenesis *in vivo.* Effect of starvation and re-feeding, and studies on cholesterol synthesis. Biochem. J. *99:* 333–340 (1966).
65 JANSEN, G. R.; ZANETTI, M. E., and HUTCHISON, C. F.: Studies on lipogenesis *in vivo.* Comparison of cholesterol and fatty acid synthesis in rats and mice. Biochem. J. *102:* 864–869 (1967).
66 JEANRENAUD, B. and HEPP, D. (ed.): Adipose tissue. Regulation and metabolic functions (Thieme, Stuttgart, and Academic Press, New York 1970).
67 JOMAIN, M. and HANSON, R. W.: Dietary protein and the control of fatty acid synthesis in rat adipose tissue. J. Lipid Res. *10:* 674–680 (1969).

68 JUNGAS, R. L. and BALL, E. G.: Studies on the metabolism of adipose tissue. XII. The effect of insulin and epinephrine on free-fatty acid and glycerol production in the presence and absence of glucose. Biochemistry *2:* 383–388 (1963).

69 JUNGAS, R. L. and BALL, E. G.: Studies on the metabolism of adipose tissue. XVII. *In vitro* effects of insulin upon the metabolism of carbohydrate and triglyceride stores of adipose tissue from fasted-refed rats. Biochemistry *3:* 1696–1702 (1964).

70 JUNGAS, R. L.: Effect of insulin on fatty acid synthesis from pyruvate, lactate, or endogenous sources in adipose tissue. Evidence for the hormonal regulation of pyruvate dehydrogenase. Endocrinology *86:* 1368–1375 (1970).

71 JUNGAS, R. L.: Hormonal regulation of pyruvate dehydrogenase. Metabolism *20:* 43–53 (1971).

72 KAHLENBERG, A.; RUBINGER, J., and KALANT, N.: Differences in the glucose metabolism and insulin responsiveness of rat and human adipose tissue. Canad. J. Biochem. *44:* 645–648 (1967).

73 KAZDOVÁ, L.; BRAUN, T.; FÁBRY, P., and POLEDNE, R.: Increased glycogen synthesis, protein synthesis and lipogenesis in rat adipose tissue induced by changed feeding frequency (in Czech). Čs. Fysiol. *15:* 522–523 (1966).

74 KAZDOVÁ, L.; BRAUN, T., and FÁBRY, P.: Increased DNA synthesis in epididymal adipose tissue of rats refed after a single fast. Metabolism *16:* 1174–1176 (1967).

75 KAZDOVÁ, L.; FÁBRY, P., and BRAUN, T.: Composition of diet and response of adipose tissue to refeeding. Physiol. Bohemoslov. *17:* 469 (1968).

76 KAZDOVÁ, L. and VRÁNA, A.: Insulin and adipose tissue cellularity. Horm. Metab. Res. *2:* 117–118 (1970).

77 KESSLER, J. I.: Effect of diabetes and insulin on the activity of myocardial and adipose tissue lipoprotein lipase of rats. J. clin. Invest. *42:* 362–367 (1963).

78 KHANADE, J. M. and NATH, M. C.: Lipogenesis in adipose tissue of rats fed different diets. Amer. J. Physiol. *201:* 1041–1043 (1961).

79 LANDAU, B. R. and KATZ, J.: Pathways of glucose metabolism; in RENOLD and CAHILL Handbook of physiology, section 5, pp. 253–272 (American Physiological Society, Washington 1965).

80 LANDAU, B. R.; KATZ, J.; BARTSCH, G. E.; WHITE, L. W., and WILLIAMS, H. R.: Hormonal regulation of glucose metabolism in adipose tissue *in vitro*. Ann. N.Y. Acad. Sci. *131:* 43–58 (1965).

81 LEVEILLE, G. A. and HANSON, R. W.: Adaptive changes in enzyme activity and metabolic pathways in adipose tissue from meal-fed rats. J. Lipid Res. *7:* 46–55 (1966).

82 LEVEILLE, G. A.: Influence of dietary fat and protein on metabolic and enzymatic activities in adipose tissue of meal-fed rats. J. Nutr. *91:* 25–34 (1967).

83 LEVEILLE, G. A.: Influence of dietary fat level on the enzymatic and lipogenic adaptations in adipose tissue of meal-fed rats. J. Nutr. *91:* 267–274 (1967).

84 LEVEILLE, G. A.; O'HEA, E. K., and CHAKRABARTY, K.: *In vivo* lipogenesis in the domestic chicken. Proc. Soc. exp. Biol. Med. *128:* 398–401 (1968).

85 LEVEILLE, G. A.: Adipose tissue metabolism. Influence of periodicity of eating and diet composition. Fed. Proc. *29:* 1294–1301 (1970).

86 LOTEN, E. G. and SNEYD, J. G. T.: An effect of insulin on adipose tissue adenosine 3′, 5′-cyclic monophosphate diesterase. Biochem. J. *120:* 187–193 (1970).

87 MACDONALD, I.: Physiological role of dietary carbohydrates. Wld. Rev. Nutr. Diet., Vol. 8, pp. 143–183 (Karger, Basel 1967).
88 MACDONALD, I.: Sucrose and blood lipids; in YUDKIN, EDELMAN and HOUGH Sugar, pp. 192–199 (Butterworths, London 1971).
89 MARTIN, D. B. and VAGELOS, P. R.: Fatty acid synthesis in adipose tissue; in RENOLD and CAHILL Handbook of physiology, section 5, pp. 211–216 (American Physiological Society, Washington 1965).
90 MAYHEW, D. A.; WRIGHT, P. H., and ASHMORE, J.: Regulation of insulin secretion. Pharmacol. Rev. *21:* 183–212 (1969).
91 MERTZ, W.: Chromium occurence and function in biological systems. Physiol. Rev. *40:* 163–239 (1969).
92 MOODY, A. J.; JEFFCOATE, S. L., and VOLUND, A.: The effects of anti-insulin serum on the disposal of an oral load of (6 ^{14}C) glucose by the tissues of the rat. Horm. Metab. Res. *2:* 193–199 (1970).
93 MOORE, R. O.: Influence of fasting and refeeding on the response of adipose tissue to insulin. Amer. J. Physiol. *205:* 222–224 (1964).
94 NAISMITH, D. J. and KHAN, N. A.: Lipoprotein lipase activity in the adipose tissue of rats fed sucrose or starch. Proc. Nutr. Soc. *30:* 12 (1970).
95 NAÏSMITH, D. J.: The hyperlipidaemic property of sucrose; in YUDKIN, EDELMAN and HOUGH Sugar, pp. 183–191 (Butterworths, London 1971).
96 NIKKILÄ, E. A. and OJALA, K.: Induction of hypertriglyceridemia by fructose in the rat. Life Sci. *4:* 937–943 (1965).
97 NIKKILÄ, E. A. and OJALA, K.: Acute effects of fructose and glucose on the concentration and removal rate of plasma triglyceride. Life Sci. *5:* 89–94 (1966).
98 OWEN, O. E. and REICHARD, G. A., jr.: Fuels consumed by man. The interplay between carbohydrates and fatty acids. Progr. biochem. Pharmacol. vol. 6, pp. 177–213 (Karger, Basel 1971).
99 PATKIN, J. K. and MASORO, E. J.: Fatty acid synthesis in normal and cold acclimated rats. Canad. J. Physiol. Pharmacol. *42:* 101–107 (1964).
100 PORTER, J. W.; KUMAR, S., and DUGAN, R. E.: Synthesis of fatty acids by enzymes of avian and mammalian species. Progr. biochem. Pharmacol., vol. 6, pp. 1–101 (Karger, Basel 1971).
101 RABEN, M. S. and HOLLENBERG, C. H.: Effect of glucose and insulin on the esterification of fatty acids by isolated adipose tissue. J. clin. Invest. *39:* 435–439 (1960).
102 RENOLD, A. E. and CAHILL, G. F., jr. (ed.): Handbook of physiology. 5. Adipose tissue (American Physiological Society, Washington 1965).
103 RIZACK, M. A.: Hormone sensitive lipolytic activity of adipose tissue; in RENOLD and CAHILL Handbook of physiology, section 5, pp. 309–312 (American Physiological Society, Washington 1965).
104 SAGGERSON, E. D. and GREENBAUM, A. L.: The effect of dietary and hormonal conditions on the activities of glycolytic enzymes in rat epididymal adipose tissue. Biochem. J. *115:* 405–417 (1969).
105 SALAMAN, M. R. and ROBINSON, D. S.: Clearing-factor lipase in adipose tissue. A medium in which the enzyme activity of adipose tissue from starved rats increases *in vitro*. Biochem. J. *99:* 640–647 (1966).
106 SALANS, L. B.; KNITTLE, J. L., and HIRSCH, J.: The role of adipose tissue cell size

and adipose tissue insulin sensitivity in the carbohydrate intolerance of human obesity. J. clin. Invest. *47:* 153–165 (1968).

107 SHAFRIR, E.; SHAPIRO, B., and WERTHEIMER, E.: Glycogen metabolism in adipose tissue; in RENOLD and CAHILL Handbook of physiology, section 5, pp. 313–318 (American Physiological Society, Washington 1965).

108 SHAFRIR, E.: Glycolysis and its reversal in adipose tissue; in MAŠEK, OŠANCOVÁ and CUTHBERTSON Nutrition. Proc. 8th Int. Congr., Prague 1969, pp. 261–267 (Excerpta Medica, Amsterdam 1970).

109 SHAPIRO, B.; CHOWERS, I., and ROSE, G.: Fatty acid uptake and esterification in adipose tissue. Biochim, biophys. Acta *23:* 115–120 (1957).

110 SHRAGO, E.; GLENNON, J. A., and GORDON, E. S.: Comparative aspects of lipogenesis in mammalian tissues. Metabolism *20:* 54–62 (1971).

111 SCHROEDER, H. A.: The role of chromium in mammalian nutrition. Amer. J. clin. Nutr. *21:* 230–244 (1968).

112 STEINBERG, D. and VAUGHAN, M.: Release of free fatty acids from a dipose tissue *in vitro* in relation to rates of triglyceride synthesis and degradation; in RENOLD and CAHILL Handbook of physiology, section 5, pp. 335–348 (American Physiological Society, Washington 1965).

113 SZANTO, S. and YUDKIN, J.: The effect of dietary sucrose on blood lipids, serum insulin, platelet adhesiveness and body weight in human volunteers. Postgrad. med. J. *45:* 602–607 (1969).

114 SZEPESI, B. and BERDANIER, C. D.: Time course of the starve-refeed response in rats. The possible role of insulin. J. Nutr. *101:* 1563–1574 (1971).

115 TANI, H.; AOYAMA, Y., and ASHIDA, K.: Depression of lipogenesis in choline-deficient guinea pigs and its possible mechanisms. J. Nutr. *94:* 516–526 (1968).

116 TEPPERMAN, J. and TEPPERMAN, H. M.: Adaptive hyperlipogenesis. Late 1964 model. Ann. N.Y. Acad. Sci. *131:* 404–411 (1965).

117 TRUEHEART, P. A. and HERRERA, M. G.: Decreased response to insulin in adipose tissue during starvation. Effect of hypophysectomy and growth hormone administration. Diabetes, N.Y. *20:* 46–50 (1971).

118 VAUGHAN, M. and STEINBERG, D.: Glyceride biosynthesis, glyceride breakdown and glycogen breakdown in adipose tissue: mechanisms and regulation; in RENOLD and CAHILL Handbook of physiology, section 5, pp. 239–252 (American Physiological Society, Washington 1965).

119 VRÁNA, A. and KAZDOVÁ, L.: Insulin sensitivity of adipose tissue and of diaphragm and adipose tissue composition in cold acclimated rats. Life Sci. *8* (I): 1103–1108 (1969).

120 VRÁNA, A. and KAZDOVÁ, L.: Metabolic differences between dietary starch and sucrose in the rat. Insulin sensitivity and composition of adipose tissue. Physiol. Bohemoslov. *18:* 518–519 (1969).

121 VRÁNA, A. and KAZDOVÁ, L.: Insulin sensitivity of adipose tissue and of diaphragm *in vitro*. Effect of the type of dietary carbohydrate (starch-sucrose). Life Sci. *9* (II): 257–265 (1970).

122 VRÁNA, A.; SLABOCHOVÁ, Z.; KAZDOVÁ, L., and FÁBRY, P.: Insulin sensitivity of adipose tissue and serum insulin concentration in rats fed sucrose or starch diets. Nutr, Rep. int. *3:* 31–37 (1971).

123 VRÁNA, A.; FÁBRY, P., and KAZDOVÁ, L.: Utilization of ^{14}C-glucose and ^{14}C-fructose in adipose tissue of rats fed a starch or sucrose diet Rev. europ. Et. clin. biol. *17:* 611–614 (1972).
124 WEBER, G.; HIRD, H. J.; STAMM, N. B., and WAGLE, D. S.: Enzymes involved in carbohydrate metabolism in adipose tissue; in RENOLD and CAHILL Handbook of physiology, section 5, pp. 225–238 (American Physiological Society, Washington 1965).
125 WEISS, L.; LÖFFLER, G.; SCHIRMANN, P., and WIELAND, O.: Control of pyruvate dehydrogenase interconversion in adipose tissue by insulin. FEBS Letters *15:* 229–231 (1971).
126 WHITE, J. E. and ENGEL, F. L.: A lipolytic action of epinephrine and norepinephrine on rat adipose tissue *in vitro*. Proc. Soc. exp. Biol. Med. *99:* 375–378 (1958).
127 WHITE, J. E. and ENGEL, F. L.: Lipolytic action of corticotropin on rat adipose tissue *in vitro*. J. clin. Invest. *37:* 1556–1563 (1958).
128 WHITNEY, J. E. and ROBERTS, S.: Influence of previous diet on hepatic glycogenesis and lipogenesis. Amer. J. Physiol. *181:* 446–450 (1955).
129 WILEY, J. H. and LEVEILLE, G. A.: Adaptive nature of glycogen synthetase in rat adipose tissue. Requirement for insulin and energy. Proc. Soc. exp. Biol. Med. *137:* 798–802 (1971).
130 WINEGRAD, A. J.: Endocrine effects on adipose tissue metabolism. Vitamins Hormones, N.Y. *20:* 141–197 (1962).
131 WINEGRAD, A. J.: Adipose tissue in diabetes; in RENOLD and CAHILL Handbook of physiology, section 5, pp. 319–330 (American Physiological Society, Washington 1965).
132 WING, D. R.; SALAMAN, M. R., and ROBINSON, D. S.: Clearing-factor lipase in adipose tissue. Factors influencing the increase in enzyme activity produced on incubation of adipose tissue from starved rats *in vitro*. Biochem. J. *99:* 648–656 (1966).
133 WOOD, F. C.; LEBOEF, B.; RENOLD, A. E., and CAHILL, G. F., jr.: Metabolism of mannose and glucose by adipose tissue and liver slices from normal and alloxan diabetic rats. J. biol. Chem. *236:* 18–21 (1961).
134 ZAKIM, D. and HERMAN, R. H.: Regulation of fatty acid synthesis. Amer. J. clin. Nutr. *22:* 200–213 (1969).

Authors' address: Dr. A. VRÁNA and Dr. P. FÁBRY, Research Centre of Metabolism and Nutrition, Institute for Clinical and Experimental Medicine, Budejovicka 800, ***Prague 4*** (Czechoslovakia)

Progr. biochem. Pharmacol., vol. 8, pp. 216–241 (Karger, Basel 1973)

Effects of Dietary Carbohydrates on Serum Lipids

I. MACDONALD

Department of Physiology, Guy's Hospital Medical School, London

Contents

I. Introduction

The concentration of some of the lipid fractions in serum has been found to be useful in the prognostication of vascular disease and for this reason considerable attention has been paid to factors which are able to vary, and more especially reduce the level of the lipid fractions. The cholesterol level in the blood has long been accepted as an important factor in forecasting vascular accidents secondary to atherosclerosis and more recently the suggestion that a raised level of triglyceride in fasting serum is associated with atherosclerosis [46] has been confirmed [21]. So far no rôle has been found for serum phospholipids or free fatty acids (FFA) in the genesis of disease. It should be pointed out, however, that the evidence relating to raised levels of cholesterol and/or triglyceride to vascular disorders is based largely on epidemiological surveys, both retrospective and prospective and, though the correlation may be high, this does not necessarily mean the two parameters are related, and even if they were shown to be related, a level in an individual is of no more value than an assessment of odds. It is not uncommon for a person with normal serum lipid concentrations to suffer from ischaemic heart disease, and conversely for raised levels to be present in a very old person, such is the multifactorial aetiology of vascular disease.

Obviously the level on any substance in the serum is the resultant of a balance between input and output and were it not for the easy and comparatively harmless procedure of obtaining blood it is doubtful whether serum would have deserved such a literature. Unlike practically every other part of the body, serum is not a site of metabolism.

The time of obtaining the sample is important in any study on the concentration of lipid in the serum. The marked increase in level of triglyceride (largely as chylomicrons) after a meal rich in fat is well-known, though the diurnal variations in the cholesterol and phospholipid levels are less striking [67]. The turnover time of the FFA is so short, and influenced by so many extraneous factors, that it is not surprising that the level of this lipid fraction undergoes violent variations in a normal person during a 24-hour period.

The usual conditions laid down for obtaining a sample of serum for an examination of its lipid profile is after a 12- to 14-hour fast. The rationale for this is that the triglyceride under these circumstances is likely to be almost entirely endogenous – the chylomicrons from the previous meal having been cleared. The endogenous triglyceride concentration under these circumstances is at its diurnal maximum.

Lipid in the serum is bound with protein to 'solubulize' it and details of the nature and immune reactions of the carrier proteins are actively under investigation [45], and beyond the scope of this review. For a thorough investigation of the serum lipid fractions ultracentrifugation and/or electrophoresis is necessary, but as the former is expensive and the latter difficult to quantify, the majority of work on the rôle of diet on the level of the various lipid fractions has used a direct estimation of cholesterol (total or free and esterified), triglyceride, phospholipid and FFA concentration. Though this direct estimation of the lipid is not entirely satisfactory – largely because it fails to distinguish which lipoprotein is involved – it is nevertheless very useful and considered by some to yield practically all the information that is usually required [101].

II. Common Dietary Carbohydrates

The commonest dietary carbohydrate to enter the portal vein is glucose. This is obtained from starch, sucrose or glucose, as such, in the diet. The next commonest carbohydrate going to the liver is fructose and the most likely source of this monosaccharide in western countries is sucrose. Other sources of fructose include honey and fruit. The only carbohydrate consumed by the very young is lactose (glucose plus galactose) and this carbohydrate is eaten to a limited extent throughout life in western societies. One carbohydrate which is consumed and cannot, by definition, have any direct metabolic rôle is that which cannot be assimilated from the gut and called variously 'fibre', 'roughage', etc. It may, however, have an indirect part to play in setting the level of serum cholesterol *(vide infra)*. Rarer dietary carbohydrates include sorbitol, manitol, xylitol, trehalose, etc. Strictly speaking, ascorbic acid and glycerol should also be included, the former is suggested to have a favourable influence on serum cholesterol levels and the latter (in increasing use in the food industry) may have sinister effects on serum triglyceride levels.

Many studies on the effect of dietary carbohydrates on serum lipids have been carried out in animals, fewer in man. As in most experiments it is unwise to interpolate from one species to another, and in the rat, for example, the absorption of fructose is active [97a] whereas in man it is passive, or at most 'facilitated' and hence large amounts give rise to an osmotic diarrhoea. It is possible, though not easy, to produce arterial disease in experimental animals – it is too easy to produce vascular disease in man, and hence the various factors, of which carbohydrate may be one, and which go to make up a pathological state in man, may not exist in experimental animals at all. Nevertheless, indications of considerable importance in the study of man have, and will continue to come from the more controlled conditions of experimental animals.

III. Immediate Effects on Serum Lipids of Carbohydrate Consumption

One means whereby dietary carbohydrate may influence serum lipid concentrations is for the carbohydrate to be converted to the lipid in question. Another mechanism that may result in altered lipid levels could be that dietary carbohydrate has no direct effect on the serum lipid levels but that it alters, in some way, a metabolic compound that secondarily changes serum lipid concentration without the dietary carbohydrate itself being in any way involved in the formation of the lipid. This indirect effect seems to be the mechanism of the immediate effect on lipids of absorbing dietary carbohydrate.

In 1957, it was reported that 100 g glucose given at hourly intervals to men lowered the serum triglyceride, with less marked falls in phospholipid and cholesterol concentrations [9]. In the following year it was shown that carbohydrate ingestion resulted in a prompt drop in FFA [48]. More detailed investigations in the rat showed that feeding carbohydrate caused a decrease in low- and high-density lipoproteins, with a more marked drop in serum triglyceride than cholesterol [18]. When glucose is given intravenously the fall in triglyceride level – maximal 30–60 min after injection – still occurs [116], thereby eliminating an absorption factor, and if insulin is added to the glucose infusion post-prandial lipaemia is further decreased [81]. The lipaemia following a meal containing 60 g of fat in healthy young men could be abolished by adding 100–250 g of glucose to the meal and the addition of glucose also caused a reduction in FFA [3]. On the other hand, single oral doses of fructose and glycerol administered to normoglyceridaemic human subjects after a fat-containing meal increased the post-prandial hyperglyceridaemia [12].

Thus, it seems that one immediate effect of ingesting carbohydrate is the reduction in the level of serum triglyceride, whether the triglyceride is endogenous or exogenous. The effect of different types of dietary carbohydrate on this triglyceride-lowering effect has recently been investigated in man and it was noted that after a formula meal containing fat the degree of lipaemia was less when the meal contained glucose than when it contained an isocaloric amount of sucrose [98]. In a comparison between fructose and glucose given intravenously to fasting non-human primates it was found that after giving fructose to the male animals no fall in triglyceride concentration occurred but in the females a significant fall took place, whereas after intravenous glucose the serum triglyceride fell in both sexes reaching a nadir about 30 min after the injection. No significant change occurred in the serum cholesterol or phospholipid concentration after the fructose or glucose injections [71]. The ^{14}C label from the fructose or glucose did not appear in the triglyceride until about 45 min after injection, thus supporting the view that the carbohydrate had lowered the endogenous triglyceride already present at the time of ingestion.

The triglyceride-lowering effects of dietary carbohydrate seen very soon after ingestion are, as will be discussed later, in contrast with the more long-term effects where dietary carbohydrate raises the triglyceride level in fasting serum. The mechanisms in both cases are therefore different. In the immediate response the effect of the dietary carbohydrate cannot be due in any way to absorption because the effect is seen whether fat accompanies the carbohydrate or not, and can also be demonstrated when carbohydrate is given intravenously.

An increase in triglyceride input obviously cannot account for the lower values and the explanation seems likely to be in an increase in the removal rate of serum triglyceride after intravenous glucose. An increase in post-heparin lipolytic activity has been found [116]. Lipoprotein lipase causes the hydrolysis of triglyceride to fatty acids and glycerol, not only of chylomicrons, but also of very low density lipoproteins [60] and lipoprotein lipase is stimulated by insulin [66, 75, 76]. Thus, the fall in serum glyceride concentration after ingesting carbohydrate could be due to the increased rate of hydrolysis resulting from increased lipoprotein lipase activity induced by insulin. Evidence to support this possibility is found in the fact that insulin given to hyperglyceridaemic patients caused a fall in serum triglyceride [129]. Further support comes from the finding that fructose or the fructose-containing disaccharide sucrose are not associated with such a marked fall in triglyceride levels as is glucose and it is known that fructose is a much less effective stimulus to insulin release than is glucose [53]. Compatible with this explanation is the less marked rise in insulin level seen when the more marked increase in alimentary lipaemia occurs after sucrose as compared to glucose ingestion [98].

A less immediate reaction of serum triglyceride levels to dietary carbohydrates is the fact on a high-carbohydrate diet the highest level of triglyceride occurs after an overnight fast [130]. The explanation for this lies in the likelihood that the serum insulin level is probably lowest at this time. If the level of the triglyceride, whether endogenous or exogenous, in the serum is important in determining arterial disease then it is worth bearing in mind that the *mean* serum triglyceride level over a 24-hour period, in a normal person, is similar whether the person is on a high-carbohydrate or high-fat containing diet, and in carbohydrate-induced hyperlipoproteinaemia the *mean* 24-hour level of serum triglyceride is higher on a high-fat diet than on a high-carbohydrate diet [130].

IV. Effects of Dietary Carbohydrate on the Level of Lipid in Fasting Serum

The knowledge that dietary carbohydrate can be converted into lipid was first demonstrated in 1850 [83] and the existence of a difference between various carbohydrates in this respect was shown in animals in 1913 [139] and in man in 1919 [61]. The existence in man, of a carbohydrate-induced hypertriglyceridaemia reported in 1961 [2] and the association between raised serum lipids and vascular disease has led to a greater interest being taken in factors which influence the extent of the 'induction' by carbohydrates of raised serum lipid concentration.

A. Triglyceride

In view of the fact that the carbohydrate from the diet can be converted to triglyceride by the adipose tissue it is not surprising that the triglyceride level in the blood is the lipid fraction most sensitive to dietary carbohydrate. Low-fat (therefore high-carbohydrate) diets lead to an increase in the lipoprotein fraction whose main lipid component is triglyceride (Sf. 20–400) [58, 68, 107] and when the estimation of triglyceride became comparatively simpler [56] these studies were extended into investigating the effects of diets containing various types of carbohydrate on triglyceride metabolism.

There are several reports that fructose, in the diet of rats, causes a more striking increase in the fasting serum triglyceride level than does glucose [15, 96, 103, 110] and fructose or sucrose both seem capable of raising the fasting serum triglyceride level compared with a normal diet in both rats [63, 133] and baboons [27, 71].

In man, the isocaloric substitution of carbohydrate for fat in the diet produced significantly higher serum triglyceride concentrations [7] and the replacement of starch by sucrose in low-fat diets in men was associated with a marked increase in fasting serum triglyceride levels [93]. In patients with carbohydrate-induced hyperglyceridaemia feeding with sucrose aggravated the hyperglyceridaemia, whereas with starch there was a fall in serum triglyceride level [82]. Striking increments in serum triglyceride concentration occurred after a high-carbohydrate diet in subjects with raised initial triglyceride levels [104]. In patients with carbohydrate-induced hypertriglyceridaemia the increases in serum triglyceride levels were more marked with sucrose and with glucose than with starch in the diet [73] whereas in a study on two hyperglyceridaemics the substitution of glucose for starch in the diet did not result in any alteration of triglyceride level [118].

Sucrose, certainly in large amounts in the diet, therefore seems to be associated with an increase in the concentration of triglyceride in fasting serum when compared with starch, and it seems that this response occurs in normoglyceridaemic as well as hyperglyceridaemic individuals. The rôle of dietary glucose is not so clear-cut, but probably in normal persons the triglyceride response is no different to that after starch, but in the carbohydrate-induced patient glucose will obviously raise the triglyceride level, but not so markedly as does sucrose.

There have been two studies reported in which the reverse procedure was adopted, namely the withdrawal of sucrose from the diet and its replacement by glucose or its polymers. In patients with ischaemic heart disease the effect of withdrawing sucrose was a fall in serum triglyceride concentration [124]. In a more recent study carried out on normal men a sucrose-free diet for 14 weeks resulted in a fall in fasting serum triglyceride concentrations only in those men whose pre-diet level of triglyceride was raised [125]. These findings substantiate those where excess sucrose was given.

The feature about sucrose that distinguishes it from other disaccharides is the presence of fructose in the molecule. It is not possible for men to consume large quantities of fructose *per se* because of the osmotic diarrhoea that it produces. However, large amounts of fructose – in sucrose – can be absorbed without intestinal hurry. In an attempt to learn whether fructose is a more hyperglyceridaemic monosaccharide than glucose, or its polymers, experimental diets containing mixtures of fructose and starch, or glucose and starch were given to healthy young men and it was found the fructose-containing diet raised the triglyceride level whereas the glucose: starch dietary mixture did not [87].

Considerable work in animals comparing the effects on lipid metabolism of fructose and glucose has been mainly directed towards the metabolic changes in the liver and these will be considered elsewhere in this volume. The lack of availability of the liver in man has limited metabolic comparisons between the lipid consequences of fructose and glucose ingestion, except as reflected in the serum. However, studies using infusions of radioactive fatty acid are compatible with the hypothesis that the formation of triglyceride fatty acid is greater after a period on a sucrose diet than on a starch diet [105]. The increased formation of serum triglyceride fatty acid after sucrose compared with starch must presumably be an effect that is not insulin-mediated because the amount of glucose incorporated into triglyceride fatty acid is increased in the presence of insulin [20, 141].

Another hypothesis for the higher serum triglyceride concentration after fructose than after glucose is that fructose increases the formation of triglyceride glycerol. Certainly significantly more radioactivity from ^{14}C-fructose than from ^{14}C-glucose is recovered in the glycerol moiety of serum triglycerides of rats [15, 102]. Glycerol when consumed either 'acutely' or 'chronically' over several weeks on an otherwise normal diet, in men, raises the serum triglyceride level [91, 111], confirming an earlier finding in rats [109].

^{14}C-sucrose is incorporated into serum triglyceride to a greater extent than ^{14}C-glucose both in the rat [96] and in the baboon [97] and following the ingestion of labelled sucrose by baboons a correlation was observed between the level of fructose in peripheral blood and the degree to which the ^{14}C from fructose appeared in the serum triglyceride: no such correlation was found with glucose [26]. This is compatible with the finding that subjects with elevated fasting serum triglyceride concentration have higher levels of serum fructose after a sucrose meal [30].

Uniformly labelled ^{14}C-fructose or ^{14}C-glucose when given either to healthy men or to those recovered from a myocardial infarct resulted in greater specific activity of the serum triglycerides after fructose than after glucose, both sugars giving higher values in the patients [89]. In the patients the increase in radioactivity of triglyceride glycerol after the fructose ingestion was 3–4 times greater than after the glucose ingestion [99].

The two hypotheses mentioned imply that an increase in the concentration of triglyceride is due to an increase in production which is not matched by an increase in removal. There is some evidence to support this as diets high in carbohydrate increase the transfer of free fatty acids into serum triglycerides in normal subjects [142] and cause both increased production and decreased utilization of this lipid [106]. In an older group an increased turnover of triglyceride fatty acid following carbohydrate feeding has been reported [123]. On the other hand, two studies on subjects with hyperlipidaemia suggested the rate of removal rather than rate of production of triglyceride fatty acid was the determining factor [126, 127]. These two opposing views of the determining factor in the elevation of serum triglyceride may have to some extent been resolved by the report that in the normal physiological lipaemia induced by carbohydrate feeding, over-production is the initiating event, whereas in patients with carbohydrate-induced hypertriglyceridaemia over-production is not implicated [35].

There are many peoples of the world who subsist on a high-carbohydrate diet and do not develop hypertriglyceridaemia [11, 62, 132] and there are several possible explanations for this. One reason is that metabolic adaptation takes place and that in experimental studies the time course is too short for this relatively slow change to occur. A rise in serum triglyceride level followed by a fall while still consuming a high-carbohydrate diet has been reported in man [12, 17] and baboons [27]. Another possible explanation is that the triglyceridogenic property of dietary carbohydrate is offset by the presence of polyunsaturated fat in the diet of many whose normal diet contains a high proportion of carbohydrate *(vide infra).*

Not all studies have found a difference in the triglyceridaemic response between sucrose and glucose (or its polymers) and the presence of fat in the diet may be one reason for this discrepancy [33]. In other reports their brevity prevents an assessment of the findings for comparison purposes [13, 84].

B. Cholesterol

As the level of serum cholesterol occupies an important rôle as a forebearer of vascular catastrophe, much effort has been expended in determining all manner of factors which can alter the serum cholesterol concentration. The success of the polyunsaturated fats in reducing cholesterol level and the failure of dietary carbohydrate to have a comparable effect has meant that most work on dietary carbohydrate: serum cholesterol relationships was done a while ago.

Diets containing a high content of carbohydrate have long been known to be associated with a fall in the serum cholesterol level in man [58, 135]. Differences between carbohydrates in their cholesterol-lowering ability were reported in 1965 when the replacement of dietary sucrose by leguminous carbohydrates caused a fall in the level of cholesterol [49, 64].

When sucrose and leguminous carbohydrates were compared for their cholesterolaemic effect, it was found that sucrose gave higher levels of cholesterol in the serum than did the complex carbohydrates [54, 65, 77] and this also appears to hold for hyperlipidaemic subjects, since in these patients the same response has been demonstrated [73]. High-sucrose diets give slightly higher levels of cholesterol than do low-sucrose diets, and the addition of lactose to the high-sucrose diet made no further difference to the serum cholesterol level. In another study, while sucrose did not alter the cholesterol level of young men, exchanging the sucrose on an isocaloric basis with starch, glucose syrup, maltose or glucose resulted in a significant fall in each case [85].

Findings similar to those found in man have also been seen in animals such as the chick [51] the rabbit [143] and the rat [4, 52, 119, 134, 136]. There is also a report that, not unlike the response of triglycerides in man, prolonged feeding of rats with a high-sucrose diet resulted first in a rise and then a fall of the cholesterol to normal values [42].

Though there are many reports in man and animals that would lead one to suppose that dietary carbohydrates – and some more than others – are an important influence in determining cholesterol level this is probably not so. In many of the animal experiments cholesterol has to be fed in order that a difference between carbohydrate effects can be produced, and antibiotics in the diet of the animals can cause differences to disappear [52, 119, 121]. The nature of the dietary protein can change the effect of dietary carbohydrate on serum cholesterol [40] and the endocrine state of the animal may markedly modify the response [131] as may the strain of animal used [137]. In man similar pitfalls await the unwary. The reduction of the amount of cholesterol in the diet, or loss of weight while on the experimental diet as well as an alteration in the amount and type of fat can all alter the serum cholesterol concentration and, bearing in mind that a change in level in one dietary component inevitably means a change in level in another component, it behoves the investigator in this area to be very careful in drawing conclusions.

Of more interest, perhaps, is the hypothesis that the fibre in the diet can lower serum cholesterol by adsorbing the cholesterol and thus allowing it to be excreted [57, 140]. Fibre (also a carbohydrate) is composed of cellulose, lignin, etc., and if it does act in a manner similar to an ion exchange resin then it would be useful to know which component is responsible. It does not appear to be the cellulose component of fibre [79, 122] but there is evidence that pectin lowers serum cholesterol in man [113] and animals [41, 43]. There is a report that the complex carbohydrates known as gums are also capable of lowering serum cholesterol in man [38].

The relationship between dietary carbohydrate and cholesterol metabolism can be via acetyl-CoA or by alteration in bile acid secretion. Radioactive carbohydrates, as sucrose or glucose, fail to appear in either the free or esterified cholesterol fractions during the first few hours after ingestion [24, 89, 97] though after xylitol the free cholesterol is labelled soon after ingestion [71]. On the other hand, there is evidence that sucrose feeding in dogs resulted in a decrease in bile acid secretion [44] and that substitution of starch for sucrose in the diet of rats increased bile acid excretion [120].

Whatever the mechanism, it is largely of academic interest because dietary carbohydrates, though they can alter serum cholesterol levels, do so to a very limited extent, in man, compared with the amount and the type of dietary fat.

C. Phospholipid

The effects of dietary carbohydrate on serum phospholipid levels are far less documented than are the effects on triglyceride or cholesterol.

The immediate effect of dietary carbohydrate on the serum phospholipid concentration is to cause a fall in man [59] and in rats [18]. In patients with atherosclerosis a small addition of sucrose to the normal diet results in an increase in the serum phospholipid level [117] and these findings were confirmed in subjects on a high-sucrose diet [10]. Fractionation of the phospholipids revealed that the increase in concentration was largely due to proportional increase in the cephalin fraction. Male baboons on a high intake of sucrose showed a significant rise in phospholipid concentration, whereas the female monkeys did not show this change [25].

The paucity of reports on the influence of dietary carbohydrates on serum phospholipids probably reflects the current status of serum phospholipid as a harbinger of disease or metabolic disorder. The substitution of sugar for starch in young men and women results in an increase of saturated fatty acid and a decrease of polyunsaturated fatty acid in serum lecithin, with reduced thrombus time [8]. There are suggestions that certain phospholipids may play a part in accelerating blood clotting and, hence, coronary thrombosis but until a more precise rôle for this lipid fraction is known it seems unlikely that much will be learned of carbohydrate: phospholipid relationships.

V. Factors Affecting Dietary Carbohydrate: Serum Lipid Relationships

The findings so far mentioned have, in the main, used large quantities of the carbohydrate under study and have used males only. Mention has been made of the state of health of the human subject, but not of the frequency of food intake. The type and amount of fat or protein have not been considered and yet these are all factors which can modify the response of serum lipids to dietary carbohydrates. In the context of the practical consequences of consuming carbohydrate it should be borne in mind that in western communities approximately equicaloric amounts of fat and carbohydrate are eaten and that a large proportion of the consumers are women in the reproductive stage of life and, as will be discussed, both these factors may override the effects of dietary carbohydrate so far discussed.

A. Dietary Fat

As dietary fats affect the lipid profile in the serum and as dietary carbohydrates can do likewise it is perhaps not surprising that a mixture of these two major components of the diet may give rise to serum lipid responses that could not be predicted. It is well documented that the amounts and type of dietary fat have marked effects on the serum cholesterol level and because of the relatively slight influence of dietary carbohydrate on this lipid parameter it would be surprising if a mixture of carbohydrate and fat in the diet in equal proportions, resulted in effects different from those predicted from a study of the fat alone. In fact, in the calculation used to predict the effect of dietary fat on serum cholesterol concentration, it has not been found necessary to take into consideration either the amount or type of dietary carbohydrate [78].

Greater interest lies in the rôle of the dietary fat on serum triglyceride levels as it is considered that this lipid fraction is, in the main, carbohydrate-related. However, it has been found that corn oil diets result in lower serum triglyceride concentrations [1] and that saturated fatty acids less than C-12 and stearic acid in the diet raise the fasting triglyceride

Table I. The reported effects of various fatty acids and of carbohydrate on the fasting serum triglyceride and cholesterol concentrations

	Carbon chain length						Carbo-hydrate
	<10:0	12:0	14:0	16:0	18:0	18:2	
Fasting serum triglyceride level	+	0	0	0	+	—	+
Fasting serum cholesterol level	0	+	+	+	0	—	0

+ = Increase; — = decrease.

level [50]. As dietary fat is not entirely exempt in the metabolism of endogenous triglyceride in the serum it is obviously necessary to learn which of the dietary constituents which had opposing effects on serum triglyceride concentration was dominant. A summary of the currently known effects of dietary fatty acids and of carbohydrate is shown in table I.

Alteration of the diet such that there was an increase in intake of unsaturated fat led to a marked reduction of serum triglyceride [36] and in short-term experimental diets the increase in serum triglyceride level associated with dietary sucrose was absent when the diet contained sunflower seed oil, but was not prevented when cream replaced the polyunsaturated fat [88]. In a similar study where cocoa butter and sucrose were given, as chocolate, no overall change occurred in the fasting serum triglyceride level, but it was noted that on this diet there was an inverse relationship between the change in triglyceride level and the control value, so that those with higher triglyceride levels showed a fall after 5 days on the chocolate diet, whereas those with very low pre-diet triglyceride levels increased this level [90]. A synergistic effect of sucrose and animal fat on the serum triglyceride of hyperlipoproteinaemic patients has been shown whereas, in contrast, starch in place of sucrose is not so uniformly hyperlipidaemic [9]. Similar findings in normolipidaemic men have also been reported [105]. In a comparison between cream and sunflower seed oil given with glucose, fructose or starch, it was shown that irrespective of the carbohydrate sunflower seed oil always resulted in a marked fall in the level of serum triglycerides. The difference between fructose and other carbohydrates on serum triglyceride level, though present, was masked by that of the polyunsaturated fat [92]. A possible explanation for the more striking effect of polyunsaturated fat on fasting serum triglyceride level could be due to the increase in activity of lipoprotein lipase as this enzyme in rats has been shown to be influenced in this manner by polyunsaturated fats [115].

Thus, it seems that though the more detailed metabolism of the interplay between the type of dietary carbohydrate and the type of dietary fat has yet to be learnt, it certainly appears that, in terms of influence in fasting serum triglyceride levels, polyunsaturated fats may be of more consequence than the type of carbohydrate.

B. Sex

The possibility that the sex of the consumer might alter the extent of the metabolic response to diet was an inevitable consideration once it was accepted that ischaemic heart disease was relatively uncommon in pre-menopausal women – assuming that diet was an aetiological factor in this disorder. There is evidence that atherosclerosis is more common in young men than in young women [138] and that this protection is less in women who have had bilateral oophorectomy [114, 144]. As stated earlier there is also evidence that would be compatible with the view that diet is an aetiological agent in atherosclerosis, and that in some persons carbohydrate can be more sinister than fat in this regard. There are also reports that carbohydrate metabolism is affected by ovarian steroids [69, 95, 146]. It is not surprising, therefore, that men on a high-carbohydrate diet have a rise in fasting serum triglyceride concentration whereas young women do not [16] and that this difference is more marked with fructose. In support of these findings that the serum lipids of pre-menopausal women behave differently to those of men in response to dietary carbohydrate, is the report that, as in men, so in post-menopausal women the fasting serum triglyceride level increases on a high-sucrose diet [86]; that young women have a similar endogenous triglyceride response to wheat starch and to sucrose [80]; and that in a diet containing a high proportion of 'simple sugars' the serum phospholipid increases more in men than in pre-menopausal women [10]. The non-human primate also shows a similar response until adaptation to a high-sucrose diet occurred in the male [27].

Further evidence of a different metabolic response to sucrose by young women is seen in that the concentration of fasting serum triglyceride is directly related to the serum level of fructose after a sucrose load in men, but not in young women [30] and after both 'acute' and 'chronic' ingestion of glycerol the rise in serum triglyceride concentration is much more marked in men than in young women [91]. In baboons the rate at which fructose is metabolized appears to be greater in male baboons than in sexually mature female baboons, while no difference between the sexes in rate of metabolism of glucose was detected and after the intravenous injection of fructose the serum triglyceride level falls in female baboons [71]. No differences between the sexes in response to intravenous xylitol was found [72].

Recent work has suggested that the difference in the response to sucrose lies in the fact that females clear serum triglycerides more rapidly than males and that a high-sucrose diet produces a equal increase in lipogenesis in both sexes [74]. It remains to be seen what effect the widespread use of oral contraceptives will have on carbohydrate: lipid interrelationships, as it is known that these agents raise the serum triglyceride level [14, 19, 47, 145] and, furthermore, have a tendency to make the glucose tolerance test more diabetic [147] largely due to oestrogen [70]. A study of triglyceride turnover during the use of oral contraceptives has shown that there is a highly significant increase in both the rate of production and efficiency of removal of serum triglyceride and that the rise in level is due to enhanced influx [108].

The apparently important rôle played by the female sex hormones in carbohydrate: lipid relationships is at the moment ill understood. There are many apparent contradictory findings, and the advent of the oestrogen-progesterone contraceptive pill would seem to make further understanding of the relationships between carbohydrates and lipids more important.

The hormonal factor (or factors) responsible for this sex difference has not been identified, though from the available evidence it seems that it is not androgens that are responsible but that an ovarian hormone prevents what would seem to be the patho-physiological increase in serum lipids with age occurring in many people. In baboons, giving testosterone to female animals on high-sucrose diets had no effect on the serum lipid levels whereas giving natural oestrogen prevented the rise associated with a high-sucrose diet in the males [27]. The increased incidence of atherosclerosis in oophorectomized women and in post-menopausal women would again suggest that an ovarian hormone(s) in some way halts the progress of atherosclerosis – perhaps by modifying serum lipid levels? Evidence against this suggestion is found in pregnancy where there is an increased concentration of serum triglyceride, despite a hypervolaemia, but in pregnancy an increase in ovarian hormones is not the only marked alteration in the hormonal pattern.

VI. Other Factors which Can Influence the Dietary Carbohydrate: Serum Lipid Relationships

A. Species

There is a difference between species and within species in the response to dietary carbohydrate. Fructose, when consumed by man in large quantities gives rise to osmotic diarrhoea, but not when the fructose is consumed in similar quantities in the disaccharide sucrose. The rat can survive for several months of diets containing over 80% fructose. This difference between man and rat is most likely due to rates of absorption of fructose [32]. It is possible that a prolonged high intake of carbohydrate can induce increased absorption as has been shown for the fructose moiety of sucrose [31]. The response to dietary carbohydrates also varies with the strain of rat used [34].

B. Dietary Protein

There are reports that the nature of the protein in the diet can affect the serum phospholipid level [6] and the serum cholesterol level [112] but no reports that the serum triglyceride concentration is affected in this way. This is perhaps surprising in view of the fact that carbohydrate:triglyceride, relationships are bound up with insulin and that some amino acids such as tryptophan and leucine are more insulinogenic [39]. However, as dietary proteins are not absorbed as amino acids [29] it could be that an effect would only be seen when amino acids are given as such. To some extent this seems to be so, in that a high-sucrose diet with an amino acid mixture causes a very striking and dramatic rise in fasting serum triglyceride levels in men whereas sucrose with other proteins such as casein (sodium and calcium salts) egg-albumen and gelatin result in increases in serum triglyceride level of a smaller order [23]. Perhaps of interest is that the very marked triglyceridogenic effect of amino acids and sucrose is not seen when the sucrose is replaced by glucose, lactose or partially hydrolyzed starch. In rats, lowering the content of protein in the diet affects the pathways of liver lipid synthesis differently, depending on the nature of the dietary carbohydrate [22]. The importance of protein in the diet in determining the serum lipid levels in rats, not dependent on the type of dietary carbohydrate, has been recently reported [102].

C. Frequency of Ingestion

There is a considerable literature on the effects of meal pattern on metabolic performance and disease [37] and on serum lipid levels [55]. Carbohydrates are often consumed frequently and in small amounts because of their easy ability to give rise to a sensation of fullness. In an experiment lasting 18 days the fasting serum triglyceride level rose early in the infrequent eating phase of a high-sucrose diet and then reached a plateau. During the 18 days when the sucrose diet was nibbled, the fasting serum triglyceride concentration rose progressively, while the phospholipid level fell significantly [94]. There are several possible explanations for this difference in triglyceride response, but in the absence of further findings such as serum insulin levels, speculation would not be profitable.

VII. Suggested Explanations for the Difference between Dietary Carbohydrates on the Response of Serum Triglyceride Level

From the evidence presented earlier it seems that the major difference between carbohydrates on serum triglyceride levels depends on the presence of fructose in the ingested carbohydrate. The inability of fructose to lower serum triglyceride levels when compared with glucose has been postulated as being due to the fact that fructose does not stimulate insulin release [128]. Some of the fructose given by mouth will be converted to glucose, most of the conversion taking place in the liver [28] so a glucose 'effect' is always seen when fructose is given but the extent of this 'effect' will be less than when equivalent amounts of glucose are given.

The failure of fructose to release insulin could account for the difference in serum triglyceride response when compared to glucose, particularly in view of the rôle of insulin in adipose triglyceride metabolism. As stated in other chapters the conversion of fructose to triglyceride is a function of the liver, whereas the conversion of glucose to triglyceride is the function of adipose tissue. The triglyceride formed from fructose in the liver is transported via the plasma and an increase in plasma concentration of this lipid reflects a disparity between input and output.

The increase in the manufacture of triglyceride in the liver after consuming fructose could be due either to an increase in the formation of the fatty acid or the formation of the glycerol moieties of the triglyceride molecule, or to both equally. Radio-isotopically labelled fructose and glucose given to fasting rats showed that there was a greater proportion of radioactivity in the serum triglyceride glycerol after fructose than glucose. When glycerol is given by mouth, a rise in serum triglyceride level takes place whether the triglyceride concentration is measured within 3 h of glycerol ingestion or in fasting serum after several days on an increased glycerol intake. Young women do not react to glycerol in this manner. Thus the evidence is compatible with the view that fructose, though it can be converted to fatty acid either directly or via its prior conversion to glucose, seems to exert its major effect on serum triglyceride metabolism through a relative increase in the glycerol moiety of triglyceride. The fact that polyunsaturated fats interfere extensively in the dietary carbohydrate: serum lipid interrelationship remains to be explained.

VIII. Conclusions

It seems, therefore, that the serum lipids, and especially the concentration of triglyceride, can be influenced by the amount and the type of dietary carbohydrate. The degree to which dietary carbohydrate can alter serum triglyceride levels depends, among other things, on the type of fat accompanying the carbohydrate, the sex of the consumer and the frequency of ingestion.

IX. References

1 AHRENS, E. H.; HIRSCH, J.; INSULL, W.; TSALTAS, T. T.; BLOMSTRAND, R., and PETERSON, M. L.: The influence of dietary fats on serum lipid levels in man. Lancet *i:* 943–953 (1957).

2 AHRENS, E. H.; HIRSCH, S.; OETTLE, K.; FARQUHAR, J. W., and STEIN, Y.: Carbohydrate-induced and fat-induced lipemia. Trans. Ass. amer. Physicians *74:* 134–146 (1961).

3 ALBRINK, M. J.; FITZGERALD, J. R., and MAN, E. B.: Reduction of alimentary lipemia by glucose. Metabolism *7:* 162–171 (1958).

4 ALLEN, R. J. and LEAHY, J. S.: Some effects of dietary dextrose, fructose, liquid glucose and sucrose in the adult male rat. Brit. J. Nutr. *20:* 339–347 (1966).

5 ANDERSON, J. T.; GRANDE, F., and KEYS, A.: Responses of various serum lipids to high fat and high carbohydrate diets in man. 5th Int. Congr. Nutr. 1960.

6 ANDERSON, J. T.; GRANDE, F., and KEYS, A.: Effect on man's serum lipids of two proteins with different amino acid composition. Amer. J. clin. Nutr. *24:* 524–530 (1971).

7 ANDERSON, J. T.; GRANDE, E. F.; MATSUMOTO, Y., and KEYS, A.: Glucose, sucrose and lactose in the diet and blood lipids in man. J. Nutr. *79:* 349–359 (1963).

8 ANTAR, M. A.: Studies of thrombus formation and fatty acid composition of serum lecithins and cephalins in man during high sugar and starch diets. J. Atheroscler. Res. *8:* 569–578 (1968).

9 ANTAR, M. A.; LITTLE, J. A.; LUCAS, P.; BUCKLEY, G. C., and CSIMA, A.: Interrelationship between the kinds of dietary carbohydrate and fat in hyperlipoproteinemic patients. 3. Synergistic effect of sucrose and animal fat on serum lipids. Atherosclerosis *11:* 191–201 (1970).

10 ANTAR, M. A. and OHLSON, M. A.: Effect of simple and complex carbohydrates upon total lipids, nonphospholipids and different fractions of phospholipids of serum of young men and women. J. Nutr. *85:* 329–337 (1965).

11 ANTONIS, A. and BERSOHN, I.: Serum triglyceride levels in South African Europeans and Bantu and in ischaemic heart disease. Lancet *i:* 998–1002 (1960).

12 ANTONIS, A. and BERSOHN, I.: The influence of diet on serum triglycerides in South African white and Bantu prisoners. Lancet *i:* 3–9 (1961).

13 ANTONIS, A.; ILES, C., and PILKINGTON, T. R. E.: The effects of dietary carbohydrate on serum lipid levels. Proc. Nutr. Soc. *27:* 2A (1968).

14 AURELL, M.; CRAMER, K., and AYBO, G.: Serum lipids and lipoproteins during long-term administration of an oral contraceptive. Lancet *i:* 291–293 (1966).

15 BAR-ON, H. and STEIN, Y.: Effect of glucose and fructose administration on lipid metabolism in the rat. J. Nutr. *94:* 95–105 (1968).

16 BEVERIDGE, J. M. R.; JAGANNATHAN, S. N., and CONNELL, W. F.: The effect of the type and amount of dietary fat on the level of plasma triglycerides in human subjects in the postabsorptive state. Canad. J. Biochem. *42:* 999–1003 (1964).

17 BIERMAN, E. L. and HAMLIN, J. T.: The hyperlipemic effect of a low-fat, high-carbohydrate diet in diabetic subjects. Diabetes, N.Y. *10:* 432–437 (1961).

18 BRAGDON, J. H.; HAVEL, R. J., and GORDON, R. S.: Effects of carbohydrate feeding on serum lipids and lipoproteins of the rat. Amer. J. Physiol. *189:* 63–67 (1957).

19 BRODY, S.; KERSTELL, J.; NILSSON, L., and SVANBORG, A.: The effects of some ovulation inhibitors on the different plasma lipid fractions. Acta med. scand. *183:* 1–7 (1968).

20 CAHILL, G. F.; LEBOEUF, B., and RENOLD, A. E.: Studies on rat adipose tissue *in vitro*. III. Synthesis of glycogen and glyceride-glycerol. J. biol. Chem. *234:* 2540–2543 (1959).

21 CARLSON, L. A. and BOTTIGER, L. E.: Ischaemic heart disease in relation to fasting values of plasma triglycerides and cholesterol. Lancet *i:* 865–868 (1972).
22 COHEN, A. M. and TEITELBAUM, A.: Effect of different levels of protein in sucrose and starch diets on lipid synthesis in the rat. Israel. J. med. Sci. *2:* 727–732 (1966).
23 COLES, B. L. and MACDONALD, I.: The influence of dietary protein on dietary carbohydrate: lipid interrelationships. Nutr. Metabol. *14:* 238–244 (1972).
24 COLTART, T. M.: The effect of hormones on the interrelationship between dietary carbohydrates and lipid metabolism; PhD Thesis London (1968).
25 COLTART, T. M.: Changes in serum phospholipids in male and female baboons on a sucrose diet. Nature, Lond. *222:* 575–576 (1969).
26 COLTART, T. M. and CROSSLEY, J. N.: Influence of dietary sucrose on glucose and fructose tolerance and triglyceride synthesis in the baboon. Clin. Sci. *38:* 427–437 (1970).
27 COLTART, T. M. and MACDONALD, I.: Effect of sex hormones on fasting serum triglycerides in baboons given high-sucrose diets. Brit. J. Nutr. *25:* 323–331 (1971).
28 COOK, G. C.: Absorption products of D(–)fructose in man. Clin. Sci. *37:* 675–687 (1969).
29 CRAFT, I. L.; GEDDES, D.; HYDE, C. W.; WISE, I. J., and MATTHEWS, D. M.: Absorption and malabsorption of glycine and glycine peptides in man. Gut *9:* 425–437 (1968).
30 CROSSLEY, J. N.: Sucrose tolerance and fasting serum glyceride concentrations in young men and women. Proc. Nutr. Soc. *26:* iii–iv (1967).
31 CROSSLEY, J. N. and MACDONALD, I.: The influence in male baboons, of a high sucrose diet on the portal and arterial levels of glucose and fructose following a sucrose meal. Nutr. Metabol. *12:* 171–178 (1970).
32 DAHLQVIST, A. and THOMPSON, D. L.: The digestion and absorption of sucrose by the intact rat. J. Physiol., Lond. *167:* 193–200 (1963).
33 DUNNIGAN, M. G.; FYFE, T.; MCKIDDIE, M. T., and CROSBIE, S. M.: The effects of isocaloric exchange of dietary starch and sucrose on glucose tolerance, plasma insulin and serum lipids of man. Clin. Sci. *38:* 1–9 (1970).
34 DURAND, A. M. A.; FISHER, M., and ADAMS, M.: The influence of type of dietary carbohydrate. Arch. Path. *85:* 318–324 (1968).
35 EATON, R. P.: Synthesis of plasma triglycerides in endogenous hypertriglyceridemia. J. Lipid Res. *12:* 491–497 (1971).
36 ENGELBERG, H.: Mechanisms involved in the reduction of serum triglycerides in man upon adding unsaturated fats to the normal diet. Metabolism *15:* 796–807 (1966).
37 FABRY, P.: Feeding patterns and nutritional adaptation (Butterworth, London 1967).
38 FAHRENBACH, M. J.; RICCARDI, B. A.; SAUNDERS, J. C.; LAURIE, I. N., and GRANT, W. C.: Comparative effect of Guar gum and pectin on human serum cholesterol levels. Circulation *32:* suppl. II, pp. 11–12 (1965).
39 FAJANS, S. S.; FLOYD, J. C.; KNOFF, R. F., and CONN, J. W.: Effect of amino acids and proteins on insulin secretion in man. Rec. Hormone Res. *23:* 617–656 (1967).

40 FARNELL, D. R. and BURNS, M. J.: Dietary starch. Effect on cholesterol levels. Metabolism *11:* 566–571 (1962).

41 FAUSCH, H. D. and ANDERSON, T. A.: Influence of citrus pectin feeding on lipid metabolism and body composition of swine. J. Nutr. *85:* 145–149 (1965).

42 FILLIOS, L. C.; NAITO, C.; ANDRUS, S. B.; PORTMAN, O. W., and MARTIN, R. S.: Variations in cardiovascular sudanophilia with changes in the dietary level of protein. Amer. J. Physiol. *194:* 275–279 (1958).

43 FISHER, H.; SILLER, W. G., and GRIMINGER, P.: The retardation by pectin of cholesterol-induced atherosclerosis in the fowl. J. Atheroscler. Res. *6:* 292–298 (1966).

44 FOSTER, M. G.; HOOPER, C. W., and WHIPPLE, G. H.: The metabolism of bile acids. IV. Endogenous and exogenous factors. J. biol. Chem. *38:* 393–411 (1919).

45 FREDRICKSON, D. S.: The regulation of plasma lipoprotein concentrations as affected in human mutants. Proc. nat. Acad. Sci., Wash. *64:* 1138–1146 (1969).

46 FREDRICKSON, D. S.; LEVY, R. I., and LEES, R. S.: Fat transport in lipoproteins. An integrated approach to mechanisms and disorders. New. Engl. J. Med. *276:* 34, 94, 148, 215, 273 (1961).

47 GERSHBERG, H.; HULSE, M., and JAVIER, Z.: Hypertriglyceridemia during treatment with oestrogen and oral contraceptives. An alteration in hepatic function? Obstet. Gynec., N.Y. *31:* 186–189 (1968).

48 GOODMAN, D. S. and GORDON, R. S.: The metabolism of plasma unesterified fatty acid. Amer. J. clin. Nutr. *6:* 669–680 (1958).

49 GRANDE, F.; ANDERSON, J. T., and KEYS, A.: Effect of carbohydrates of leguminous seeds, wheat and potatoes on serum cholesterol concentration in man. J. Nutr. *86:* 313–317 (1965).

50 GRANDE, F.; ANDERSON, J. T., and KEYS, A.: Diets of different fatty acid composition producing identical serum cholesterol levels in man. Amer. J. clin. Nutr. *25:* 53–60 (1972).

51 GRANT, W. C. and FAHRENBACH, M. J.: Influence of sucrose and casein on plasma cholesterol in animals on purified diets. Fed. Proc. *16:* 50 (1957).

52 GRANT, W. C. and FAHRENBACH, M. J.: Effect of dietary sucrose and glucose on plasma cholesterol in chicks and rabbits. Proc. Soc. exp. Biol. Med. *100:* 250–252 (1959).

53 GRODSKY, G. M.; BATTS, A. A.; BENNETT, L. L.; VCELLA, C.; MCWILLIAMS, N. B., and SMITH, D.: Effects of carbohydrates on secretion of insulin from isolated rat pancreas. Amer. J. Physiol. *205:* 638–644 (1963).

54 GROEN, J. J.: Effect of bread in the diet on the serum cholesterol. Amer. J. clin. Nutr. *20:* 191–197 (1967).

55 GWINUP, G.; BYRON, R. C.; ROUSH, W. H.; KRUGER, F. A., and HAMWI, G. J.: Effect of nibbling *versus* gorging on serum lipids in man. Amer. J. clin. Nutr. *13:* 209–213 (1963).

56 HANDEL, E. V. and ZILVERSMITH, D.: Micromethod for the direct determination of serum triglycerides. J. Lab. clin. Med. *50:* 152–157 (1957).

57 HARDINGE, M. G.; CHAMBERS, A. C.; CROOKS, H., and STARE, F. J.: Nutritional studies of vegetarians. III. Dietary levels of fiber. Amer. J. clin. Nutr. *6:* 523–525 (1958).

58 HATCH, F. T.; ABELL, L. L., and KENDALL, F. E.: Effects of restriction of dietary fat and cholesterol upon serum lipids and lipoproteins in patients with hypertension. Amer. J. Med. *19:* 48–60 (1955).

59 HAVEL, R. J.: Early effects of fasting and of carbohydrate ingestion on lipids and lipoproteins of serum in man. J. clin. Invest. *36:* 855–859 (1957).

60 HAVEL, R. J.: Adipose tissue; in Handbook of physiology, section 5, pp. 499–507 (American Physiological Society, Washington 1965).

61 HIGGINS, H. L.: The rapidity with which alcohol and some sugars may serve as nutrient. Amer. J. Physiol. *41:* 258–265 (1919).

62 HIGGINSON, J. and PEPLER, W. J.: Fat intake, serum cholesterol concentration, and atherosclerosis in South African Bantu. II. Atherosclerosis and coronary artery disease. J. clin. Invest. *33:* 1366–1371 (1954).

63 HILL, P.: Effect of fructose on rat lipids. Lipids *5:* 621–627 (1970).

64 HODGES, R. E. and KREHL, W. A.: The role of carbohydrates in lipid metabolism. Amer. J. clin. Nutr. *17:* 334–346 (1965).

65 HODGES, R. E.; KREHL, W. A.; STONE, D. B., and LOPEZ, A.: Dietary carbohydrates and low cholesterol diets. Effects on serum lipids of man. Amer. J. clin. Nutr. *20:* 198–208 (1967).

66 HOLLENBERG, C. H.: Effect of nutrition on activity and release of lipase from rat adipose tissue. Amer. J. Physiol. *197:* 667–670 (1959).

67 HOLLISTER, L. E. and WRIGHT, A.: Diurnal variation of serum lipids. J. Atheroscler. Res. *5:* 445–450 (1965).

68 HORLICK, L.: Further observations on dietary modification of serum cholesterol levels. Canad. med. Ass. J. *85:* 1127–1131 (1961).

69 HOUSSAY, B. A.; FOGLIA, V. G., and RODRIGUEZ, R. R.: Production or prevention of some types of experimental diabetes by estrogens or corticosteroids. Acta endocrin., Kbh. *17:* 146–164 (1954).

70 JAVIER, Z.; GERSHBERG, H., and HULSE, M.: Ovulatory suppressants, estrogens and carbohydrate metabolism. Metabolism *17:* 443–456 (1968).

71 JOURDAN, M. H.: The effect of a sucrose-enriched diet on the metabolism of intravenously administered fructose in baboons. Nutr. Metab. *14:* 28–37 (1972).

72 JOURDAN, M. H., HENDERSON, J. R., and MACDONALD, I.: The effect of C^{14} xylitol given intravenously on the serum glucose, insulin and lipid concentrations of male and female baboons. Nutr. Metabol. *14:* 92–99 (1972).

73 KAUFMANN, N. A.; POZNANSKI, R.; BLONDHEIM, S. A., and STEIN, Y.: Changes in serum lipid levels of hyperlipemic patients following the feeding of starch, sucrose and glucose. Amer. J. clin. Nutr. *18:* 261–269 (1966).

74 KEKKI, M. and NIKKILA, E. A.: Plasma triglyceride turnover during use of oral contraceptives. Metabolism *20:* 878–879 (1971).

75 KESSLER, J. I.: Effect of insulin on the release of plasma lipolytic activity and clearing of emulsified fat intravenously administered to pancreatectomised and alloxanised dogs. J. Lab. clin. Med. *60:* 747–755 (1962).

76 KESSLER, J. I.: Effect of diabetes and insulin on the activity of myocardial and adipose tissue lipoprotein lipase of rats. J. clin. Invest. *42:* 362–367 (1963).

77 KEYS, A.; ANDERSON, J. T., and GRANDE, F.: Diet-type (fats constant) and blood lipids in man. J. Nutr. *70:* 257–266 (1960).

78 KEYS, A.; ANDERSON, J. T., and GRANDE, F.: Serum cholesterol response to changes in the diet. IV. Particular saturated fatty acids in the diet. Metabolism *14:* 776–787 (1965).

79 KEYS, A.; GRANDE, F., and ANDERSON, J. T.: Fiber and pectin in the diet and serum cholesterol concentration in man. Proc. Soc. exp. Biol. Med. *106:* 555–558 (1961).

80 KLUGH, C. A. and IRWIN, M. I.: Serum levels of young women as related to source of dietary carbohydrate. Fed. Proc. *25:* 672 (1966).

81 KRUT, L. H. and BARSKY, R. F.: Effect of enhanced glucose utilisation on post-prandial lipemia in ischaemic heart disease. Lancet *ii:* 1136–1138 (1964).

82 KUO, P. T. and BASSETT, D. R.: Dietary sugar in the production of hyper-glyceridemia. Ann. intern. Med. *62:* 1199–1212 (1965).

83 LAWES, J. B. and GILBERT, J. H.: Composition of foods in relation to respiration and feeding of animals. Report 323 (British Association for Advanced Science, London 1850).

84 LEES, R. S. and FREDRICKSON, D. S.: Carbohydrate induction of hyperlipemia in normal men. Clin. Res. *13:* 327 (1965).

85 MACDONALD, I.: The effects of various dietary carbohydrates on serum lipids during a five day regimen. Clin. Sci. *29:* 193–197 (1965).

86 MACDONALD, I.: The lipid response of post-menopausal women to dietary carbohydrates. Amer. J. clin. Nutr. *18:* 86–90 (1966).

87 MACDONALD, I.: Influence of fructose and glucose on serum lipid levels in men and pre- and post-menopausal women. Amer. J. clin. Nutr. *18:* 369–372 (1966).

88 MACDONALD, I.: Interrelationship between the influences of dietary carbohydrates and fats on fasting serum lipids. Amer. J. clin. Nutr. *20:* 345–351 (1967).

89 MACDONALD, I.: Ingested glucose and fructose in serum lipids in healthy men and after myocardial infarction. Amer. J. clin. Nutr. *21:* 1366–1373 (1968).

90 MACDONALD, I.: Effects of a skimmed milk and chocolate diet on serum and skin lipids. J. Sci. Food Agric. *19:* 270–272 (1968).

91 MACDONALD, I.: Effects of dietary glycerol on the serum glyceride level of men and women. Brit. J. Nutr. *24:* 537–543 (1970).

92 MACDONALD, I.: Relationship between dietary carbohydrates and fats in their influence on serum lipid levels. Clin. Sci. *43:* 265–274 (1972).

93 MACDONALD, I. and BRAITHWAITE, D. M.: The influence of dietary carbohydrates on the lipid pattern in serum and in adipose tissue. Clin. Sci. *27:* 23–30 (1964).

94 MACDONALD, I.; COLES, B. L.; BRICE, J., and JOURDAN, M. H.: The influence of frequency of sucrose intake on serum lipid, and carbohydrate levels. Brit. J. Nutr. *24:* 413–423 (1970).

95 MACDONALD, I. and CROSSLEY, J. N.: Glucose tolerance during the menstrual cycle. Diabetes, N.Y. *19:* 450–452 (1970).

96 MACDONALD, I. and ROBERTS, J. B.: The incorporation of various ^{14}C dietary carbohydrates into serum and liver lipids. Metabolism *14:* 991–999 (1965).

97 MACDONALD, I. and ROBERTS, J. B.: The serum lipid response of baboons to various carbohydrate meals. Metabolism *16:* 572–579 (1967).

97a MACRAE, A. R, and NEUDOERFFER, T. S.: Support for the existence of an active transport mechanism of fructose in the rat. Biochim. biophys. Acta *288* (1): 137–144 (1972).

98 MANN, J. I.; TRUSWELL, A. S., and PIMSTONE, B. L.: The different effects of oral sucrose and glucose on alimentary lipemia. Clin. Sci. *41:* 123–129 (1971).

99 MARUHAMA, Y.: Conversion of ingested carbohydrate ^{14}C into glycerol and fatty acids of serum triglyceride in patients with myocardial infarction. Metabolism *19:* 1085–1093 (1970).

100 MARUHAMA, Y. and MACDONALD, I.: Some changes in the triglyceride metabolism of rats on high fructose or glucose diets. Metabolism *21:* 835–842 (1972).

101 MASAREI, J. R.; SUMMERS, M.; CURNOW, D. H.; CULLEN, K. J.; MCCALL, M. G.; STENHOUSE, N. S., and WELBOURN, T. A.: Lipoprotein electrophoretic patterns, serum lipids and coronary heart disease. Brit. med. J. *i:* 78–82 (1971).

102 MCGREGOR, D.: The effects of some dietary changes upon the concentrations of serum lipids in rats. Brit. J. Nutr. *25:* 213–224 (1971).

103 MUKHERJEE, S.; BASO, M., and TRIVEDI, K.: Effect of low dietary levels of glucose, fructose and sucrose on rat lipid metabolism. J. Atheroscler. Res. *10:* 261–272 (1969).

104 NESTEL, P. J.: Carbohydrate-induced hypertriglyceridemia and glucose utilization in ischaemic heart disease. Metabolism *15:* 787–795 (1966).

105 NESTEL, P. J.; CARROLL, K. F., and HAVENSTEIN, N.: Plasma triglyceride response to carbohydrates, fats and calorie intake. Metabolism *19:* 1–18 (1970).

106 NESTEL, P. J. and HIRSCH, E. Z.: Triglyceride turnover after diets rich in carbohydrate or animal fat. Austr. Ann. Med. *14:* 265–269 (1965).

107 NICHOLLS, A. V.; DOBBIN, V., and GOFMAN, J. W.: Influence of dietary factors upon human serum lipoprotein concentrations. Geriatrics *12:* 7–17 (1957).

108 NIKKILA, E. A. and KEKKI, M.: Polymorphism of plasma triglyceride kinetics in normal human adult subjects. Acta med. scand. *190:* 49–59 (1971).

109 NIKKILA, E. A. and OJALA, K.: Hyperglyceridemia induced by glycerol feeding. Life Sci. *3:* 1021–1023 (1964).

110 NIKKILA, E. A. and OJALA, K.: Induction of hyperglyceridemia by fructose in the rat. Life Sci. *4:* 937–943 (1965).

111 NIKKILA, E. A. and PELKONEN, R.: Enhancement of alimentary hyperglyceridemia by fructose and glycerol in man. Proc. Soc. exp. Biol. Med. *123:* 91–94 (1966).

112 OLSON, R. E.; NICHAMAN, M. Z.; NITTKA, J., and EAGLES, J. A.: Effect of amino acid diets upon serum lipids in man. Amer. J. clin. Nutr. *23:* 1614–1625 (1970).

113 PALMER, G. H. and DIXON, D. G.: Effect of pectin dose on serum cholesterol levels. Amer. J. clin. Nutr. *18:* 437–442 (1966).

114 PARRISH, H. M.; CARR, C. A.; HALL, D. G., and KING, J. M.: Time interval from castration in premenopausal women to development of excessive coronary atherosclerosis. Amer. J. Obstet. Gynec. *99:* 155–162 (1967).

115 PAWAR, S. S. and TIDWELL, H. C.: Effect of ingestion of unsaturated fat on lipolytic activity of rat tissues. J. Lipid Res. *9:* 334–336 (1968).

116 PERRY, W. F. and CORBETT, B. N.: Changes in plasma triglyceride concentration following the intravenous administration of glucose. Canad. J. Physiol. Pharmacol. *42:* 353–356 (1964).

117 PLESHKOV, A. M.: Effect of prolonged use of easily absorbable carbohydrates (sugar) on blood lipid levels in patients with atherosclerosis. Ter. Arkh. *35:* 66–70 (1963); translation in Fed. Proc. *23:* 334–336 (1964).
118 PORTE, D.; BIERMAN, E. L., and BAGDADE, J. D.: Substitution of dietary starch for dextrose in hyperlipemic subjects. Proc. Soc. exp. Biol. Med. *123:* 814–816 (1966).
119 PORTMAN, O. W.; LAWRY, E. Y., and BRUNO, D.: Effect of dietary carbohydrates on experimentally induced hypercholesterolemia and hyperbetalipoproteinemia in rats. Proc. Soc. exp. Biol. Med. *91:* 321–323 (1956).
120 PORTMAN, O. W.; MANN, G. V., and WYSOCKI, A.: Bile acid excretion by the rat. Nutritional effects. Arch. Biochem. *59:* 224–232 (1955).
121 PORTMAN, O. W. and STARE, F. J.: Dietary regulation of serum cholesterol levels. Physiol. Rev. *39* (3): 407–442 (1959).
122 PRATHER, E. S.: Effect of cellulose on serum lipids in young women. J. amer. diet. Ass. *45:* 230–233 (1964).
123 REAVEN, G. M.; HILL, D. B.; GROSS, R. C., and FARQUHAR, J. W.: Kinetics of triglyceride turnover of very low density lipoproteins of human plasma. J. clin. Invest. *44:* 1826–1833 (1965).
124 RIFKIND, B. M.; LAWSON, D. H., and GALE, M.: Effects of short-term sucrose restriction on serum lipid levels. Lancet *ii:* 1379–1381 (1966).
125 ROBERTS, A. M.: Some effects of a sucrose-free diet on fasting serum lipid levels. Proc. Nutr. Soc. *30:* 71–72 (1971).
126 RYAN, W. G. and SCHWARTZ, T. B.: Dynamics of plasma triglyceride turnover in man. Metabolism *14:* 1243–1254 (1965).
127 SAILER, S.; SANDHOFER, F. und BRAUNSTEINER, H.: Umsatzraten für freie Fettsäuren und Triglyceride in Plasma bei essentieller Hyperlipämie. Klin. Wschr. *44:* 1032–1036 (1966).
128 SAMOLS, E. and DORMANDY, T. L.: Insulin response to fructose and galactose. Lancet *i:* 478–479 (1963).
129 SCHLIERF, G. and KINSELL, R.: Effect of insulin in hypertriglyceridemia. Proc. Soc. exp. Biol. Med. *120:* 272–274 (1965).
130 SCHLIERF, G.; REINHEIMER, W., and STOSSBERG, V.: Diurnal patterns of plasma triglycerides and free fatty acids in normal subjects and in patients with endogenous (type IV) hyperlipoproteinemia. Nutr. Metabol. *13:* 80–91 (1971).
131 SCHULTZ, A. L. and GRANDE, F.: Effects of starch and sucrose on the serum lipids of dogs before and after thyroidectomy. J. Nutr. *94:* 71–73 (1968).
132 SCHWARTZ, M. J.; ROSENSWEIG, B.; TOOR, M., and LEWITUS, Z.: Lipid metabolism and arteriosclerotic heart disease in Israelites of Bedouin, Yemenite and European origin. Amer. J. Cardiol. *12:* 157–168 (1963).
133 SHIFF, T. S.; ROHEIM, P. S., and EDER, H. A.: Effects of high sucrose diets and 4-aminopyrazolo-pyrimidine on serum lipids and lipoproteins in the rat. J. lipid Res. *12:* 596–603 (1971).
134 SIMKO, V. und CHORUATAHOVA, V.: Der Fettstoffwechsel von Ratten unter dem Einfluss der Fütterung mit verschiedenen Kohlenhydraten. Ernährungsforschung *13:* 361–368 (1968).
135 STARKE, H.: Effect of the rice diet on the serum cholesterol fractions of one

hundred and fifty four patients with hypertensive vascular disease. Amer. J. Med. *9:* 494–499 (1950).
136 STAUB, H. W. and THIESSEN, R.: Dietary carbohydrate and serum cholesterol in rats. J. Nutr. *95:* 633–638 (1968).
137 TAYLOR, D. D.; CONWAY, E. S.; SCHUSTER, E. M., and ADAMS, M.: Influence of dietary carbohydrates on liver content and on serum lipids in relation to age and strain of rat. J. Nutr. *91:* 275–282 (1967).
138 TAYLOR, R. D.; CORCORAN, A. C., and PAGE, I. H.: Menopausal hypertension. Critical study. Amer. J. med. Sci. *213:* 475–476 (1947).
139 TOGEL, O.; BREZINA, E. und DURIG, A.: Ueber die kohlenhydratsparende Wirkung des Alkohols. Biochem. Z. *1:* 296–345 (1913).
140 TROWELL, H.: Crude fiber, dietary fiber and atherosclerosis. Atherosclerosis *16:* 138–140 (1972).
141 VAUGHAN, M.: Effect of hormones on glucose metabolism in adipose tissue. J. biol. Chem. *236:* 2196–2199 (1961).
142 WATERHOUSE, C.; KEMPERMAN, J. H., and STORMANT, J. H.: Alterations in triglyceride metabolism as produced by dietary change. J. Lab. clin. Med. *63:* 605–620 (1964).
143 WELLS, W. W. and ANDERSON, S. C.: The increased severity of atherosclerosis in rabbits on a lactose-containig diet. J. Nutr. *68* (4): 541–549 (1959).
144 WUEST, J. H.; DRY, T. J., and EDWARDS, J. E.: Degree of coronary atherosclerosis in bilaterally oophorectomized women. Circulation *7:* 801–809 (1953).
145 WYNN, V. and DOAR, J. W. H.: Some effects of oral contraceptives on carbohydrate metabolism. Lancet *ii:* 715–719 (1966).
146 WYNN, V. and DOAR, J. W. H.: Some effects of oral contraceptives on carbohydrate metabolism. Lancet *ii:* 761–766 (1969).
147 WYNN, V.; DOAR, J. W. H., and MILLS, G. L.: Some effects of oral contraceptives on serum lipid and lipoprotein levels. Lancet *ii:* 720–723 (1966).

Author's address: Dr. IAN MACDONALD, Department of Physiology, Guy's Hospital Medical School, *London SE1 9RT* (England)

Progr. biochem. Pharmacol., vol. 8, pp. 242–270 (Karger, Basel 1973)

Dietary Carbohydrates in Lipid Disorders in Man[1]

MARGARET J. ALBRINK

Department of Medicine, West Virginia University School of Medicine, Morgantown, W.Va.

Contents

1 Some of the studies reported herein were supported in part by USPHS Grants AM 09252 and K6 HE 486.

I. Introduction

Interest in dietary carbohydrate and lipid metabolism in man stems from its possible rôle in atherosclerosis. Vital statistics show the relentless persistance of arteriosclerotic cardiovascular disease (ASCVD) as the major source of death in affluent nations [1]. Most efforts concerning the etiology and treatment of ASCVD have centered about dietary saturated fat and cholesterol as the agents responsible for hypercholesterolemia and the formation of the atherosclerotic plaque, and the latter as the agent responsible for high death rate from ASCVD.

Recent evidence points to other factors, however; factors in which obesity, diabetes and hypertriglyceridemia may play a rôle. Because of the possible important relationship of dietary carbohydrate to these risk factors, a brief review of pertinent risk factors will first be presented.

II. Atherogenesis

A. The Hyperlipidemias

All lipids circulate as constituents of lipoprotein molecules. The hyperlipidemias have been classified by FREDRICKSON *et al.* into 5 types of hyperlipoproteinemia according to electrophoretic pattern [2]. The 5 types can be condensed into two basic groups, according to whether the problem is primarily one of elevated cholesterol (type II or pure hypercholesterolemia), or whether it is one of elevated triglycerides with and without hypercholesterolemia (types I, III and IV). Hypertriglyceridemia in turn can be divided into two groups, exogenous or fat-induced (type I), and endogenous or carbohydrate-induced (types III and IV). Type V has elements of both type IV and type I. Type I is extremely rare and is not associated with increased incidence of atherosclerosis, diabetes or obesity. Type II carries with it premature ASCVD. Types III, IV and V are associated with ASCVD, obesity and diabetes. The hyperlipoproteinemias may be primary or may be secondary to other disorders. The primary hyperlipoproteinemias in turn may be familial or acquired, various cultural influences being responsible in the latter instance. Whether the inherited abnormality is concerned primarily with the metabolism of the protein or with the lipid part of the molecule is unknown. In either event diet has a strong influence on the concentrations of circulating lipids. For practical purposes the hyperlipoproteinemias usually encountered are types II or IV. The associated lipid abnormality of concern is thus cholesterol or triglyceride. In the discussion of hypertriglyceridemia in this chapter reference to endogenous hypertriglyceridemia is assumed regardless of specific type. While dietary carbohydrate may have a minor influence on cholesterol, its chief influence is on triglycerides and related metabolic abnormalities.

B. Rôle of Cholesterol

The correlation between ASCVD, serum cholesterol and dietary saturated fat has been repeatedly documented. The subject has been extensively reviewed and on the basis of findings recommendations have been made to decrease the intake of saturated fats in Americans [3]. Yet, the correlation between saturated fat intake and serum cholesterol within the USA is difficult to document [4]. People respond to a reduction of dietary saturated fat with a fall in serum cholesterol concentration but the reduction is of the order of only about 10% in highly motivated free-living people when the effect of associated weight loss is excluded [5]. Such a change in diet may lead to a decrease in cardiovascular mortality in institutionalized persons [6].

C. Triglycerides

Plasma triglycerides, though less well studied, also constitute a risk factor for ASCVD [3, 7–9]. The increased risk may be explained in part by associated hypercholesterolemia [9]. The statistical association between serum triglycerides and serum cholesterol, while not very strong, is real and probably reflects their interrelated fate. The clearance of triglyceride particles from the plasma probably involves first their association with certain serum lipoprotein apoproteins and with cholesterol and phospholipids. Through activation of one or more lipoprotein lipases [10, 11] the triglyceride part of the complex is removed, leaving a residue which is even richer in cholesterol. The final cholesterol-rich remnant is probably removed by the liver and destroyed, or recirculated with a new triglyceride particle. If the removal of triglyceride is impaired at some stage the particle will be rich or poor in triglyceride depending on the stage of degradation of the triglyceride part of the molecule [12]. The fate of cholesterol is, thus, intimately related to the metabolism of triglyceride. The cause of some, perhaps most, of the cases of hypercholesterolemia may be found in the metabolism not of cholesterol but of triglyceride.

The response of pure hypercholesterolemia to dietary manipulation may differ from the response of hypercholesterolemia secondary to hypertriglyceridemia (see below). Hypertriglyceridemia probably exerts an atherogenic effect independent of coexisting hypercholesterolemia [8, 13a, 13b]. Some of the risk may be exerted through a mechanism other than atheroma formation.

III. Nonatherogenic Risk Factors

In spite of increased public awareness of potential harm from ingestion of saturated fat, the incidence of myocardial infarction in white US males did not decline in the years 1960–1967 [14], and in one study home care produced better results than hospital care [15]. One possible explanation is that years, probably decades will be required for an effect of diet to lead to a decrease in atherosclerosis. However, whether atherosclerosis is fatal or immediately symptomatic evidently depends on factors other than the degree of atherosclerosis.

A. Sudden Death

Recently attention has been turned to other factors which may influence the outcome of ASCVD. PAUL has emphasized the importance of sudden unexpected death (SUD) in the lack of statistical benefit from efforts to reduce ASCVD mortality [16], SUD being defined in these studies as death within 24 h. According to KULLER, 60% of all persons dying of ASCVD die suddenly [17]. Thus, even with sophisticated coronary care units to cope with acute problems a large number of patients probably does not survive to reach the hospital.

The cause of SUD of ASCVD patients is almost totally unknown. Electrical abnormalities of the heart with death from cardiac arrythmia have been suggested. Increased blood viscosity, increased platelet stickiness, platelet emboli and impaired fibrinolysis have all been suggested. Impairment of diffusion of oxygen from the blood stream by high concentrations of circulating polymers, proteins and perhaps other substances has been proposed as an anoxia-producing abnormality which may influence the course of ASCVD [18].

B. Rôle of Obesity

The epidemiological factors which influence sudden death are evidently different from those which cause the underlying atherosclerosis. In the Framingham Study, SUD, angina pectoris and preinfarction angina were all found to be related to obesity, best measured by scapular skinfold thickness, and not to serum cholesterol concentrations or to diastolic hypertension, factors which are thought to increase atherosclerosis [19]. Obesity was found not to be related to degree of atherosclerosis in a multi-country pathologic study of atherosclerosis [20]. Its relation with the symptomatic expression of ASCVD rather than its cause would explain many of the puzzling inconsistencies concerning obesity as an ASCVD risk factor.

C. Rôle of Triglycerides

The mechanism by which obesity causes such symptomatic expression of ASCVD is unknown. All the factors which might influence electrical irritability of heart, oxygen transfer and blood viscosity mentioned above are suspect. Obesity is known to be associated statistically with a cluster of metabolic abnormalities, one or more of which could be responsible for the observed increased SUD. These include hypertriglyceridemia, mild diabetes and hyperinsulinism, all of which may be atherogenic [21–26] but may also have an additional rôle in precipitating symptomatic expresion of the disease through some unknown blood flow factor. The finding of the Newcastle and Scottish trials that Clofibrate, a drug which almost specifically lowers plasma triglycerides, decreased the mortality from sudden death, particularly in patients who had angina pectoris, lends support to a rôle for triglycerides in such a 'blood flow factor' [27]. Hypertriglyceridemia, when present in type III hyperlipidemia, is often associated with impaired blood flow to the extremities. The impairment is corrected by correction of the hypertriglyceridemia, in a time too short to be accounted for by resorption of atheroma [28a]. On admittedly very tenuous grounds it might be postulated that symptomatology of ASCVD, including angina pectoris and SUD, is related to a blood flow factor or factors which when superimposed on atherosclerosis further compromise the delivery of oxygen to tissues. The blood flow factor in turn may be related to some aspect of hypertriglyceridemia, obesity and diabetes.

In a study of US factory workers, the parents of men with elevated triglycerides, when compared to parents of men with normal triglycerides, were found to have more ASCVD but less non-ASCVD illness. If they did not die of ASCVD they were characterized by unusual longevity [28b]. Perhaps elevation of circulating triglycerides may be a manifestation of a genetic trait conferring increased ability to cope with physiologic emergencies. The thrombogenic effect of impaired fibrinolysis reported in patients with hypertriglyceridemia may be a case in point [29]. Such a trait would protect against hemorrhage and, because of interrelations among thrombosis, antigen-antibody reactions and the kallikrein system, might confer more general survival value. Readiness to cope with physiologic emergencies may carry the penalty of impaired removal of circulating triglycerides and of increased susceptibility to thromboembolic disease in the setting of modern civilized life.

IV. Rôle of Dietary Carbohydrate in ASCVD

Not only serum cholesterol but also the other risk factors of obesity, elevation of circulating triglycerides, mild diabetes and hyperinsulinemia, may be influenced by diet. Atherosclerotic disease has reached epidemic proportions only in recent times. Modern deviations from the dietary environment in which man evolved have been proposed as causes of the epidemic of ASCVD. The evolution of man's eating habits will be briefly reviewed in search of changes which might have promoted each of the above factors.

A. Evolution of Man's Diet

1. Man the Hunter

Ancient preagricultural man lived by hunting, fishing and gathering such vegetable foods as he could find. Evidence concerning the details of his diet is extremely fragmentary. According to anthropologists, man had fully evolved to his present somatic and presumably enzymatic state 40,000 years ago but most of the grains, such as corn, wheat and rice, which form such an important part of the diet of modern man, were not to be cultivated from wild grasses for another 30,000 years. The impact of the change brought by agriculture could be better evaluated if we knew more of the eating habits of ancient man. A picture of ancient man as an omnivore with strong carnivorous tendencies has been drawn by COON [30].

The few remaining hunting and gathering societies, while they are not necessarily representative of ancient man, provide additional insight into the eating habits of preagricultural man. The study of such societies first in their natural setting and then as they become 'civilized' provides invaluable information concerning the impact of modern eating habits on the so-called degenerative diseases.

The eating patterns of existing hunting and gathering societies vary widely and are dependent upon geographic location. The closer such peoples live to the tropics where edible plants are abundant, the more do they rely on the gathering of nonmeat foods for subsistence. With increasing latitude above the equator, as edible vegetation becomes scarce, fishing becomes more prominent. In the most northern lattitudes hunting becomes the dominant and almost the only source of food. The percent of calories derived from meat increases from a low of 33% in the tropics to a high of nearly 100% in the Arctic [31]. Thus, such societies might be largely vegetarian or largely carnivorous but not both.

Thus, prehistoric man, if he resembled modern nonagricultural man, was exposed to food varying from the fruits and roots of the jungle to the flesh of animals and fish during the time he was evolving to his present form. He should, therefore, have the enzymatic machinery for coping with a wide variety of food, but not necessarily with the quantity in which it occurs in civilized countries.

2. The Advent of Agriculture

YUDKIN has called attention to the tremendous food revolution caused by the advent of agriculture about 8,000 years ago [32]. With it, large quantities of starchy grain foods became available and man's diet was probably suddenly higher in carbohydrate than even the relatively high carbohydrate diet of the tropical hunters and gatherers. Another product of agriculture, refined sugar, has become widely available, particularly in the past 50 years [32].

3. The Industrial Revolution

Among the technological developments made possible by the industrial revolution is the mass production and distribution of food. In affluent societies food of all sorts, of both animal and vegetable origin, is available with little physical effort. One of the results, evident in all peoples as they 'emerge' into technological life, is the simultaneous emergence of obesity. The reason is not entirely clear. Availability of food *per se* does not seem to be the only answer, for availability of food is not necessarily the limiting factor in the food intake of hunting and gathering societies.

An interesting aspect of food intake of modern hunting and gathering societies is, indeed, the limitation of food intake. Of many edible plants and animals available, the hunting and gathering peoples utilize only about 20–30% of available edible species, an adaptation that would protect them from fluctuations in supply [31] and probably also from overnutrition. In these societies where procuring of food requires labor by most or all of its members, the effort man is willing to put forth is thus limited to the easiest 20–30% of available food. Little effort is put into the more arduous pursuit of hunting unless the more easily procured plant foods are not available.

We can guess that such natural constraints on caloric intake were also imposed on pretechnological man. As soon as the delicate balance between desire for food and the effort required to procure it was upset by the advent of the industrial revolution and its application to the food industry, obesity appeared and with it diabetes and ASCVD.

Another mechanism whereby industrial food processing might cause obesity is the production of foods of higher caloric density than occur in nature. There are wide individual differences in the ability to recognize differences in caloric density [33]. Whether effortless availability of all food, whether increased caloric density or whether some specific food such as sugar is responsible for obesity is difficult to determine. Whether specific types of food have an effect on risk factors independent of the effect of obesity is even more difficult to ascertain.

B. Epidemiologic Studies of the Effect of Dietary Carbohydrate

1. Relationship between Intake of Complex Carbohydrate and Diabetes

Epidemiologic evidence may help discern whether the nonspecific factor of quantity of food relative to caloric expenditure or whether some specific food is responsible for various ASCVD risk factors.

The distinction has been made between complex carbohydrate, i.e. starchy food which requires digestion before it can be absorbed as its component sugars, and simple carbohydrate or sugars which require little or no digestion and are therefore more rapidly absorbed. Since the agricultural revolution resulted in availability of large quantities of complex carbohydrate as grains or tubers, the question must be asked whether such high intake of carbohydrate, even in complex form, might cause a carbohydrate disposal problem manifest as diabetes.

That a high-carbohydrate diet causes diabetes and obesity is a widely held belief but is difficult to prove [34]. Whether in lean persons a high-carbohydrate diet causes more diabetes than a low-carbohydrate high protein and fat Arctic diet is hard to say. The Eskimos and Athebaskan Indians of Alaska who habitually eat a very low carbohydrate diet have almost no diabetes and no obesity [35]. Diabetes does exist in lean high-carbohydrate consuming peoples, such as the Japanese, according to some sources, as frequently as in the US [36], although arteriosclerotic complications are less frequent. According to WEST and KALBFLEISCH, on the other hand, obesity is the chief determinant of diabetes in world-wide studies [37]. CLEAVE *et al.* have studied many populations and have concluded that a high-carbohydrate diet, if the carbohydrate is in the form of starch rather than sugar, is evidently not associated with obesity; thus, the Zulu diet contains 85% carbohydrate while the neighboring Masai diet is very low in carbohydrate, yet both are lean [38].

While genetic differences may also be a factor, evidence is suggestive but not conclusive that the Eskimo low-carbohydrate diet is least conducive to diabetes, that a high-carbohydrate diet may be associated with more diabetes but with few detrimental complications unless obesity is also present.

2. Relationship between Sugar Intake, Diabetes and ASCVD

Many studies have suggested an interrelation between dietary sugar[2] (sucrose) and ASCVD. YUDKIN was the first to draw attention to sugar as a

2 The term 'sugar' is used synonymously with 'sucrose'.

risk factor [39]. COHEN [40] suggested that high sucrose intake was responsible for the increased incidence of arteriosclerosis and diabetes in Yeminite Jews. ANTAR *et al.* [41] and LOPEZ *et al.* [42] reported a high correlation between increasing simple carbohydrate intake and increasing arteriosclerotic death rate in the US and Europe in recent decades. The increase in arteriosclerosis of Durban Indians has been related to high sugar intake and diabetes [38]. CLEAVE *et al.* hypothesized that after 20 years of high sugar intake, 70 lb *per capita* per year, a society would begin to show increased incidence of diabetes and heart disease [38]. YUDKIN and RODDY [43] found that patients with coronary or peripheral vascular disease ingested twice as much sugar as control groups.

On the other hand, other studies have shown only a low-grade association if any between sugar intake and vascular disease [44, 45]. PAPP *et al.* [46] found no difference in sucrose consumption between coronary patients and controls. Another report suggested that the low-grade association between sugar intake and arteriosclerosis was merely a manifestation of the association between sugar intake and smoking [47]. The disagreement between studies may be explained by as yet unknown differences in the baseline characteristics of the various populations under study. In all studies the sugar intake cannot be separated from the effect of associated obesity which almost invariably accompanies high sugar intake.

If the relation between high sugar intake, diabetes and ASCVD is mediated by the effect of sugar on obesity then the mechanism by which sugar causes obesity must be sought. Sugar, because of high palatability and high caloric density, might increase the total amount of carbohydrate ingested and thus increase total caloric intake, but statistics show rather that sugar has tended to replace complex carbohydrate rather than to supplement it [41]. Sugar might act merely as a nonspecific marker of affluence and the sum total of the many aspects of affluence which might increase ASCVD. Finally, the effect of sugar might be due to some peculiar property of a specific carbohydrate such as fructose and its effect on serum lipids and glucose tolerance either directly or through some influence on obesity.

C. Effect of Various Carbohydrates on Glucose Tolerance

The effect of dietary carbohydrate on obesity, lipids and glucose tolerance depends on whether the carbohydrate is simple or complex, and on the constituent sugar.

1. Complex Carbohydrate

Theoretically ingestion of the complex carbohydrate starch, its partial breakdown product dextrins or the disaccharide maltose, all of which are rapidly degraded into glucose in the intestinal tract, should have the same effect on plasma insulin and glucose as an equivalent amount of ingested glucose. In actuality ingested starch has slightly less effect on plasma insulin and glucose than does an equivalent amount of glucose [48]. The slower absorption of starch, owing to the time required to digest the starch, could account for the lower insulin response. Some other property of the starch, perhaps related to the protein or perhaps even to the uncertainty of estimating the glucose equivalent of a given amount of starch, could be important.

Simple carbohydrates consist chiefly of sucrose, glucose and fructose of fruits, and refined sugar (sucrose). Since sucrose is broken down in the gastrointestinal (GI) tract into glucose and fructose, differences between sucrose and starch could be attributed to the fructose content of sugar as well as to its refined state.

2. Metabolic Differences between Glucose and Fructose

a) Liver

Fructose is able to bypass several enzymatic steps required for glucose to enter the glycolytic pathway in liver. Its uptake by liver is, therefore, much more rapid than that of glucose. For these reasons in man as well as in animals fructose is a much better precursor in liver of both fatty acids and glyceride-glycerol than is glucose [49] as explained elsewhere [ZAKIM, this volume].

b) Other Tissues

Orally administered fructose in man is removed largely by the liver, in contrast to glucose which is disposed of mainly in the periphery [50]. Much of the fructose is converted into glucose by liver [51]. Fructose is so completely extracted that it causes almost no increase in peripheral circulating fructose and only a slight increase in glucose and insulin [48, 52]. Fructose causes little or no insulin secretion from isolated rat islets [53].

Fructose can be removed by most peripheral tissues except brain. Since its removal is not dependent on insulin its removal is not impaired by factors which impair removal of glucose such as cortisone [54].

3. Effect of Fructose and Glucose on Glucose Tolerance

The immediate effect of a carbohydrate meal on blood glucose depends on the source and amount of carbohydrate in the meal [48, 55]. In general, glucose causes a greater rise than starch or sucrose, and fructose the least rise. Protein causes no rise [55].

Fructose causes only a minimal rise in circulating insulin in intact man [48]. Glucose causes the greatest rise, followed by starch. Sucrose provokes less rise than starch (see below).

The long-range effect of fructose on glucose tolerance has been the subject of debate for many years. Because of its ability to bypass insulin-dependent limitations to entry into liver metabolic pathways, fructose might be expected to be a useful form of carbohydrate for diabetics. Furthermore, since it has little insulin-stimulating effect, fructose, if substitued for glucose or glucose-yielding polysaccharides, would avoid the rapid increase in blood glucose and resulting insulin stimulation. Dietary fructose would be expected to 'spare' insulin and thus prevent islet cell exhaustion and resultant diabetes. Neither expectation has been borne out [52]. HILL and CHAIKOFF [56], quite the contrary, found that prolonged fructose feeding in dogs caused intolerance to glucose. They later demonstrated decreased hepatic glucokinase in such animals and postulated that decrease in this enzyme was responsible for the impaired glucose tolerance.

COHEN *et al.* [57] compared glucose tolerance curves of human subjects subsisting for 5 weeks alternately on diets in which either sucrose or bread constituted the main source of carbohydrate. The glucose tolerance curves were significantly higher on the sucrose diet.

CLEAVE *et al.* advanced a theory for a diabetogenic effect of sugar (sucrose). They hypothesized that dietary sugar intake by causing earlier or greater output of insulin would, after some 20 years of excessive intake, lead to eventual pancreatic exhaustion and diabetes [38]. While there is much epidemiological evidence in favor of his hypothesis the diabetes could as well be caused by the associated obesity. CLEAVE *et al.* have indeed commented on the regularity of concurrence of diabetes and obesity, as though both were caused by the same factor, sugar.

Against sugar-induced hyperinsulinism in all of the above studies as the cause of diabetes, through pancreatic exhaustion, is the fact that sucrose causes much less output of insulin than does glucose or starch [48, 53]. The insulin response to 100 g glucose, sucrose or equivalent amount of starch in 9 healthy young men is shown in figure 1 [48]. Glucose stimulated the largest output of insulin, starch the next largest, and sucrose the least. Fructose caused less output of insulin even than sucrose (not shown). Sucrose, being half glucose and half fructose, thus caused less of an increase in insulin than did glucose, owing to the fact that fructose stimulates little or no insulin output from the pancreas.

Some, perhaps all, of the diabetogenic effect of high-fructose, low-glucose diets could be explained by decrease in glucose-dependent enzymes such as glucokinase, and by decreased pancreatic responsiveness. To this extent the diabetogenic effect of fructose would be an adaptation similar to that of any other form of glucose deprivation, and would be evident only when a glucose meal interrupted the period of glucose deprivation.

Perhaps the most satisfactory explanation for the common occurence together of high sugar intake, obesity and diabetes is the more rapid entry of fructose than glucose into metabolic pathways resulting in fat synthesis. Once obesity is established the well-known insulin resistance of the obese could account for the diabetes and the hyperinsulinism. The latter could, as hypothesized by CLEAVE *et al.* [38], lead eventually to pancreatic exhaustion and insulin-deficient diabetes, as discussed in the next section. The argument for a causal rôle for sugar in the pathogenesis of obesity and diabetes is strong but not proven.

V. Metabolic Correlates of Obesity

Many of the epidemiologic associations between sugar intake, ASCVD and diabetes cannot be separated from the effects of related obesity. Knowledge of the metabolic characteristics of the obese may help to explain the above relationships.

Obesity frequently is associated with hypertriglyceridemia, impaired glucose tolerance and atherosclerosis [21–24]. The association is not complete, however, for there are many thin persons with high triglycerides and many obese persons with normal triglycerides.

A. Insulin Resistance

Hyperinsulinism and peripheral resistance or antagonism to the action of insulin have been widely reported in obese persons and animals. The relative ineffectiveness of insulin in obese diabetics is an old clinical observation. Resistance to insulin has also been shown in hypertriglyceridemic and obese persons as a failure of plasma glucose to drop as far or as rapidly after intravenous administration of insulin as in lean persons [58]. Using the isolated forearm technique, RABINOWITZ and ZIERLER showed insensitivity to the effect of infused insulin on glucose uptake by both muscle and

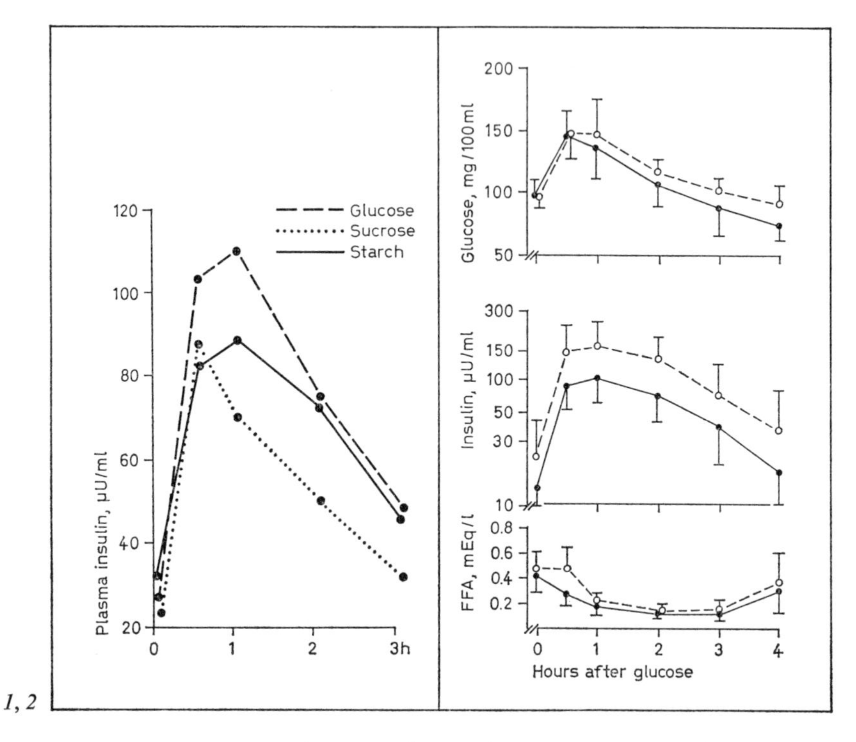

1, 2

Fig. 1. Insulin response to ingestion of 100 g glucose, sucrose or the glucose equivalent of starch in 9 healthy young men. The tests were carried out in random order about 1 week apart. From SWAN *et al.* [48].

Fig. 2. Response of plasma glucose, insulin and free fatty acids (FFA) to ingestion of 100 g glucose in 14 very obese persons (o------o) (ponderal index less than 11) and 67 thin persons (•———•) (ponderal index greater than 12.5). Insulin is shown on a logarithmic scale.

adipose tissue [59]. Mild diabetes is frequently present in obese persons owing, in part, to the relative ineffectiveness of insulin in the obese. The elevation of basal and postcibal insulin usually observed in nondiabetic obese persons is thought to be a manifestation of insulin antagonism or ineffectiveness in the peripheral tissues [21].

SALANS *et al.* demonstrated insulin insensitivity *in vitro* in adipose tissue from obese humans and related the insensitivity to increased adipose cell size, rather than to increased numbers of adipose cells [60]. The insulin levels and the insulin insensitivity were restored to normal by weight loss and were thus secondary to the obese state.

SIMS and HORTON confirmed the secondary nature of the insulin antagonism by inducing obesity in thin men. With weight gain, plasma triglycerides rose and insulin also increased. Both variables were restored to normal by subsequent weight loss [61].

The antagonism to insulin has been variously attributed to an abnormality in circulating insulin, to a circulating insulin antagonist or to some property of the overstuffed adipose cells. None of these possibilities has thus far been confirmed. It seems likely that the ineffectiveness of insulin affects all tissue, not only the large adipose cells, and that it is in some way related to the general metabolic condition of impaired carbohydrate combustion or accelerated fat metabolism. The elevation of circulating amino acids in obese persons is an indication that amino acids share with glucose insensitivity to the action of insulin [62].

The nature of the signal to the pancreas that the peripheral tissues are insulin-insensitive has been the subject of some debate. The elevated concentrations of the substrates themselves, i.e. amino acids or glucose, seem reason enough for the increased insulin response. If the elevated basal insulin reflects increased basal need for insulin, then this need must be taken into consideration in evaluating the additional insulin need imposed by a glucose meal. Reference to basal status can be incorporated by expressing the postcibal insulin response as percent of baseline [63] or by logarithmic conversion of all insulin values. When insulin response to a glucose challenge is plotted on a logarithmic scale (fig. 2) the rise above fasting is seen to be of the same magnitude as that for nonobese persons. The same log difference observed basally between obese and thin persons is maintained throughout the test.

The higher insulin response of nondiabetic obese persons can thus be explained entirely by the higher basal insulin levels, as stated previously by BIERMAN and PORTE [63]. While the insulin response with reference to baseline values is not excessive, the insulin concentration in arithmetic units necessary to achieve a normal increase above baseline is greatly increased by even modest increases in basal insulin. Each carbohydrate meal calls upon the pancreas for a much greater insulin output when basal insulin is elevated than when it is normal.

If the prolonged overstimulation of the pancreas leads to eventual pancreatic exhaustion and insulin-deficient diabetes [64], a high-carbohydrate diet might, therefore, be potentially diabetogenic for persons with elevated basal insulin levels. Factors which raise the basal insulin, by imposing need for greater postcibal rise, could also be considered diabetogenic. Obesity qualifies as such a factor.

B. Effect of Diet on Plasma Insulin

The effect of habitual long-range diet on basal plasma insulin concentration is not known. High intake of carbohydrate was reported by BRUNZELL *et al.* to be associated with adaptive improvement in glucose tolerance and, as a consequence, with lower basal and postglucose insulin concentrations [65]. However, GREY and KIPNIS [66] found that a low-carbohydrate diet in obese persons was associated with improved glucose tolerance and lower plasma insulin concentrations. SIMS *et al.* found that whether a high-carbohydrate diet raised or lowered the plasma insulin concentration depended on degree of obesity. Normal subjects who voluntarily greatly expanded their fat body mass failed to show the decrease in plasma insulin following a very high carbohydrate diet which they had shown while on the same diet at normal weight, and actually showed increased insulin and glucose [67]. A high-carbohydrate diet thus may not be desirable in all persons.

VI. Metabolic Characteristics of Hypertriglyceridemia

A. Insulin Resistance

The same metabolic abnormalities which have been reported in obesity also occur in hypertriglyceridemia, i.e. impaired glucose tolerance, increased concentration of basal and postcibal circulating insulin, and insulin resistance [58]. Since hypertriglyceridemia is often associated with obesity, the insulin resistance of hypertriglyceridemia may be, in part, a manifestation of associated obesity [22]. However, we have found higher insulin levels in obese patients with hypertriglyceridemia than equally obese patients without it [unpublished observations]. In all likelihood the combination of hypertriglyceridemia and obesity imposes more of a burden on the pancreas than either alone.

B. Effect of Very High Carbohydrate Diet on Serum Lipids

1. Carbohydrate-Induced Lipemia

AHRENS coined the term 'carbohydrate-induced lipemia' to signify the type of hypertriglyceridemia which became more intense following several days of a very high carbohydrate diet, to contrast this type with the rare fat-induced type [68]. In contrast to alimentary lipemia, in which triglyceride concentration is greatest 3–5 h after a fat meal, in carbohydrate-induced lipemia the triglyceride concentration reaches its highest concentration after an overnight fast. Since the triglycerides under these circumstances are endogenously synthesized from carbohydrate the term 'endogenous lipemia' is preferred by some [63]. The phenomenon of elevation of triglycerides following a high-carbohydrate diet has since been studied by many workers. These have been reviewed by KAUFMANN [69].

By and large such a diet in which carbohydrate constitutes 70–80% of the calories causes an approximate doubling of the serum triglycerides [63, 70]. The phenomenon is thus nonspecific and is seen in all persons. A possible exception is the patient with chylomicronemia in whom the decrease in circulating exogenously derived triglycerides upon being fed a low-fat, very high carbohydrate diet may more than compensate for the increase in endogenously derived triglycerides [70].

Since plasma triglycerides are as a rule doubled after a few days of high-carbohydrate feeding, persons with initially elevated triglycerides will show a greater absolute increase in triglycerides than will persons with normal triglycerides. The excess triglyceride is demonstrable as increased pre-β-lipoprotein, an indication of its endogenous source [2]. While some studies suggested that the carbohydrate-induced increase of triglyceride was derived from fatty acids newly synthetized from carbohydrate by the liver [71], others indicate that the fatty acid component of the carbohydrate-induced increment in triglyceride synthesis is derived from circulating fatty acids [72, 73]. The excess carbohydrate probably does play a rôle in the synthesis of excess α-glycerolphosphate which, in turn, is an obligatory receptor of free fatty acids, diverting them from other fates to triglyceride synthesis.

2. Effect of Type of Carbohydrate on Triglycerides

The influence of type of carbohydrate on triglycerides or on carbohydrate-induced lipemia has been the subject of some debate. Epidemiologic studies suggest that lifelong intake of high-carbohydrate diet is not associated with elevated triglycerides [74, 75], though associated thinness makes interpretation difficult [23]. Because of epidemiologic evidence suggesting that intake of sugar was associated with vascular disease [39, 40, 76] attention was turned to the effect of type of carbohydrate on serum lipids.

Many authors have demonstrated a probable lowering effect of complex carbohydrate and probable raising effect of sugar, especially fructose or sucrose, on serum lipids [77] [see MACDONALD, this volume]. MCGANDY *et al.* [78] reviewing this field concluded that, as far as cholesterol is concerned, sucrose or purified carbohydrate in quantities in which carbohydrate is usually ingested had only a minimal cholesterol-raising effect. Other authors failed to confirm the increased cholesterol in sucrose-fed man [79]. GRANDE *et al.* [80] found varying influence of starch on lipids according to the source of the starch. In general, simple carbohydrate has a greater effect on triglycerides than on cholesterol.

A greater increase in triglycerides when fructose or sucrose rather than starch is the source of carbohydrate has been reported by some [81, 82], as has an increase in alimentary lipemia [83]. Increased incorporation of fructose into serum lipids is the probable mechanism, owing to the fact that fructose is a better precursor of fat than is glucose [49].

The converse of sucrose restriction in treating hypertriglyceridemic post-myocardial infarction patients is difficult to evaluate because of the concomitant weight loss [84]. Some authors have found that when weight is strictly controlled the source of carbohydrate has no effect on the degree of carbohydrate-induced lipemia. KAUFMANN [69] stressed the extreme variability of response.

In summary, the type of carbohydrate may have a small effect on cholesterol in the direction of a slight increase in cholesterol on high-sucrose diet and a decrease on a high-starch diet, but only when the diet is extremely high in carbohydrate. The type of carbohydrate may have a greater effect on triglyceride, sucrose and fructose having a more marked triglyceride raising effect than starch but with great individual variability. Few comparisons have been made of starch, dextrins and glucose, all of which should yield only glucose on digestion and should, except for a slight delay needed for hydrolysis of starch in the GI tract, have comparable effect on plasma glucose and theoretically on plasma triglycerides. Starches, however, are not simple substances – they cannot easily be separated from constituent proteins. Various types of starchy foods may have differing effects on serum lipids of man as suggested by GRANDE *et al.* [80]. Indeed, the diabetes literature of the preinsulin era abounds with the references to the desirability of some sources of starch over others in the control of glycosuria in diabetes.

The marked metabolic differences between glucose and fructose are consistent with a greater serum triglyceride-raising effect of fructose. This effect has not been proven with assurance, however, because of difficulty of separating the dietary effect from the many other variables which influence responsiveness: sex, age, and other variables.

3. Mechanism of Carbohydrate-Induced Lipemia

Most workers agree that the increase in triglycerides following a very high carbohydrate diet, carbohydrate-induced lipemia, is caused by increased hepatic synthesis of triglyceride which is subsequently delivered to the circulation in very low density lipoproteins [73, 85]. There is little evidence that the high-carbohydrate diet results in a decrease in fractional removal rate. While transient in some persons, carbohydrate-induced lipemia may be permanent in others [75].

If one accepts the thesis that a high-carbohydrate diet causes a doubling of circulating triglyceride concentration because of a nonspecific increase in triglyceride synthesis, then carbohydrate-induced lipemia can be explained as a normal adaptation to a high-carbohydrate diet. The peripheral tissues are limited in the amount of carbohydrate they can utilize or store as glycogen. Excess carbohydrate is converted to triglyceride for storage in adipose tissue. In some species, including man, adipose tissue is evidently not a very active site of fatty acid synthesis from carbohydrate [60]; rather excess glucose is converted in the liver into fatty acids and glycerol, the latter two combining to form triglyceride. The triglyceride is secreted into the circulation when glucose metabolism is active, subsequently removed by adipose tissue and stored. The mechanism whereby glucose is diverted from the glucose-surfeited peripheral tissues to the liver is not known. OLEFSKY *et al.* believe that peripheral resistance or impedence to the action of insulin causes an increase in circulating insulin, the latter acting on the liver to cause increased triglyceride synthesis [86]. Others have found a decrease rather than an increase in insulin following some days on a high-carbohydrate diet [65]. The debate concerning the effect of diet on plasma insulin has been discussed above. The induction of certain hepatic enzymes, including glucokinase, may promote the uptake of large amounts of glucose by liver. If such enzymes are chiefly responsible for diversion of excess glucose to liver the rôle of insulin is secondary, for glucokinase depends on normal insulin secretion for its induction by a high-carbohydrate diet [87]. On theoretical grounds fructose, because of its rapid entry into fat synthesis pathways in the liver [88], might be converted into triglyceride at a lower level of carbohydrate surfeit than is glucose.

4. Mechanism of Basal Hypertriglyceridemia

The mechanism of hypertriglyceridemia in patients receiving a normal diet is less clear. A genetic component to such basal hypertriglyceridemia is suggested by a number of studies, but the nature of the genetic trait is not known [13a]. Some evidence suggests that, as is the case following a very high carbohydrate diet, the increased hepatic synthesis of triglycerides under the influence of hyperinsulinism is responsible [24]. Thus REAVEN *et al.* proposed that on a normal diet as well as on a high-carbohydrate diet impedence to peripheral action of insulin on glucose leads to a rise in circulating insulin and that the latter acts upon the liver to increase the synthesis of triglycerides and their subsequent secretion into the circulation as very low density lipoproteins [24, 86]. According to these workers, then, increased synthesis rather than decreased removal is responsible for hypertriglyceridemia whether the diet is normal or high in carbohydrate.

Other evidence suggests, however, that the basic lesion is one of impaired removal of triglycerides by the peripheral tissues, with little evidence for increased synthesis [72, 73]. A very high carbohydrate diet, by causing a nonspecific increase in triglyceride production, merely magnifies the preexistent lipemia, the latter usually being approximately doubled [63, 70]. Unless there is preceding impaired removal, therefore, the absolute increase caused by a very high carbohydrate diet will be small, according to this theory.

Since obesity and hypertriglyceridemia may exist independently of each other, the hypertriglyceridemia trait is probably distinct from obesity but is influenced by it. Whether obesity exerts its hypertriglyceridemic effect through increased synthesis or impaired removal is unknown. On theoretical grounds obesity, by causing insulin antagonism, could impair the peripheral uptake of glucose, causing an increase in circulating insulin which might act in the same manner as a high-carbohydrate diet to cause increased triglyceride synthesis. On the other hand, if basal hypertriglyceridemia is caused by impaired removal of triglycerides and if obesity is a cause of basal hypertriglyceridemia, then obesity, perhaps by causing insulin resistance, might interfere with the removal of triglycerides by adipose tissue, a process which is insulin-dependent [89]. Some evidence exists for such impairment [71].

There is lack of certainty regarding the effect of obesity on plasma triglycerides. In the general population the correlation between obesity and triglycerides is very low [23]. As stated above, not only is there disagreement regarding the importance of increased plasma insulin in causing increased triglyceride synthesis by liver but there is, furthermore, lack of agreement regarding the effect of obesity on plasma insulin, REAVEN *et al.* finding no correlation [24], BIERMAN and PORTE finding an important correlation with basal insulin [63]. Lack of a precise method for estimating obesity may be in part responsible for the disagreement. A number of studies suggest that acquired obesity, i.e. obesity acquired after maturity, is more likely to be associated with large adipose cells [61], with hypertriglyceridemia [90] and with insulin resistance and hyperinsulinism [60]. Even here there is disagreement. Not all workers can demonstrate the insulin resistance of large adipose cells [91]. Adequate prior feeding is important in demonstrating normal insulin sensitivity in adipose tissue of humans [92] as is adequate intake of carbohydrate [61].

A final and perhaps most important reason for confusion is lack of a method for distinguishing between the genetic trait for hypertriglyceridemia, if such there be, and acquired or nutritional hypertriglyceridemia [13a]. Hypertriglyceridemia may be sporadic or may be inherited, probably as a dominant trait. Overwelght is not necessary for manifestation of the genetic trait, since it may occur in thin people. While changes in obesity may cause corresponding changes in triglycerides in both the sporadic and genetic variety, the mechanism of obesity may be different from the mechanism of the inherited trait. Until it is possible to separate the effects of one from the effects of the other there will probably continue to be disagreement in studies of the effect of obesity on hypertriglyceridemia.

The mild diabetes frequently observed with hypertriglyceridemia is probably a manifestation of the association of both diabetes and hypertriglyceridemia with obesity. Any association between diabetes and impaired removal of triglycerides is probably mediated by associated obesity and is not a direct effect [22].

The issue of the relative rôle of increased synthesis and decreased removal in the genesis of hypertriglyceridemia is unsolved and is important in assessing the rôle of diet in the genesis of hypertriglyceridemia. The level of dietary carbohydrate at which hypertriglyceridemia occurs must vary with the severity of impairment of removal, and even if not primary may become an important determinant of triglyceride concentration in persons with severely impaired removal.

If the basic lesion is one of impaired removal, the specific rôle of dietary carbohydrate in its etiology is limited to the rôle of sugar in promoting obesity, of sucrose and fructose in being more effective inducers of obesity and of lipemia during very high carbohydrate diets. If, on the other hand, the lesion is one of increased synthesis for a given level of dietary carbohydrate then dietary carbohydrate has a much more direct causal relationship to hypertriglyceridemia. Until these issues are settled the treatment of hypertriglyceridemia is largely empirical.

C. Dietary Treatment of Hypertriglyceridemia

1. Weight Loss

Whether the relationship between hypertriglyceridemia and obesity is causal or is coincidental there is little doubt that obesity causes an exaggeration of hypertriglyceridemia. Short-range weight loss is almost always followed by a reduction of triglycerides though not necessarily to normal concentrations. Owing to the difficulty in inducing and maintaining weight loss the long-range effect of weight loss on hypertriglyceridemia is unknown. In epidemiologic studies hypertriglyceridemia is unusual in societies characterized by leanness [23].

2. Carbohydrate Restrictions

A low-carbohydrate isocaloric diet also promotes reduction of triglycerides. It is not certain whether the reduction reflects improved removal or decreased synthesis, probably the latter. The few reports of long-range effects of low-carbohydrate diets on triglycerides have not separated the effects of concomitant weight loss from the effects of the carbohydrate restriction [84].

While weight loss *per se* may lower the triglycerides regardless of composition of the diet, the reduction may be greater with a low-carbohydrate than with a high-carbohydrate diet, at least in some persons. Figure 3 compares the effect of low- and high-carbohydrate diets of a patient of the author's who had type III hyperlipidemia. Triglycerides were clearly higher during high carbohydrate intake despite caloric restriction and weight loss. In this particular patient the cholesterol too was higher during high carbohydrate intake (fig. 4), an indication that the elevation of cholesterol in type III disease, at least, is secondary to the elevation of triglycerides [12]. While, in general, cholesterol more often decreases than increases upon institution of a low-fat, high-carbohydrate diet there are many exceptions. No rules from which to predict the effect of a low-fat diet on the cholesterol of a given patient are yet available. In any patient with marked hypertriglyceridemia the likelihood that cholesterol will increase as a result of such a diet must be kept in mind. Accoding to WILSON and LEES, changes in cholesterol concentration of certain lipoprotein fractions may be more important than the relatively minor effects of diet on serum total cholesterol [93]. Upon induction of hypertriglyceridemia by a very high carbohydrate diet, cholesterol may shift out of the low-density lipoproteins into the very low density lipoproteins, while the opposite shift of cholesterol from very

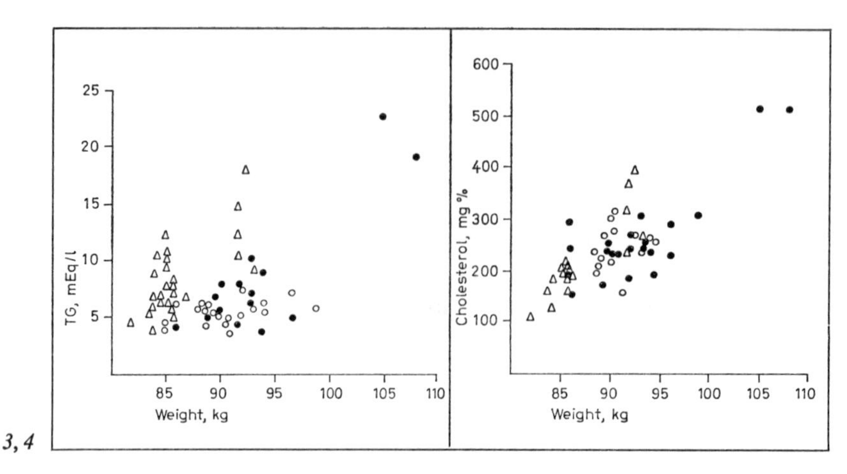

Fig. 3. Effect of weight loss and composition of diet on serum triglycerides of a 50-year-old man (T.T.) with type III hyperlipidemia. Triglycerides (TG) are expressed as mEq/l of triglyceride fatty acids (5.5 upper limit of normal). The two closed circles in the right upper corner represent the first and last determinations. The remaining symbols indicate triglycerides during weight loss as he was treated with a low-calorie diet which was, variously, low in carbohydrate (about 50 g daily) or high in carbohydrate (300–400 g), or a balanced diet (no diet). After achieving maximal weight loss he failed to return to the clinic for 1 year at which time both weight and triglycerides were restored to their original high values. o = Low carbohydrate; Δ = low fat, high carbohydrate; • = no diet.

Fig. 4. Effect of weight loss and composition of the diet on serum cholesterol of the same 50-year-old man (T.T.) shown in figure 3. See legend of figure 3 for details. o = Low carbohydrate; Δ = low fat; • = no diet.

low density lipoproteins into low-density lipoproteins takes place when triglycerides are reduced by weight loss. Whether the latter change is transient or permanent or whether it is undesirable is not known.

Whether or not the decision to limit total carbohydrate intake of any particular hypertriglyceridemic patient is made, the possibility that sugar is more hypertriglyceridemic than starch seems definite enough to suggest that the dietary treatment of hypertriglyceridemia at least includes elimination of refined sugar from the diet.

3. Dietary Fiber

Reference has been made to the cholesterol-lowering effect of leguminous seeds, pectins, fruit and leafy vegetables [80]. The suggestion has recently been made that dietary fiber, especially cereal fiber, has a cholesterol-lowering effect in man [94]. Similar conclusions have been drawn for animal experiments. A possible mechanism in rats is the binding of bile acids to non-digestible residue, with resultant accelerated conversion of cholesterol to bile acids [95]. The possible lesser incidence of carbohydrate-induced lipemia, when carbohydrate is administered in complex form rather than as sugar, may also eventually be explained by some unknown action of dietary fiber on carbohydrate digestion.

VII. Summary

The possible rôle that dietary carbohydrate plays in ASCVD is considered from the point of view or both atherogenic and nonatherogenic causes of death in this disease. Obesity, diabetes and hypertriglyceridemia are associated with increased risk of death from ASCVD. Epidemiologic evidence suggests a rôle for dietary sugar in the etiology of all three, but it is impossible to separate the effects of sugar from the effects of associated obesity.

Clinical and laboratory evidence support a rôle for obesity and related insulin resistance in the genesis or aggravation of diabetes and hypertriglyceridemia. Diet may also be a factor. A very high carbohydrate diet causes an approximate doubling of triglycerides, but the effect on diabetes is variable. Sucrose may have a greater hypertriglyceridemic effect than starch. The long-range effect of more physiologic variation in amount and type of carbohydrate on serum triglycerides is not known.

VIII. References

1 National Heart and Lung Institute: Arteriosclerosis. A report by the Task Force on Arteriosclerosis, vol. II. DHEW Publication No. (NIH) 72–219 (National Institutes of Health, Washington 1971).

2 Fredrickson, D. S.; Levy, R. I., and Lees, R. S.: Fat transport in lipoproteins. An integrated approach to mechanisms and disorders. New Engl. J. Med. *296:* 34, 94, 148, 215, 273 (1967).

3 Inter-Society Commission for Heart Disease Resources: Primary prevention of the atherosclerotic disease. Circulation *42:* 55–95 (1970).

4 Kannel, W. B. and Gordon, T.: The Framingham study. An epidemiological investigation of cardiovascular disease (US Government Printing Office, Washington D.C.)

5 American Heart Association: The national diet heart study. Final report. Monograph No. 18 (American Heart Association, New York 1968).

6 Dayton, S.; Pearce, M. L.; Hashimoto, S.; Dixon, W. J., and Tomiyasu, U.: A controlled clinical trial of a diet high in unsaturated fat in preventing complications of atherosclerosis. American Heart Association Monograph No. 25 (American Heart Association, New York 1969).

7 ROSENMAN, R. H.; FRIEDMAN, M.; STRAUS, R.; JENKINS, C. D.; YZANSKI, S. J., and WURM, M.: Coronary heart disease in the western collaborative group study. A follow-up experience of 4½ years. J. chron. Dis. *23:* 173–190 (1970).

8 ALBRINK, M. J.; MEIGS, J. W., and MAN, E. B.: Serum lipids, hypertension, and coronary artery disease. Amer. J. Med. *31:* 4–23 (1961).

9 KANNEL, W. B.; CASTELLI, W. P.; GORDON, T., and MCNAMARA, P. M.: Serum cholesterol, lipoproteins, and the risk of coronary heart disease. The Framingham study. Ann. intern. Med. *74:* 1–12 (1971).

10 HAVEL, R. J.; SHORE, V. G.; SHORE, B., and BIER, D. M.: Role of specific glycoproteins of human serum lipoproteins in the activation of lipoprotein lipase. Circulat. Res. *27:* 595–600 (1970).

11 GRETEN, H.; LEVY, R. I.; FALES, H., and FREDRICKSON, D. S.: Hydrolysis of diglyceride and glyceryl monoester diethers with 'lipoprotein lipase'. Biochim. biophys. Acta *210:* 39–45 (1970).

12 HAZZARD, W. R.; PORTE, D., and BIERMAN, E. L.: Abnormal lipid composition of chylomicrons in broad-β disease (type III hyperlipoproteinemia). J. clin. Invest. *49:* 1853–1881 (1970).

13a GOLDSTEIN, J. L.; HAZZARD, W. R.; SCHROTT, H. G.; BIERMAN, E. L., and MOTULSKY, A. G.: Genetics of hyperlipdemia in coronary heart disease. Trans. Ass. amer. Physicians (in press).

13b CARLSON, L. A. and BOTTIGER, L. E.: Ischaemic heart-disease in relation to fasting values of plasma triglycerides and cholesterol. Stockholm prospective study. Lancet *i:* 865–868 (1972).

14 National Vital Statistics System: Leading component of upturn in mortality, 1952–1967. DHEW Publication No. 72-1008, series 20, No. 11 (Washington, 1971).

15 MATHER, H. G.; PEARSON, N. G.; READ, K. L.; SHAW, D. B.; STEED, G. R.; THORNE, M. G.; JONES, S.; GUERRIER, C. V.; ERANT, C. D.; MCHUGH, P. M.; CHOWDHURY, N. R.; JAXARY, M. H., and WALLACE, T. J.: Acute myocardial infarction. Home and hospital treatment. Brit. med. J. *iii:* 334–338 (1971).

16 OGLESBY, P.: Myocardial infarction and sudden death. Hosp. Pract. *6:* 91–103 (1971).

17 KULLER, L.: Sudden death in arteriosclerotic heart disease. The case for preventive medicine. Amer. J. Cardiol. *24:* 617–628 (1969).

18 NAVARI, R. M.; GAINER, J. L., and HALL, K. R.: Effect of plasma constituents on oxygen diffusivity; in HERSHEY Blood oxygenation, p. 243 (Plenum Press, New York 1970).

19 DAMON, A.; DAMON, S. T.; HARPENDING, H. C., and KANNEL, W. B.: Predicting coronary heart disease from body measurements of Framingham males. J. chron. Dis. *21:* 781–802 (1969).

20 MONTENEGRO, M. R. and SOLBERG, L. A.: Obesity, body weight, body length and atherosclerosis. Lab. Invest. *18:* 594–603 (1968).

21 KARAM, J. H.; GRODSKY, G. M., and FORSHAM, P. H.: Insulin secretion in obesity: pseudodiabetes. Amer. J. clin. Nutr. *21:* 1445–1454 (1968).

22 BAGDADE, J. D.; BIERMAN, E. L., and PORTE, D.: Influence of obesity on the relationship between insulin and triglyceride levels in endogenous hypertriglyceridemia. Diabetes, N.Y. *20:* 664–672 (1971).

23 ALBRINK, M. J. and MEIGS, J. W.: Serum lipids, skin-fold thickness, body bulk and body weight of native Cape Verdeans, New England Cape Verdeans, and United States factory workers. Amer. J. clin. Nutr. *24:* 344–352 (1971).

24 REAVEN, G. M.; LERNER, R. L.; STERN, M. P., and FARQUHAR, J. W.: Role of insulin in endogenous hypertriglyceridemia. J. clin. Invest. *46:* 1756–1767 (1967).

25 OSTRANDER, L. D.; NEFF, B. J.; BLOCK, W. D.; FRANCIS, T., and EPSTEIN, F. H.: Hyperglycemia and hypertriglyceridemia among persons with coronary heart disease. Ann. intern. Med. *67:* 34–41 (1967).

26 HEINLE, R. A.; LEVY, R. I.; FREDRICKSON, D. S., and GORLIN, R.: Lipid and carbohydrate abnormalities in patients with angiographically documented coronary artery disease. Amer. J. Cardiol. *24:* 178–186 (1969).

27 DEWAR, H. A. and OLIVER, M. F.: Secondary prevention trials using Clofibrate. A joint commentary on the Newcastle and Scottish trials. Brit. med. J. *4:* 778–786 (1971).

28a ZELIS, R.; MASON, D. T.; BRAUNWALD, E., and LEVY, R.: Effects of hyperlipoproteinemia and their treatment on the peripheral circulation. J. clin. Invest. *49:* 1007–1015 (1970).

28b MEIGS, J. W.; CALDWELL, T. B., and ALBRINK, M. J.: Epidemiology of coronary disease in industrial workers. Arch. environm. Hlth *10:* 467–474 (1965).

29 STEVENSON, M.; HARPER, L., and ALBRINK, M.: Effect of low and high carbohydrate diets on fibrinolysis. Circulation *33* (42): III–24 (1970).

30 COON, C. S.: The origin of races, p. 307 (Knopf, New York 1963).

31 LEE, R. B.: What hunters do for a living, or, How to make out on scarce resources; in LEE and DEVORE Man the hunter, pp. 30–43 (Aldine, Chicago 1968).

32 YUDKIN, J.: Evolutionary and historical changes in dietary carbohydrates. Amer. J. clin. Nutr. *20:* 108–115 (1967).

33 CAMPBELL, R. G.; HASHIM, S. A., and ITALLIE, T. B.: Studies of food intake regulation in man. Responses to variations in nutritive density in lean and obese subjects. New Engl. J. Med. *285:* 1402–1407 (1971).

34 ALBRINK, M. J. and DAVIDSON, P. C.: Dietary therapy and prophylaxis of vascular disease in diabetics. Med. Clin. N. Amer. *55:* 877–887 (1971).

35 MOURETOFF, G. T.; CARROLL, N. V., and SCOTT, E. M.: Diabetes mellitus in Athebasken Indians in Alaska. Diabetes, N.Y. *18:* 29–32 (1969).

36 BLACKARD, W. G.; OMORI, Y., and FREEDMAN, L. R.: Epidemiology of diabetes mellitus in Japan. J. chron. Dis. *18:* 415–427 (1965).

37 WEST, K. M. and KALBFLEISCH, J. M.: Influence of nutritional factors on prevalence of diabetes. Diabetes, N.Y. *20:* 99–108 (1971).

38 CLEAVE, T. L.; CAMPBELL, G. D., and PAINTER, N. S.: Diabetes, coronary thrombosis and the saccharine disease (Wright, Bristol 1969).

39 YUDKIN, J.: Diet and coronary thrombosis. Hypothesis and fact. Lancet *ii:* 155–162 (1957).

40 COHEN, A. M.: Fats and carbohydrate as factors in atherosclerosis and diabetes in Yeminite Jews. Amer. Heart J. *65:* 291–293 (1963).

41 ANTAR, M. A.; OHLSON, M. A., and HODGES, R. E.: Changes in retail market food supplies in the U.S. in the last seventy years in relation to the incidence of coronary heart disease, with special reference to dietary carbohydrates and essential fatty acids. Amer. J. clin. Nutr. *14:* 169–178 (1964).

42 LOPEZ, S. A.; KREHL, W. A.; HODGES, R. E., and GOOD, E. I.: Relationship between food consumption and mortality from atherosclerotic heart disease in Europe. Amer. J. clin. Nutr. *19:* 361–367 (1966).

43 YUDKIN, J. and RODDY, J.: Levels of dietary sucrose in patients with occlusive atherosclerotic disease. Lancet *ii:* 6–8 (1964).

44 PAUL, O.; MACMILLAN, A., and PARK, H.: Sucrose intake and coronary heart disease. Lancet *ii:* 1049–1051 (1968).

45 ELWOOD, P. C.; WATER, W. E.; MOORE, S., and SWEETNAM, P.: Sucrose consumption and ischaemic heart disease in the community. Lancet *i:* 1014–1016 (1970).

46 PAPP, O. A.; PADILLA, L., and JOHNSON, A. L.: Dietary intake in patients with and without myocardial infarction. Lancet *ii:* 259–261 (1965).

47 Medical Research Council Working-Party: Dietary sugar intake in men with myocardial infarction. Lancet *ii:* 1265–1271 (1970).

48 SWAN, D. C.; DAVIDSON, P., and ALBRINK, M.: Effect of simple and complex carbohydrates on plasma non-esterified fatty acids, plasma sugar and plasma insulin during oral carbohydrate tolerance tests. Lancet *i:* 60–63 (1966).

49 ZAKIM, D. and HERMAN, R. H.: Fructose metabolism. II. Regulatory control to the triose level. Amer. J. clin. Nutr. *21:* 315–319 (1968).

50 BUTTERFIELD, W. J. H.; SARGEANT, B. M., and WHICHELOW, M. J.: The metabolism of human forearm tissues after ingestion of glucose, fructose, sucrose or liquid glucose. Lancet *i:* 574–577 (1964).

51 ATWELL, M. E. and WATERHOUSE, C.: Glucose production from fructose. Diabetes, N.Y. *20:* 193–199 (1971).

52 MOORHOUSE, J. A. and KARK, R. M.: Fructose and diabetes. Amer. J. Med. *23:* 46–58 (1957).

53 LACEY, P. E.; YOUNG, D. A., and FINK, J.: Studies on insulin secretion *in vitro* from isolated islets of the rat pancreas. Endocrinology *83:* 1155–1161 (1968).

54 PAPPER, S.: Effects of cortisone, fructose and glucose tolerance tests on normal men. J. Lab. clin. Med. *48:* 13–19 (1956).

55 CONN, L. W. and NEWBURGH, C. H.: Glycemic response to isoglucogenic quantities of protein and carbohydrate. J. clin. Invest. *15:* 665–671 (1936).

56 HILL, R. and CHAIKOFF, I. L.: Loss and repair of glucose-disposal mechanism in dog fed fructose as sole dietary carbohydrate. Proc. Soc. exp. Biol. Med. *91:* 265–267 (1956).

57 COHEN, A. M.; TEITLEBAUM, A.; BALOGH, M., and GROEN, J.: Effect of interchanging bread and sucrose as main source of carbohydrate in a low fat diet on the glucose tolerance curve of healthy volunteer subjects. Amer. J. clin. Nutr. *19:* 59–62 (1966).

58 DAVIDSON, P. C. and ALBRINK, M. J.: Insulin resistance in hyperglyceridemia. Metabolism *14:* 1059–1070 (1965).

59 RABINOWITZ, D. and ZIERLER, K. L.: Forearm metabolism in obesity and its response to intra-arterial insulin. Characterization of insulin resistance and evidence for adaptive hyperinsulinism. J. clin. Invest. *41:* 2173–2181 (1962).

60 SALANS, L. B.; KNITTLE, J. L., and HIRSCH, J.: The role of adipose cell size and adipose tissue insulin sensitivity in the carbohydrate intolerance of human obesity. J. clin. Invest. *47:* 153–165 (1968).

61 SIMS, E. A. H. and HORTON, E. S.: Endocrine and metabolic adaptation to obesity and starvation. Amer. J. clin. Nutr. *21:* 1455–1470 (1968).

62 Felig, P.; Marliss, E., and Cahill, G. F.: Plasma amino acid levels and insulin secretion in obesity. New Engl. J. Med. *281:* 811–816 (1969).

63 Bierman, E. L. and Porte, D., jr.: Carbohydrate intolerance and lipemia. Ann. intern. Med. *68:* 926–933 (1968).

64 Ogilvie, R. F.: Sugar tolerance in obese subjects. A review of sixty-five cases. Quart. J. Med. *4:* 345–358 (1935).

65 Brunzell, J. D.; Lerner, R. L.; Hazzard, W. R.; Porte, D., and Bierman, E. L.: Improved glucose tolerance with high carbohydrate feeding in mild diabetes. New Engl. J. Med. *284:* 521–524 (1971).

66 Grey, N. and Kipnis, D. M.: Effect of diet composition on the hyperinsulinemia of obesity. New Engl. J. Med. *285:* 827–831 (1971).

67 Sims, E. A. H.; Bray, G. A.; Danforth, E. J.; Glennon, J. A.; Horton, E. S.; Salans, L. B., and O'Connell, M.: Experimental obesity in man. VI. The effect of variations in intake of carbohydrate, lipid and cortisol metabolism; in Levine and Pfeiffer Hormone and metabolic research (in press).

68 Knittle, J. L. and Ahrens, E. H.: Carbohydrate metabolism in two forms of hyperglyceridemia. J. clin. Invest. *43:* 485–495 (1964).

69 Kaufmann, N. A.: Carbohydrate induced lipemia. Atherosclerosis *11:* 365–367 (1970).

70 Glueck, C. J.; Levy, R. I., and Fredrickson, D. S.: Immunoreactive insulin, glucose tolerance, and carbohydrate inducibility in types II, III, IV and V hyperlipoproteinemia. Diabetes, N.Y. *18:* 739–747 (1969).

71 Nestel, P. J.: Metabolism of linoleate and palmitate in patients with hypertriglyceridemia and heart disease. Metabolism *14:* 1–9 (1965).

72 Havel, R. J.; Kane, J.; Balasse, E. O.; Segal, N., and Basso, L. V.: Splanchnic metabolism of free fatty acids and production of triglycerides of very low density lipoproteins in normotriglyceridemic and hypertriglyceridemic humans. J. clin. Invest. *49:* 2017–2035 (1970).

73 Quarfordt, S. H.; Frank, A.; Shames, D. M.; Berman, M., and Steinburg, D.: Very low density lipoprotein triglyceride transport in type IV hyperlipoproteinemia and the effects of carbohydrate-rich diets. J. clin. Invest. *49:* 2281–2297 (1970).

74 Kirkeby, K.: Plasma lipids on a moderately low fat high carbohydrate diet, rich in polyunsaturated fatty acids. Acta med. scand. *180:* 767–776 (1966).

75 Antonis, A. and Bersohn, I.: The influence of diet on serum triglycerides in South African and Bantu prisoners. Lancet *i:* 3–9 (1961).

76 Yudkin, J.: Sweet and dangerous (Wyden, New York 1972).

77 Macdonald, I. and Braithwaite, D. M.: The influence of dietary carbohydrate on the lipid pattern in serum and in adipose tissue. Clin. Sci. *27:* 23–30 (1964).

78 McGandy, R. B.; Hegsted, D. M., and Stare, F. J.: Dietary fats, carbohydrates and atherosclerotic vascular disease. New Engl. J. Med. *277:* 186–192, 242–247 (1967).

79 Shamma'a, M. and Al-Khaldi, U.: Dietary carbohydrates and serum cholesterol in man. Amer. J. clin. Nutr. *13:* 194–196 (1963).

80 Grande, F.; Anderson, J. T., and Keys, A.: Effect of carbohydrates of leguminous seeds, wheat and potatoes on serum cholesterol concentrations in man. J. Nutr. *86:* 313–317 (1965).

81 HODGES, R. E.; KREHL, W. A.; STONE, D. R., and LOPEZ, A.: Dietary carbohydrate and low cholesterol diets. Effects on serum lipids of man. Amer. J. clin. Nutr. *20:* 198–208 (1967).

82 KUO, P. T.: Dietary sugar in the production of hypertriglyceridemia in patients with hyperlipemia and atherosclerosis. Trans. Ass. amer. Physicians *78:* 97–115 (1965).

83 NIKKILA, E. A. and PELKONEN, R.: Enhancement of alimentary hyperglyceridemia by fructose and glycerol in man. Proc. Soc. exp. Biol. Med. *123:* 91–94 (1966).

84 RIFKIND, B. M.; LAWSON, D. H., and GALE, M.: Effect of short-term sucrose restriction on serum-lipid levels. Lancet *ii:* 1379–1381 (1966).

85 EATON, R. P.: Synthesis of plasma triglycerides in endogenous hypertriglyceridemia. J. Lipid Res. *12:* 491–497 (1971).

86 OLEFSKY, J.; REAVEN, G., and FARQUHAR, J.: Cause of endogenous hypertriglyceridemia in man. J. clin. Res. *20:* 552 (1972).

87 BROWN, J.; MILLER, D. M., and LEVE, G. D.: Hexokinase isoenzyme in liver and adipose tissue of man and dog. Science *155:* 205–207 (1967).

88 ZAKIM, D.; HERMAN, R. H., and GORDON, C., jr.: The conversion of glucose and fructose to fatty acids in the human liver. Biochem. Med. *2:* 427–437 (1969).

89 ROBINSON, D. S. and WING, D. R.: Regulation of adipose tissue clearing factor lipase activity. In: Adipose tissue regulation and metabolic functions, pp. 41–46 (Academic Press, New York 1970).

90 ALBRINK, M. J. and MEIGS, J. W.: Interrelationship between skinfold thickness, serum lipids and blood sugar in normal men. Amer. J. clin. Nutr. *15:* 255–261 (1964).

91 GOLDRICK, R. B. and MCLOUGHLIN, G. M.: Lipolysis and lipogenesis from glucose in human fat cells of different sizes. Effect of insulin, epinephrine, and theophylline. J. clin. Invest. *49:* 1213–1223 (1970).

92 BRAY, G. A.: Metabolic obesity in rats and man; in JEANRENAUD and HEPP Adipose tissue regulation and metabolic functions, pp. 175–180 (Academic Press, New York 1970).

93 WILSON, D. E. and LEES, R. S.: Metabolic relationships among the plasma lipoproteins. J. clin. Invest. *51:* 1051–1057 (1972).

94 TROWELL, H.: Ischemic Heart disease and dietary fiber. Amer. J. clin. Nutr. *25:* 926–932 (1972).

95 BALMER, J. and ZILVERSMIT. D. B.: Cholesterol absorption and turnover in rats fed laboratory chow or semi-synthetic diets. Fed. Proc. *32:* 934 (1973).

Author's address: Prof. MARGARET J. ALBRINK, MD, Department of Medicine, West Virginia University School of Medicine, *Morgantown, WV 26506* (USA)

Progr. biochem. Pharmacol., vol. 8, pp. 271–299 (Karger, Basel 1973)

Triglyceride Metabolism in Diabetes mellitus

E. A. NIKKILÄ

Third Department of Medicine, University of Helsinki Central Hospital, Helsinki

Contents

I. Introduction

Diabetes represents a clinical disease or syndrome which is difficult to define in unequivocal terms. It still seems justified to use the diagnosis as synonymous to elevated fasting blood glucose value irrespective of whether the cause for this increase is known or not. Analogously, a chemical or latent diabetes is a condition in which fasting blood glucose is normal but the degree of hyperglycemia after any type of glucose load is abnormally high or delayed. These definitions do not include any statement on insulin secretion and, accordingly, the diabetes may be associated with high, normal or low insulin secretion. In fact, much controversy still exists as to whether diabetes primarily involves some defect in the output of insulin from the β-cells or 'peripheral' resistance in the assimilation of glucose, which is secondarily compensated by an increased secretion of insulin. It is becoming increasingly evident that diabetes, as defined in the above manner, is a heterogeneous group of diseases which may have common features in addition to the increased blood glucose and which, therefore, are not easily distinguished from each other. One of the diabetes-associated abnormalities involves triglyceride metabolism, which is the subject of this review. As a rather comprehensive paper on the relationship of glucose and insulin and triglycerides has appeared recently [92] and as some aspects are being dealt with in other articles of this volume, the presentation here has been made rather short without any attempt towards a complete bibliography of the extensive field.

The factors which may influence triglyceride metabolism in different types of diabetes are many and, therefore, the occurrence and degree of hyperglyceridemia and fatty liver are highly variable. In addition to high blood glucose itself the triglyceride synthesis and breakdown in different tissues may be influenced by lack of insulin and excess of insulin. Other factors active in fat mobilization include growth hormone, adrenaline and glucagon, all of which may be elevated in diabetes. Things are further complicated by the frequent association of diabetes and obesity and by the fact that a mild diabetes frequently accompanies different forms of familial hypertriglyceridemia. This makes it often difficult to decide in an individual case whether the primary disorder is diabetes or hyperlipemia. These complex relationships are made important by their association with coronary heart disease and other manifestations of atherosclerosis.

With reference to the two basic types of human diabetes the alterations of triglyceride metabolism are here discussed separately for insulin deficiency and hyperinsulinism. The former state is present in untreated or poorly controlled juvenile-type human diabetes and in experimental diabetes induced by alloxan, streptozotocin, mannoheptulose or anti-insulin serum.

Hyperinsulinism, on the other hand, is a characteristic feature of the major part of obese adult-onset diabetics. Its animal counterpart is represented by the hereditary obesity-hyperglycemic syndrome of mice and by the fat Zucker rat.

II. Prevalence of Hypertriglyceridemia in Human Diabetes

The association of hyperlipemia with uncontrolled juvenile-type diabetes was already well recognized in the preinsulin era [23] and the frequent occurrence of elevated plasma triglyceride level in insulin deficiency has been repeatedly confirmed [1, 6, 38, 76, 91, 94, 148]. Exact data on the prevalence of hypertriglyceridemia in diabetics as compared to matched nondiabetics are not available, however, and the reasons for this are quite obvious. Firstly, the triglyceride level in the general population is highly variable, being dependent on age, sex, diet and degree of obesity. Thus, diabetics should be compared to control subjects who have a normal glucose metabolism and are matched for the above parameters. Secondly, the triglyceride level in a diabetic patient is variable, being influenced by the degree of diabetic control and by the extent of vascular complications. Many reports on plasma triglycerides in diabetics do not provide individual data but record only mean values and, thus, the prevalence of hyperglyceridemia remains unknown. Much material also suffers from pooling of diabetics with different types of the disease and different age and obesity.

In the majority of reported series diabetics of any type show significantly higher mean plasma triglyceride levels than nondiabetics even though many patients have their triglyceride within a normal range. *Obesity* influences triglyceride concentration in diabetics more than in nondiabetics [26, 76, 119]. Thus, BRAUNSTEINER *et al.* [26] found a mean increase of 8 mg/100 ml per every 10-kg increment of body weight in nondiabetic subjects while the corresponding figures for diabetics treated with tolbutamide or insulin were 28 and 39 mg/100 ml, respectively. The plasma triglyceride does not show a significant correlation to *fasting blood glucose* level but, taken as a group, patients with poor control of diabetes (fasting blood sugar above 150 mg/100 ml) have a higher degree of hyperglyceridemia than well balanced patients [1, 76, 91, 115].

Duration of diabetes, on the other hand, does not influence plasma lipid values more than predicted by age regression [91, 115]. Apart from the hypertriglyceridemia associated with severe renal involvement, the plasma triglyceride values are similar in diabetics with and without *microangiopathy* [5, 76, 115]. On the other hand, diabetics with clinical *coronary disease* show a high prevalence of hypertriglyceridemia [5, 76, 115] but it is not clear from the published material whether the plasma lipid levels are different in diabetic and nondiabetic patients with overt coronary heart disease. An impaired glucose tolerance (subclinical diabetes) may be associated with coronary atherosclerosis irrespective of the plasma triglyceride level [96].

Dietary influences on plasma lipid levels in different types of diabetes have been intensely discussed throughout the years, already beginning in the preinsulin era. The numerous early studies have been reviewed by BLIX in 1926 [23]. It is well-known that a diet high in *carbohydrates* and particularly in sugar will elevate the endogenous plasma triglyceride level in nondiabetic subjects even though the individual response is highly variable [92]. A similar response has been reported to occur in patients with diabetes [20, 23]. However, with a moderate increase of dietary carbohydrate content and decrease of fat calories ERNEST *et al.* [47] like STONE and CONNOR [134] could not observe any change of plasma triglyceride level in well controlled insulin-dependent diabetics, and VAN ECK [45] even obtained a fall of triglycerides by reducing dietary fat and increasing carbohydrate intake. Exchange of fructose (80 g daily) for starch does not influence plasma triglycerides in insulin-dependent adult diabetics [110] or diabetic children [3].

Dietary fat was formerly believed to be responsible for the visible lipemia (i.e., a more severe hypertriglyceridemia) at times associated with diabetes [7]. The concept of a fat-induced diabetic lipemia has been revived recently by BAGDADE *et al.* [12]. They described 5 patients presenting with severe hyperglycemia, moderate ketosis and gross hyperlipemia. In one of these patients elimination of fat from the diet decreased plasma triglycerides but did not bring them to a normal range. The circulating triglycerides were mainly in the form of chylomicrons, and plasma postheparin lipolytic activity was slightly decreased until insulin treatment was started. There seems to be no doubt that much of the excess plasma triglyceride in these cases was really of dietary origin. Also it has been recently shown by PARKER *et al.* [107] that the lipids in diabetic eruptive xanthomas originate from chylomicrons. However, it still seems uncertain whether the 'diabetic' hypertriglyceridemia was primarily induced by dietary fat in a similar manner as in familial type I hyperlipidemia. Since the exogenous and endogenous plasma triglycerides have a common saturable removal system the fractional removal of both diminishes with increasing size of the circulating triglyceride pool. At a certain triglyceride level the elimination of alimentary fat from the blood is so slow that an overnight fast is not sufficient for removal of dietary fat, and an exogenous hyperlipemia is associated with a primarily endogenous one. This concept fits in well with the observation of BAGDADE *et al.* [12] that a fat-free regimen reduces the plasma triglycerides but does not abolish the hyperlipemia. Thus, even though there may be some impairment in the removal of exogenous triglycerides from the blood in diabetes (see below), dietary fat contributes to the postabsorptive hypertriglyceridemia only in cases with extreme increase of endogenous plasma triglycerides. In the vast majority of diabetic patients, basal triglyceride values are lower during a high-fat intake than during an ordinary diet [5].

Attempts have been made to separate *definite syndromes presenting with diabetes and hyperlipemia* but the field remains in a state of confusion. THANNHAUSER [137], in his classic review, described the combination of mild diabetes and gross hyperlipemia and similar cases were included in an account of idiopathic hyperlipemia presented by LEVER *et al.* [80]. ADLERSBERG and WANG [2] outlined a syndrome of 'idiopathic hyperlipemia, mild diabetes mellitus and severe vascular damage' by describing 5 cases with extreme elevation of plasma neutral fat, hypercholesterolemia, obesity, eruptive xanthomatosis, nonketotic diabetes and arterial disease. In fact, this combination of biochemical abnormalities in less striking degrees is rather common and a whole spectrum of cases can be found with diabetes varying from a latent one to a severe ketoacidosis and the degree of hyperlipidemia extending from a moderate hypertriglyceridemia to a gross hyperlipemia with much elevated plasma chylomicron and very low density lipoprotein-triglyceride and cholesterol concentrations. Since many cases with primary familial types III, IIB or IV (or any type between these) are also associated with some disturbance of glucose and insulin metabolism it seems that any separation of defined syndromes is not possible before more is known on the biochemical background of elevated plasma triglyceride and lipoprotein levels.

Hypertriglyceridemia regularly accompanies the uncommon but curious disease called lipoatrophic diabetes [113]. An association of diabetes, acromegaly and hyperlipemia has been described [34] and LYNCH *et al.* [82] have reported a family with coexistence of diabetes mellitus, hyperlipemia, short stature and hypogonadism.

III. Plasma FFA Concentration and Turnover in Insulin Deficiency

A marked elevation of basal plasma free fatty acid (FFA) is one of the characteristic biochemical changes of human insulin-deficient diabetes [19, 79]. On acute deprivation of insulin by injection of anti-insulin serum the rise of FFA occurs rapidly and parallels the increase of blood glucose [135, 136, 151].

Untreated juvenile diabetics also show an exaggerated response of plasma FFA to exercise [36]. There is much evidence that the rise of plasma FFA is entirely accounted for by an increased mobilization from adipose tissue. Thus, an increased release of FFA occurs *in vitro* from adipose tissue of diabetic rats [104, 135, 144] and man [30, 35]. All estimations of the turnover rate of plasma FFA in diabetes are also consistent with an increased inflow of FFA into plasma. As a general rule, the turnover rate and concentration of plasma FFA show a good correlation with each other and the regression is similar for diabetics and nondiabetics [42, 120]. In depancreatized dogs the net transport of plasma FFA is increased 2- to 4-fold over the values of control dogs [15]. In rats given a constant infusion of labeled FFA the specific activity of plasma FFA is decreased on injection of anti-insulin serum, a finding which indicates that the release of FFA into plasma is accelerated [16]. It should be noted, however, that none of these studies excludes the possibility that diabetes could also impair the assimilation of FFA by some individual tissues. In fact, this has been suggested to be the case in forearm muscle preparation [32].

The exact mechanism of the increased liberation of FFA from fat cells in diabetes is not fully established. It is obvious that lipolysis is increased and re-esterification is decreased. A stimulation of the hormone-sensitive lipase activity is expected since, in addition to the fall of the antilipolytic activity exerted by insulin, diabetes is associated with increased plasma concentrations of several lipolytic hormones like catecholamines [40], growth hormone [55] and glucagon [141]. It is also known that in insulin-deficient rats both hypophysectomy and adrenalectomy decrease the release of FFA and glycerol by adipose tissue [49, 51]. The greater rise of plasma FFA and glycerol on exercise in insulin-deficient diabetics might also be due to the exaggerated response of growth hormone to muscular work [55].

The plasma glycerol level is also elevated in diabetes [37, 109, 142], apparently due to increased release from adipose tissue [35, 105].

IV. Hepatic Metabolism of Fatty Acids and Triglycerides in Insulin Deficiency

A. Uptake and Esterification of FFA

The liver is extremely active and rapid in the extraction of FFA from the blood and in their incorporation into lipid esters [93, 127]. A fraction of 30–40 % of albumin-bound FFA is removed during a single passage through the liver [61, 87] and the hepatic clearance represents about one-third of the total plasma FFA outflow transport [25, 102, 132]. This fractional removal of FFA by the liver is not influenced by the concentration of FFA in the inflowing blood [87], which means that the absolute amount of FFA taken up is in direct correlation to the FFA level in portal blood. This correlation is not changed by any of the nutritional or hormonal factors studied so far. Thus, there is considerable agreement on the fact that fractional hepatic FFA uptake is not altered in diabetes [14, 15, 130–132, 151]. In only a few instances has an increased [58] or decreased [77] relative FFA extraction been reported but these latter studies have used less reliable techniques. It thus seems likely that the absolute FFA uptake of the liver in diabetes parallels the concentration of plasma FFA.

Esterification of FFA occurs immediately on their uptake [46]. In diabetes the activation of fatty acids to CoA and carnitine derivatives is obviously not impaired since the levels of both acyl-CoA [140, 145] and of long-chain acylcarnitines [24] are increased in the liver of alloxan-diabetic rats. The glycerol-3-phosphate concentration in the liver is reduced in acute insulin deficiency produced by anti-insulin serum [69] but is not changed in the chronic diabetic state induced with alloxan [41] or streptozotocin [78]. Diabetes does not influence the hepatic uptake of glycerol *in vitro* [9] or *in vivo* [14] or the glycerol kinase activity of the liver [41], and as the glycerol influx from plasma to liver is increased it is not likely that the synthesis of glycerol phosphate is defective.

Direct assays of the enzymes of triglyceride synthesis have not been carried out in the diabetic liver, but CORDER and KALKHOFF [41] have demonstrated that in rat-liver supernatant fraction supplemented with necessary cofactors the incorporation of both palmitate and glycerol-3-phosphate into triglyceride is significantly increased in alloxan diabetes. Stimulation of esterification by addition of glycerol-3-phosphate was also higher in the livers of diabetic rats than of controls. These results suggest that the activity of 'triglyceride synthetase' enzyme complex is increased in diabetic liver. However, studies made *in vivo* in rats or by liver perfusion with labeled precursors have not supported the view that lack of insulin, as such, favors the synthesis of triglycerides in the liver. Thus, in perfused liver isolated from rat with an acute alloxan diabetes the incorporation of radioactive FFA into liver triglyceride is not different from corresponding controls, whereas the synthesis of phospholipids seems to be markedly depressed [59]. In livers taken from rats with a chronic alloxan diabetes under insulin deprivation the conversion of perfusate FFA label into all lipid esters of the liver is impaired [56]. On the other hand, when radioactive FFA is injected *in vivo* to anesthetized rats the proportion of the dose recovered in liver triglycerides is similar in normal and diabetic animals [97]. This finding is consistent with earlier studies demonstrating that in liver slices the incorporation of ^{14}C-fructose and of ^{14}C-glucose into tissue glyceride-glycerol is identical in pancreatectomized and normal rats [39].

On the basis of the available evidence listed above it seems justified to conclude that an acute or chronic diabetic state with insulin deficiency does not, in principle, interfere with the ability of the liver to take up FFA and glycerol and to synthetize triglycerides. It follows from this that the extent of liver triglyceride synthesis in diabetes is determined by the plasma FFA and glycerol concentrations and by the competitive pathways of fatty acid metabolism, i.e. ketogenesis, oxidation and incorporation into phospholipids or other lipid esters. It is well-known that diabetes diverts more fatty acids to ketone bodies and to the oxidative pathway, but in the face of elevated FFA uptake from plasma the absolute amount of triglyceride formed in the liver from plasma FFA is likely to be increased. On the other hand, this may be more or less compensated by the decreased availability of endogenous fatty acids because of the block of lipogenesis, and the net result of these two opposing factors is unknown and perhaps variable. Moreover, the blood glucose level may be an important factor which influences the hepatic fatty acid and triglyceride metabolism independently of insulin, and hyperglycemia can partly counterbalance the effects of insulin lack as is, for example, suggested by the data of HAFT [54]. All these considerations help us to understand why the liver triglyceride concentration and metabolism, as a whole, do not behave in a uniform manner in all individuals with diabetes. Furthermore, all the studies made so far on fat metabolism in diabetes have dealt with liver triglyceride as a single pool even though it is well recognized that the synthesis and turnover and their regulation may be highly dissimilar in the membrane, storage and secretory compartments of liver cell triglyceride. It is easy to realize that our current knowledge of these biochemical changes is at a very 'macroscopic' stage.

B. Hepatic Synthesis of Fatty Acids

That hepatic fatty acid synthesis is markedly depressed in insulin deficiency diabetes of experimental animals was shown by DRURY [44] and by STETTEN and BOXER [133]. During the subsequent 30 years this finding has been well confirmed and much work has been done to elucidate the mechanism of this inhibition. The whole topic has been recently reviewed [92] and this makes it unnecessary to repeat any details here particularly as new essential information does not seem to have accumulated in the meantime. Whether a similar block of hepatic fatty acid synthesis exists also in insulin-deficient man has not been established and should, therefore, not be taken as a fact. It may be important to recognize that most of the evidence for absent or low lipogenesis in diabetes is derived from animals with absolute lack of insulin. In the human disease some insulin is usually present and the simultaneous hyperglycemia may oppose the effects of insulin deficiency. Thus, in perfused rat-liver an increase of glucose concentration in the medium enhances lipogenesis even in the absence of insulin [54]. Consequently, the extent of inhibition of hepatic fatty acid synthesis may be highly variable in human diabetes and the possibility of normal or even increased lipogenesis is not ruled out by present evidence.

It is possible that in a diabetic organism the hepatic FFA uptake and endogenous fatty acid synthesis have a reciprocal relationship and thus mutually compensate each other's contribution to liver and plasma triglyceride synthesis. In the absence of insulin the FFA mobilization is much stimulated while the synthesis of fatty acids in the liver is markedly depressed and all plasma and liver triglyceride fatty acid is derived from plasma FFA fraction. When insulin is present the lipolysis will be inhibited at concentrations which still are insufficient to prevent hyperglycemia. Under these conditions the hepatic lipogenesis may become supernormal and stimulate the formation of liver and plasma triglycerides from endogenous substrate. Additional contribution in the same direction is made by exogenous insulin which often corrects the defect in lipolysis but does not prevent hyperglycemia. A high rate of plasma triglyceride synthesis may, therefore, be expected in juvenile-type diabetics treated with insufficient amounts of insulin.

V. Secretion of Triglycerides by the Diabetic Liver

Estimation of the rate of output of triglycerides from the liver has appeared to be difficult and therefore the values reported for different conditions are very divergent. The transport is too slow to make the concentration gradient across the liver large enough for any exact chemical measurement. Incorporation of a radioactive precursor into plasma triglycerides may occur through complex pathways with different flow rates and, moreover, only part of the secreted triglyceride molecules may be derived from circulating precursors. Under experimental conditions an isolated perfused liver is theoretically ideal, since no triglyceride is lost from the system and the net synthesis can be calculated, but it is quite obvious that the conditions even at their best are not equivalent to those occurring *in vivo*. Thus, the triglyceride concentration of the liver may rapidly decrease during perfusion lasting a few hours without much output of triglyceride into perfusate [59]. It is likely that lipolysis and release of FFA occur from perfused liver in contrast to *in vivo* conditions, in which a significant liberation of FFA from the liver has not been established [15].

The output of triglycerides into the medium is markedly reduced in the isolated perfused liver of an alloxan-diabetic rat [56, 58, 59]. The secretion is stimulated by increasing medium FFA concentration but at all FFA levels the net hepatic release of triglycerides is less in diabetic animals than in controls [59]. Also, the incorporation of medium ^{14}C-palmitic acid into the secreted triglyceride is subnormal in the perfused livers of diabetic rats [58]. The triglyceride secretion is increased by pretreatment of the animals with insulin [59], but is not influenced by insulin added to the medium [60]. In acute insulin deficiency produced by anti-insulin serum the triglyceride secretion by the liver remains unchanged for several hours but begins to deteriorate thereafter and ceases after a 12-hour lack of insulin [151]. The secretory block subsides only slowly after administration of insulin and is not corrected by insulin added to the perfusion medium [151]. Similar results pointing to a decreased secretion of triglycerides by diabetic liver have been obtained in some *in vivo* studies.

Thus, SAILER *et al.* [120] found that relatively less of the FFA transported in plasma appeared in plasma triglyceride in insulin-deficient diabetics as compared to normals. Likewise, in pancreatectomized dogs BASSO and HAVEL [15] demonstrated that the conversion of plasma radioactive FFA into plasma VLDL-triglyceride was either zero or extremely low. In contrast to these observations, SPITZER *et al.* [132] could not detect a significant difference between alloxan-diabetic and control dogs in the fractional conversion of plasma FFA into plasma triglyceride. As the total hepatic FFA influx was much increased in the diabetic animals due to high plasma FFA transport, the absolute rate of triglyceride secretion (from plasma FFA) was increased. Similar findings were recently reported by BALASSE *et al.* [14] in dogs after injection of anti-insulin serum. The fraction of plasma palmitate taken up by the liver and converted to plasma triglyceride increased from 11 to 26% after anti-insulin serum. As the total FFA turnover was simultaneously accelerated the plasma triglyceride production (from FFA) increased 5-fold on deprivation of insulin. The flux rates given in these two studies on dogs are very similar for both control and diabetic animals even though the diabetic state was chronic in the former and very acute in the latter. There is, so far, no explanation for the contradictory results obtained in chronic and acutely diabetic dogs in HAVEL's laboratory [14, 15]. Electron micrographs taken from the liver of anti-insulin serum treated rats show accumulation of lipoprotein particles at endoplasmic reticulum, Golgi apparatus and the space of Disse, indicating an increased secretion of lipids [65]. This change is closely related to the increase of plasma FFA since inhibition of the FFA mobilization by nicotinic acid effectively prevents accumulation of lipoprotein particles in the liver [65] as well as the increase of plasma triglyceride after injection of anti-insulin serum [53].

Since protein synthesis is known to be depressed in diabetes, it has been thought that the secretion of triglyceride is blocked by the inability of the liver to produce enough apolipoproteins. This possibility could, then, also account for the difference between acute and chronic insulin deficiency and for the discrepancy between observations made *in vivo* and those obtained in liver perfusion studies, where the supply of substrate for *de novo* protein synthesis is limited. Actually, WILCOX *et al.* [146] have shown that the incorporation of radioactive amino acids into the major classes of plasma lipoproteins is depressed in perfused liver of alloxan-diabetic rats. However, this defect could not be corrected by insulin treatment while the secretion of triglycerides increased. Protein synthesis is depressed to a similar degree in alloxan and anti-insulin serum diabetic rats [139] but it is possible that there is a time-lag after the injection of anti-insulin serum when the protein synthesis is still maintained at a normal level but FFA is already being mobilized at an increased rate.

In human juvenile-type diabetics with ketoacidosis, the secretion of endogenous triglycerides into circulation is regularly above normal and there is a significant correlation between plasma triglyceride level and production rate [94, 95]. The fractional removal rate of triglycerides is in the low normal range or decreased, but this may be accounted for by the increased concentration [94]. Also in poorly controlled insulin-dependent diabetics without ketoacidosis the plasma triglyceride influx is moderately elevated or at the upper normal range [94]. It seems from these results that the hypertriglyceridemia associated with insulin-deficient diabetes of man is mainly caused by an increased release of triglycerides into blood rather than by a deficient removal system. This postulate is not compatible with the results of the perfusion studies of rat-liver or with the conclusions of SAILER *et al.* [120]. At the present time the reasons for this discrepancy are not established.

VI. Removal of Plasma Triglycerides in Insulin Deficiency

Many investigators in the preinsulin era believed that the lipemia often seen in diabetics was due to a deficient removal of dietary fat from the blood [7]. This view was later challenged, however, by a number of dietary studies indicating that even poorly controlled insulin-dependent diabetics can usually consume large amounts of fat without increase of the *fasting* triglyceride level [23, 83, 134]. BLIX [23] had already noticed that there are rare exceptions to this rule and it is obvious that these are cases with gross elevation of VLDL-triglycerides and accumulation of chylomicrons (type V diabetic lipemia) described in the literature [12]. Even though the actual fat-induced hyperlipidemia is uncommon in diabetics many studies have shown that the elimination of single oral fat loads is often delayed in patients with insulin-deficient diabetes [62, 70, 123]. Of particular interest is the finding of KALLIO [70] that nondiabetic normoglyceridemic relatives of diabetic index patients often have an impaired oral fat tolerance. Evidence for the rôle of exogenous fat in the hypertriglyceridemia of alloxan-diabetic rats is conflicting [18, 33, 84, 118].

In spite of a number of studies it is still not clear whether the removal efficiency of plasma triglycerides is primarily decreased in insulin deficiency or whether the lowered fractional clearance of fat from the circulation is accounted for by the increased pool size. Following insulin withdrawal, the basal plasma triglyceride concentration rises and the fractional removal rate of injected emulsified fat, chylomicrons and tracer triglycerides decreases in diabetic animals [27, 52, 58, 71, 143] and man [13]. In one study only did diabetic rats clear chylomicrons at the same rate before and after insulin deprivation [28]. In order to dissociate the effects of diabetes and hypertriglyceridemia (= increased endogenous triglyceride pool) on the clearance rate of exogenous fat, we carried out the intravenous Intralipid® test in a number of diabetic and nondiabetic patients. The results indicated that the fractional rate constant (K) of Intralipid-triglyceride disappearance was either normal or decreased in diabetic subjects but when the K value was plotted against fasting (endogenous) triglyceride concentration an identical regression was obtained in diabetics and nondiabetics [94]. From this result it appears that insulin deficiency does not impair the removal efficiency of exogenous plasma triglycerides apart from the influence through enlarged pool size. On the basis of two published turnover studies [94, 120] it also seems that endogenous plasma triglycerides are cleared at a normal rate in human insulin-deficient diabetes.

VII. Alterations of Lipoprotein Lipase Activity in Diabetes

There is already circumstantial evidence that lipoprotein lipase (clearing factor lipase), located at the capillary endothelial cells, is the enzyme responsible for the uptake of circulating chylomicron and VLDL-triglycerides [22, 116]. Accordingly, variations in the activity of this enzyme might be expected to influence the removal efficiency and concentration of plasma triglycerides. However, evidence for this association is so far only fragmentary and far from unequivocal [116]. This applies also to diabetes where conflicting results have been reported on the activity of lipoprotein lipase in tissues and in postheparin plasma.

Requirement of insulin for the synthesis and maintenance of lipoprotein lipase in *adipose tissue* is well established. In rats made insulin-deficient with alloxan [27, 72, 125], streptozotocin [114] or mannoheptulose [99] the adipose tissue lipoprotein lipase activity is much reduced and addition of insulin *in vitro* can induce synthesis of lipoprotein lipase depressed by starvation [63, 99, 122]. The mechanism of this action of insulin is not clear; increased concentration of intracellular FFA [98] or cyclic AMP [150] and decreased formation of ATP [108] have been suggested as possible mechanisms of the enzyme repression in insulin deficiency.

The *myocardial* lipoprotein lipase activity was found to be supernormal in diabetic rats [72]. Recently this result has been challenged, however, by careful perfusion studies of alloxan-diabetic rat hearts [4] indicating a normal 'acute' release of lipoprotein lipase but a depressed liberation during prolonged perfusion. The authors presented an attractive hypothesis on two pools of the enzyme, one being rapidly released and located at the vascular bed while the other is at the site of actual synthesis and comes out slowly on perfusion with heparin. The former is normal in diabetes while the synthesis of myocardial lipoprotein lipase is reduced. Uptake of chylomicron triglycerides by the myocardium has been reported to be decreased [75] or increased [52] in diabetes.

Studies on the influence of insulin deficiency on *postheparin plasma* lipoprotein lipase activity have yielded highly controversial results. In alloxan-diabetic or pancreatectomized rats and dogs the values are subnormal and increase on administration of insulin [4, 71, 85]. In rabbits, on the other hand, diabetes does not influence the postheparin plasma lipolytic activity [101]. Also, in human diabetes most reported postheparin plasma lipoprotein lipase values are not different from normal [43, 67, 111, 129]. Even in ketoacidosis with clearly elevated plasma triglyceride levels, the postheparin plasma lipoprotein lipase values remain normal [67]. However, when postheparin plasma lipoprotein lipase is determined in the same individual before and after insulin withdrawal a definite decrease of the activity is observed [13]. In diabetic hyperlipemia subnormal postheparin plasma lipoprotein lipase values have been consistently recorded [10, 12, 147].

A single injection of heparin presumably releases only part of the lipoprotein lipase present in the vascular bed and it is doubtful whether the activity measured in plasma 10–20 min after the heparin injection reflects the amount of enzyme in tissues. In order to get a better estimate of the total tissue lipoprotein lipase pool and probably also of the synthesis of the enzyme, BIERMAN *et al.* [17] have used a 4-hour heparin infusion technique and determined the steady-state level of plasma postheparin lipoprotein lipase. Their preliminary data [29] suggest that the rapidly releasable activity is normal in diabetes but, in contrast to nondiabetic subjects the diabetics show a progressive decrease of postheparin plasma lipoprotein lipase on continued heparin administration. This has been interpreted as a sign of a reduced synthesis pool of lipoprotein lipase in diabetes. SHIGETA *et al.* [129] had previously noted that the 20-min postheparin lipase activity is decreased in diabetics while the 10-min value is normal, a finding which is in good agreement with BIERMAN's data. These results are also strikingly consistent with the two-pool model and reduction of synthesis pool in diabetes as suggested by AKTIN and MENG [4] (see above) for myocardial lipoprotein lipase. It remains for future work to explore whether the amount and activity of the enzyme in the synthesis pool has any significance in the removal of triglycerides by different tissues.

The contribution of different tissues to the postheparin plasma lipoprotein lipase activity is unknown. As there seems to be considerable differences in the hormonal and nutritional control of this enzyme in various organs and changes in these may occur, even in opposite directions, the plasma lipolytic activity may remain constant in spite of wide variations in the lipoprotein lipase activity of an individual tissue. Under these conditions it is likely that the overall removal efficiency of plasma triglycerides is not influenced by a reduction of lipoprotein lipase activity of, for example, adipose tissue but the relative distribution of triglycerides between the sites of assimilation is altered. Evidence for this possibility in diabetes is provided by observations that uptake of chylomicrons is diminished in adipose tissue [27, 52, 88] but increased in skeletal muscle [88] and in liver [58] in alloxan-diabetic rats. As new methods have recently been developed to separate the hepatic lipoprotein lipase from other lipolytic activities of postheparin plasma [74] it will be of interest to know whether this enzyme is influenced by insulin deficiency.

VIII. Triglyceride Metabolism in Diabetes with Hyperinsulinism

In the adult-onset type of diabetes the plasma insulin level is not subnormal in absolute terms but is either in the normal range or elevated. It is certainly less than in a normal subject when related to blood glucose level but from the metabolic point of view the absolute insulin concentration in blood may be more important than the relative value expressed as insulin : glucose ratio. In these cases the liver is exposed to high-insulin and high-glucose concentrations simultaneously, and this influences the triglyceride metabolism in a different manner than insulin deficiency even though the end-result is similar in both conditions, namely elevation of endogenous triglycerides in plasma and liver. It is difficult to decide which of the two basic abnormalities, hyperinsulinemia or hyperglycemia, is mainly responsible for the accumulation of triglycerides since the two are closely inter-related to each other and, in addition, to the usually coexisting obesity which also influences fat metabolism. An animal model of this type of diabetes is represented by the hereditary obese hyperglycemic mice (New Zealand obese, Bar Harbor obob and others).

A. Effect of Insulin, Glucose and Obesity on Plasma FFA Metabolism

Insulin is the most potent antilipolytic agent which inhibits the action of the 'hormone-sensitive' lipase of adipose tissue at concentrations physiologically present in peripheral blood. The mechanism of this action is not yet clear but it is obviously mediated by a rapid decrease of cyclic AMP in the fat cell [31]. Opinions differ whether this lowering is caused by activation of cyclic AMP breakdown by phosphodiesterase [81] or by inhibition of adenylate cyclase [64, 68]. It is also possible that the antilipolytic activity of insulin is unrelated to its effects on cyclic AMP [48]. Whatever the mechanism may appear to be, the lipolysis is much more sensitive to insulin than is the glucose transport.

Hyperglycemia also inhibits the production of FFA from adipose tissue. Obviously this occurs through providing more glycerol-3-phosphate and ATP for esterification of intracellular fatty acids before their release from the fat cell.

Both insulin and glucose thus depress the production (turnover) rate of FFA and since they do not influence the removal of FFA from the blood [8, 19, 42] the concentration of plasma FFA is decreased. The fraction of plasma FFA transport taken up by the liver is not changed by insulin [50, 59, 138] and, consequently, the absolute hepatic FFA uptake is diminished by high plasma insulin level.

As one would expect, obese subjects have a higher total circulating FFA turnover rate than normal-weight persons [21, 86, 90]. Expressed per unit body weight or fat tissue weight the FFA turnover is independent of relative body weight [90] but in relation to the lean body mass obese subjects produce more FFA than persons of normal weight [21]. This means that the FFA production by a given amount of adipose tissue is constant. Thus, in obese patients the FFA influx into the liver is increased over that of normal-weight subjects and this increment may be even higher than estimated on the basis of total body FFA turnover rate, since a considerable part of the adipose tissue drains directly into the portal circulation. It is not known whether the fractional hepatic FFA uptake is influenced by obesity.

Obesity may thus counterbalance the reduction of FFA release produced by hyperglyceridemia and hyperinsulinism.

B. Effect of Insulin, Hyperglycemia and Obesity on Hepatic Triglyceride Synthesis and Release

When insulin or glucose are administered to a nondiabetic subject the plasma triglyceride concentration is reduced within a few hours. This acute effect is obviously related to the decreased influx of FFA and glycerol into the liver and to activation of lipoprotein lipase [92]. It is transient, however, and chronic hyperglycemia and hyperinsulinism have a hyperglyceridemic effect, as illustrated by the induction of hyperlipemia by dietary sugars.

The question of the possible direct effects of insulin on hepatic fatty acid and triglyceride synthesis and secretion is still open. A number of both *in vitro* and *in vivo* studies favors the view that insulin stimulates hepatic lipogenesis [92] by some unknown mechanism. A similar effect may also be obtained by perfusing the liver with high-glucose medium and under these circumstances insulin does not further increase the fatty acid synthesis [54, 138]. Esterification of added FFA in rat-liver slices is not influenced by insulin [117]. A similar negative result has been obtained with human liver slices taken from patients with insulin-producing tumors [124]. The incorporation of labeled glycerol and fructose into glyceride glycerol and fatty acid was not different from normal. On the other hand, in perfused rat-liver, raising the insulin concentration in the medium increases the esterification and decreases the oxidation of FFA [138]. Also glucose stimulates the conversion of palmitate into triglycerides in isolated liver cells [103].

Controversial results have also been reported on the rôle of insulin in regulation of hepatic triglyceride secretion. Earlier, Heimberg *et al.* [58] observed no difference in the output of triglycerides from perfused liver between control and insulin-treated normal rats. Recently, Topping and Mayes [138] demonstrated that addition of insulin to perfusion medium significantly increased the secretion of VLDL-triglycerides and the conversion of medium FFA to triglycerides. It seems from their results that insulin stimulated the formation of plasma VLD-triglyceride both from the medium FFA (direct secretory pathway) and from the fatty acid synthetized in the liver. It is confusing, however, that in several studies in man the conversion of plasma FFA into plasma triglyceride (rate of appearance of radioactivity from infused FFA to plasma triglyceride) was not influenced at all [42, 89] or was even decreased on infusing insulin for several hours [66].

Induction of hyperglycemia and hyperinsulinemia by glucose infusion likewise decreases the production of plasma triglyceride from plasma FFA in man [121]. It may well be, however, that the chronic effects of hyperinsulinemia and its associated hyperglycemia cannot be reproduced in acute experiments, where the time is too short for any enzyme inductions. That these may be essential in the pathogenesis of hyperlipidemia is shown by the biochemical changes noted in obese hyperglycemic mice. These animals have a chronic hyperinsulinism and elevated blood glucose level in addition to obesity. The hepatic lipogenesis and the triglyceride synthesis are increased many-fold over the corresponding values of non-obese littermates [149].

Increase of adipose tissue mass with its associated rise of FFA production may appear to be a more important causative factor of hypertriglyceridemia and fatty liver in adult-onset type diabetes than is hyperglycemia or hyperinsulinism alone. Insulin secretion and plasma insulin levels are not different in obese patients with or without hypertriglyceridemia [11, 100]. It is not known to what extent diabetes influences the alterations of plasma FFA and triglyceride turnover present in obesity. Kinetic studies have indicated that the secretion of triglycerides into plasma is often increased in adult-onset type diabetics while the fractional removal and efficiency of elimination are not different from normal [94]. However, an adequate comparison between diabetics and nondiabetics with matched relative body weight has not been made in this respect.

IX. Lipoatrophic Diabetes

Hypertriglyceridemia, severe atrophy of adipose tissue, insulin-resistant hyperglycemia and absence of ketosis form the characteristic metabolic syndrome of a rare disorder called lipoatrophic diabetes or generalized lipodystrophy. The disease may be either congenital or of late-onset type and a variable number of other abnormalities is frequently found in these patients [113]. It is not known how the different metabolic anomalies are linked to each other and what the primary disturbance is. As the scarcity or almost total absence of adipose tissue is common to all patients, many believe that the basic error is either a defect in the development of fat cells or a metabolic error in triglyceride synthesis. A further possibility would be an unusually rapid mobilization of fat from the adipose tissue but this is not supported by the available evidence [57, 106]. Plasma FFA level is usually normal and it cannot be raised much by catecholamines [57, 73], a finding which is expected on account of the extremely small fat stores. The presence of an excessive amount of insulin antagonists or of pituitary fat mobilizing factor has been suggested but never proved.

Plasma triglyceride concentration is generally elevated but may occasionally be normal [57]. Much evidence favors the view that defective removal is an important pathogenetic factor behind the hyperlipemia, and this possibility would also fit in with the common view that adipose tissue is an important site of plasma triglyceride elimination. Postheparin plasma lipolytic activity is subnormal in most of the cases studied [57, 73, 106, 128]. The patients show a decreased clearance of alimentary fat [57] and of intravenous fat emulsion [106] and they may develop a fasting chylomicronemia. The lipoprotein pattern is either of type IV (increased pre-β) or type V (increased pre-β and chylo). It is possible that the synthesis and release of endogenous plasma triglycerides is increased and this also contributes to the development of hyperlipemia. Some evidence for this possibility has been presented earlier [126] and is consistent with the hyperglycemia and hyperinsulinemia associated with lipoatrophic diabetes. It is plausible, then, that the rate of hepatic lipogenesis is high and plasma triglycerides are synthetized from the endogenous fatty acids of the liver rather than from plasma FFA. It is pertinent that in partial lipodystrophy the plasma triglycerides may be markedly elevated in spite of a normal or only slightly decreased postheparin lipolytic activity while the plasma insulin response to glucose is constantly exaggerated [112].

X. Summary

Diabetes represents a heterogeneous group of disorders, which are associated with multiple hormonal and metabolic disturbances. Insulin secretion may be low, normal or increased and abnormalities in the secretory patterns of growth hormone, glucagon and catecholamines are often present. Plasma triglyceride level may be normal but usually, diabetes is accompanied by slight to moderate hypertriglyceridemia. Gross hyperlipemia with or without skin xanthomas is occasionally seen.

In insulin deficiency (untreated juvenile-type diabetes, experimental diabetes) the release of FFA and glycerol from adipose tissue is enhanced and their increased flow into the liver stimulates the synthesis and secretion of plasma triglycerides. Simultaneously, the removal of triglycerides from the circulation may be reduced due to decreased synthesis and inhibition of the activity of lipoprotein lipase by FFA in fat cells. In chronic and severe hypoinsulinism, the hepatic triglyceride secretion becomes depressed apparently due to a block in apoprotein peptide synthesis.

In adult-onset type diabetes, the plasma insulin levels are usually increased above normal and, together with elevated blood glucose, this stimulates the hepatic fatty acid synthesis. The production of VLDL-triglycerides from these fatty acids is enhanced, and the total hepatic synthesis of plasma triglycerides may be further increased by the associated obesity and concomitant accelerated FFA turnover. Thus, an overproduction hyperglyceridemia is common in these patients. Occasionally, a removal defect is also present and leads to marked elevations of plasma VLDL particles and chylomicrons.

Dietary influences on the diabetes-associated hyperlipidemia are not established. In most cases, a high carbohydrate diet increases plasma triglyceride level while a low carbohydrate-high fat diet or reduction of calory intake will relieve the hyperlipidemia. In cases with gross hyperlipemia, elimination of dietary fat is often followed by a marked decrease of plasma triglyceride and disappearance of xanthomas. Response to hypolipidemic drugs is variable.

XI. Acknowledgements

This work was aided by a grant from the Finnish State Medical Research Council (Academy of Finland). The assistance of Mrs. ULLA MATILAINEN in the preparation of the manuscript is appreciated.

XII. References

1 ADLERSBERG, D. and EISLER, L.: Circulating lipids in diabetes mellitus. J. amer. med. Ass. *170:* 1261–1265 (1959).

2 ADLERSBERG, D. and WANG, C.-I.: Syndrome of idiopathic hyperlipemia, mild diabetes mellitus, and severe vascular damage. Diabetes, N.Y. *4:* 210–218 (1955).

3 ÅKERBLOM, H. K.; SILTANEN, I., and KALLIO, A.-K.: Does dietary fructose affect the control of diabetes in children? Acta med. scand., suppl. 542, pp. 195–202 (1972).

4 AKTIN, E. and MENG, H. C.: Release of clearing factor lipase (lipoprotein lipase) *in vivo* and from isolated perfused hearts of alloxan diabetic rats. Diabetes, N.Y. *21:* 149–156 (1972).

5 ALBRINK, M. J.; LAVIETES, P. H., and MAN, E. B.: Vascular disease and serum lipids in diabetes mellitus. Observations over thirty years (1931–1961). Ann. intern. Med. *58:* 305–323 (1963).

6 ALBRINK, M. J. and MAN, E. B.: Serum triglycerides in health and diabetes. Diabetes, N.Y. *7:* 194–201 (1958).

7 ALLEN, F. M.: The role of fat in diabetes. Amer. J. med. Sci. *143:* 313–371 (1917).

8 ARMSTRONG, D. T.; STEELE, R.; ALTSZULER, N.; DUNN, A.; BISHOP, J. S., and BODO, R. C. DE: Regulation of plasma free fatty acid turnover. Amer. J. Physiol. *201:* 9–15 (1961).

9 ASHMORE, J.; RENOLD, A. E.; NESBETT, F. B., and HASTINGS, A. B.: Studies on carbohydrate metabolism in rat liver slices. V. Glycerol metabolism in relation to other substrates in normal and diabetic tissue. J. biol. Chem. *215:* 153–161 (1955).

10 BAGDADE, J. D.: Diabetic lipaemia complicating acute pancreatitis. Lancet *ii:* 1041–1043 (1969).

11 BAGDADE, J. D.; BIERMAN, E. L., and PORTE, D., jr.: Influence of obesity on the relationship between insulin and triglyceride levels in endogenous hypertriglyceridemia. Diabetes, N.Y. *20:* 664–672 (1971).

12 BAGDADE, J. D.; PORTE, D., jr., and BIERMAN, E. L.: Diabetic lipemia. New Engl. J. Med. *276:* 427–433 (1967).

13 BAGDADE, J. D.; PORTE, D., jr., and BIERMAN, E. L.: Acute insulin withdrawal and the regulation of plasma triglyceride removal in diabetic subjects. Diabetes, N.Y. *17:* 127–132 (1968).

14 BALASSE, E. O.; BIER, D. M., and HAVEL, R. J.: Early effects of anti-insulin serum on hepatic metabolism of plasma free fatty acids in dogs. Diabetes, N.Y. *21:* 280–288 (1972).

15 BASSO, L. V. and HAVEL, R. J.: Hepatic metabolism of free fatty acids in normal and diabetic dogs. J. clin. Invest. *49:* 537–547 (1970).

16 BIEBERDORF, F. A.; CHERNICK, S. S., and SCOW, R. O.: Effect of insulin and acute diabetes on plasma FFA and ketone bodies in the fasting rat. J. clin. invest. *49:* 1685–1693 (1970).

17 BIERMAN, E. L.: Personal communication (1972).

18 BIERMAN, E. L.; AMARAL, J. A. P., and BELKNAP, B. H.: Hyperlipemia and diabetes mellitus. Diabetes, N.Y. *15:* 675–679 (1966).

19 BIERMAN, E. L.; DOLE, V. P., and ROBERTS, T. N.: An abnormality of nonesterified fatty acid metabolism in diabetes mellitus. Diabetes, N.Y. *6:* 475–479 (1957).

20 BIERMAN, E. L. and HAMLIN, J. T., III: The hyperlipemic effect of a low-fat, high carbohydrate diet in diabetic subjects. Diabetes, N.Y. *10:* 432–437 (1961).

21 BIRKENHÄGER, J. C. and TJABBES, T.: Turnover rate of plasma FFA and rate of esterification of plasma FFA to plasma triglycerides in obese humans before and after weight reduction. Metabolism *18:* 18–32 (1969).

22 BLANCHETTE-MACKIE, J. and SCOW, R. O.: Sites of lipoprotein lipase activity in adipose tissue perfused with chylomicrons. Electron microscope cytochemical study. J. Cell Biol. *51:* 1–25 (1971).

23 Blix, G.: Studies on diabetic lipemia. I. Acta med. scand. *64:* 142–174 (1926).
24 Bøhmer, T.; Norum, K. R., and Bremer, J.: The relative amounts of long-chain acylcarnitine, acetylcarnitine, and free carnitine in organs of rats in different nutritional states and with alloxan diabetes. Biochim. biophys. Acta *125:* 244–251 (1966).
25 Bragdon, J. H. and Gordon, R. S., jr.: Tissue distribution of C^{14} after the intravenous injection of labeled chylomicrons and unesterified fatty acids in the rat. J. clin. Invest. *37:* 574–578 (1958).
26 Braunsteiner, H.; Sailer, S. und Sandhofer, F.: Plasmalipide bei Patienten mit Diabetes mellitus. Klin. Wschr. *44:* 116–119 (1966).
27 Brown, D. F.: Triglyceride metabolism in the alloxan-diabetic rat. Diabetes, N.Y. *16:* 90–95 (1967).
28 Brown, D. F. and Olivecrona, T.: The effect of glucose availability and utilization on chylomicron metabolism in the rat. Acta physiol. scand. *66:* 9–18 (1966).
29 Brunzell, J. D.; Smith, N. D.; Porte, D., jr., and Bierman, E. L.: Evidence for multiphasic release of postheparin lipolytic activity. J. clin. Invest. *51:* 16a (1972).
30 Buckle, R. M.: Mobilization of free fatty acids from adipose tissue from normal and diabetic subjects. Influence of glucose and insulin. Diabetes, N.Y. *12:* 133–140 (1963).
31 Butcher, R. W.; Baird, C. E., and Sutherland, E. W.: Effects of lipolytic and antilipolytic substances on adenosine 3′,5′-monophosphate levels in isolated fat cells. J. biol. Chem. *243:* 1705–1712 (1968).
32 Butterfield, W. J. H. and Schless, G.: Observations on the peripheral metabolism of nonesterified fatty acids. Diabetes, N.Y. *8:* 450–454 (1959).
33 Cagan, R. N.; Nichols, R., and Loewe, L.: Serum lipids in diabetic and non-diabetic rats. Effect of varying lipid diets. Diabetes, N.Y. *5:* 112–115 (1956).
34 Camitta, F. D. and Gray, T. K.: A study in the pathogenesis of hyperlipemia in a patient with acromegaly and diabetes mellitus. Diabetes, N.Y. *18:* 44–50 (1969).
35 Carlson, L. A. and Östman, J.: *In vitro* studies on the glucose uptake and fatty acid metabolism of human adipose tissue in diabetes mellitus. Acta med. scand. *174:* 215–218 (1963).
36 Carlström, S.: Studies on fatty acid metabolism in diabetics during exercise. I. Plasma free fatty acid concentration in juvenile, newly diagnosed diabetics during exercise. Acta med. scand. *181:* 609–621 (1967).
37 Carlström, S.: Studies on fatty acid metabolism in diabetics during exercise. V. Plasma concentration of free fatty acids and glycerol in newly diagnosed, adult diabetics during exercise. Acta med. scand. *182:* 363–376 (1967).
38 Chaikoff, I. L.; Smyth, F. S., and Gibbs, G. E.: The blood lipids of diabetic children. J. clin. Invest. *15:* 627–631 (1936).
39 Chernick, S. S. and Scow, R. O.: Synthesis *in vitro* of glyceride-glycerol by the liver of normal and pancreatectomized rats. J. biol. Chem. *239:* 2416–2419 (1964).
40 Christensen, N. J.: Plasma catecholamines in long-term diabetics with and

without neuropathy and in hypophysectomized subjects. J. clin. Invest. *51:* 779–787 (1972).

41 CORDER, C. N. and KALKHOFF, R. K.: Hepatic lipid metabolism in alloxan diabetic rats. J. Lab. clin. Med. *73:* 551–563 (1969).

42 CSORBA, T. R.; MATSUDA, I., and KALANT, N.: Effects of insulin and diabetes on flux rates of plasma glucose and free fatty acids. Metabolism *15:* 262–270 (1966).

43 DENBOROUGH, M. A. and PATERSON, B.: Clearing factor, fibrinolysis and blood lipids in diabetes mellitus. Clin. Sci. *23:* 485–488 (1962).

44 DRURY, D. R.: Role of insulin in carbohydrate metabolism. Amer. J. Physiol. *131:* 536–543 (1940).

45 ECK, W. S. VAN: The effect of a low fat diet on the serum lipids in diabetes and its significance in diabetic retinopathy. Amer. J. Med. *27:* 196–211 (1959).

46 ELOVSON, J.: Immediate fate of albumin-bound (1-^{14}C) stearic acid following its intraportal injection into carbohydrate refed rats. Early course of desaturation and esterification in the liver. Biochim. biophys. Acta *106:* 480–494 (1965).

47 ERNEST, I. E. B.; HALLGREN, B., and SVANBORG, A.: Short term study of effect of different isocaloric diets in diabetes. Metabolism *11:* 912–919 (1962).

48 FAIN, J. N. and ROSENBERG, L.: Antilipolytic action of insulin on fat cells. Diabetes, N.Y. *21:* 414–425 (1972).

49 FAIN, J. N. and SCOW, R. O.: Effect of hypophysectomy on lipid metabolism in pancreatectomized rats. Endocrinology *77:* 547–552 (1965).

50 FINE, M. B. and WILLIAMS, R. H.: Effect of fasting, epinephrine and glucose and insulin on hepatic uptake of nonesterified fatty acids. Amer. J. Physiol. *199:* 403–406 (1960).

51 GARLAND, P. B. and RANDLE, P. J.: Regulation of glucose uptake by muscle. X. Effects of alloxan-diabetes, starvation hypophysectomy and adrenalectomy and of fatty acids, ketone bodies and pyruvate on the glycerol output and concentrations of free fatty acids, long-chain fatty acyl-coenzyme A, glycerol phosphate and citrate-cycle intermediates in rat heart and diaphragm muscles. Biochem. J. *93:* 678–687 (1964).

52 GRIES, F. A.; POTTHOFF, S., and JAHNKE, K.: The effect of insulin on the uptake of radioactive labelled plasma triglycerides by rat tissue *in vivo*. Diabetologia *3:* 311–317 (1967).

53 GROSS, C. and CARLSON, L. A.: Metabolic effects of nicotinic acid in acute insulin deficiency in the rat. Diabetes, N.Y. *17:* 353–361 (1968).

54 HAFT, D. E.: Effects of insulin on glucose metabolism by the perfused normal rat liver. Amer. J. Physiol. *213:* 219–230 (1967).

55 HANSEN, A. P.: Abnormal serum growth hormones response to exercise in juvenile diabetics. J. clin. Invest. *49:* 1467–1478 (1970).

56 HARKEN, D. R. VAN; BROWN, T. O., and HEIMBERG, M.: Hepatic lipid metabolism in experimental diabetes. III. Synthesis and utilization of triglycerides. Lipids *2:* 231–238 (1967).

57 HAVEL, R. J.; BASSO, L. V., and KANE, J. P.: Mobilization and storage of fat in congenital and late-onset form of 'total lipodystrophy' (abstract). J. clin. Invest. *47:* 1068 (1967).

58 HEIMBERG, M.; DUNKERLEY, A., and BROWN, T. O.: Hepatic lipid metabolism in experimental diabetes. I. Release and uptake of triglycerides by perfused livers from normal and alloxan-diabetic rats. Biochim. biophys. Acta *125:* 252–264 (1966).

59 HEIMBERG, M.; HARKEN, D. R. VAN, and BROWN, T. O.: Hepatic lipid metabolism in experimental diabetes. II. Incorporation of (I-^{14}C) palmitate into lipids of the liver and of the d $<$ 1.020 perfusate lipoproteins. Biochim. biophys. Acta *137:* 435–445 (1967).

60 HEIMBERG, M.; WEINSTEIN, I., and KOHOUT, M.: The effects of glucagon, dibutyryl cyclic adenosine 3′,5′-monophosphate and concentration of free fatty acid on hepatic lipid metabolism. J. biol. Chem. *244:* 5131–5139 (1969).

61 HILLYARD, L. A.; CORNELIUS, C. E., and CHAIKOFF, I. L.: Removal by the isolated rat liver of palmitate-1-C^{14} bound to albumin and of palmitate-1-C^{14} and cholesterol-4-C^{14} in chylomicrons from perfusion fluid. J. biol. Chem. *234:* 2240–2245 (1959).

62 HIRSCH, E. F.; PHIBBS, B. P., and CARBONARO, L.: Parallel relation of hyperglycemia and hyperlipemia (esterified fatty acids) in diabetes. Arch. intern. Med. *91:* 106–117 (1953).

63 HOLLENBERG, C. H.: Effect of nutrition on activity and release of lipase from rat adipose tissue. Amer. J. Physiol. *197:* 667–670 (1959).

64 ILLIANO, G. and CUATRECASAS, P.: Modulation of adenylate cyclase activity in liver and fat cell membranes by insulin. Science *175:* 906–908 (1972).

65 JONES, A. L.; RUDERMAN, N. B., and EMANS, J. B.: An electron microscopic study of hepatic lipoprotein synthesis in the rat following anti-insulin serum, nicotinic acid and puromycin administration. Gastroenterology *56:* 402 (1969).

66 JONES, D. P. and ARKY, R. A.: Effects of insulin on triglyceride and free fatty acid metabolism in man. Metabolism *14:* 1287–1293 (1965).

67 JONES, D. P.; PLOTKIN, G. R., and ARKY, R. A.: Lipoprotein lipase activity in patients with diabetes mellitus, with and without hyperlipemia. Diabetes, N.Y. *15:* 565–570 (1966).

68 JUNGAS, R. L.: Role of cyclic-3′,5′-AMP in the response of adipose tissue to insulin. Proc. nat. Acad. Sci., Wash. *56:* 757–763 (1966).

69 KALKHOFF, R. K.; HORNBROOK, K. R.; BURCH, H. B., and KIPNIS, D. M.: Studies of the metabolic effects of acute insulin deficiency. II. Changes in hepatic glycolytic and Krebs-cycle intermediates and pyridine nucleotides. Diabetes, N.Y. *15:* 451–456 (1966).

70 KALLIO, V.: Fat loading test in males with diabetic heredity, as compared to controls, diabetics and coronary patients. Acta med. scand., suppl. 467, pp. 1–67 (1967).

71 KESSLER, J. I.: Effect of insulin on release of plasma lipolytic activity and clearing of emulsified fat intravenously administered to pancreatectomized and alloxanized dogs. J. Lab. clin. Med. *60:* 747–755 (1962).

72 KESSLER, J. I.: Effect of diabetes and insulin on the activity of myocardial and adipose tissue lipoprotein lipase of rats. J. clin. Invest. *42:* 362–367 (1963).

73 KIKKAWA, R.; HOSHI, M.; SHIGETA, Y., and IZUMI, K.: Lack of ketosis in lipoatrophic diabetes. Diabetes, N.Y. *21:* 827–831 (1972).

74 KRAUSS, R.; LEVY, R.; WINDMUELLER, H.; MILLER, L., and FREDRICKSON, D.: Selective measurement of lipoprotein lipase and hepatic triglyceride lipase in postheparin plasma (abstract). J. clin. Invest. *51:* 52a (1972).

75 KREISBERG, R. A.: Effect of diabetes and starvation on myocardial triglyceride and free fatty acid utilization. Amer. J. Physiol. *210:* 379–384 (1966).

76 KUDO, H.: Serum triglyceride levels of untreated diabetics in relation to vascular lesions and obesity. Tohoku J. exp. Med. *97:* 47–56 (1969).

77 KUHFAHL, E.; MUELLER, F., and DETTMER, D.: Studies on the uptake of fatty acids by the diabetic liver *in vivo*. Acta biol. med. germ. *18:* 563–571 (1967).

78 KUUSISTO, A. and NIKKILÄ, E.: Manuscript in preparation (1972).

79 LAURELL, S.: Plasma free fatty acids in diabetic acidosis and starvation. Scand. J. clin. Lab. Invest. *8:* 81–82 (1956).

80 LEVER, W. F.; SMITH, P. A. J., and HURLEY, H. A.: Idiopathic hyperlipemia and primary hypercholesterolemic xanthomatosis. (Clinical data and analysis of plasma lipids.) Invest. Derm. *22:* 33 (1954).

81 LOTEN, E. G. and SNEYD, J. G. T.: An effect of insulin on adipose-tissue adenosine 3′,5′-cyclic monophosphate phosphodiesterase. Biochem. J. *120:* 187–194 (1970).

82 LYNCH, H. T.; KAPLAN, A. R.; HENN, M. J., and KRUSH, A. J.: Familial co-existence of diabetes mellitus, hyperlipemia, short stature, and hypogonadism. Amer. J. med. Sci. *252:* 323–330 (1966).

83 MARSH, P. L. and WALLER, H. G.: The relation between ingested fat and the lipemia of diabetes mellitus. Arch. intern. Med. *31:* 63–75 (1923).

84 MARUHAMA, Y.: Diet and blood lipids in normal and diabetic rats. Metabolism *14:* 78–87 (1965).

85 MENG, H. C. and GOLDFARB, J. L.: Heparin-induced lipemia clearing factor in rats. Role of the pancreas in its production. Diabetes, N.Y. *8:* 211–217 (1959).

86 MILLER, H. I.; BORTZ, W. M., and DURHAM, B. C.: The rate of appearance of FFA in plasma triglyceride of normal and obese subjects. Metabolism *17:* 515–521 (1968).

87 MORRIS, B.: Some factors affecting the metabolism of free fatty acids and chylomicron triglycerides by the perfused rat's liver. J. Physiol., Lond. *168:* 584–598 (1963).

88 NAIDOO, S. S.; LOSSOW, W. J., and CHAIKOFF, I. L.: Inverse relationship between adipose tissue and skeletal muscle in the uptake of injected triglyceride fatty acid of chyle lipoproteins by diabetic rats. Experientia *23:* 829–830 (1967).

89 NESTEL, P. J.: Relationship between FFA flux and TGFA influx in plasma before and during the infusion of insulin. Metabolism *16:* 1123–1132 (1967).

90 NESTEL, P. J. and WHYTE, H. M.: Plasma free fatty acid and triglyceride turnover in obesity. Metabolism *17:* 1122–1128 (1968).

91 NEW, M. I.; ROBERTS, T. N.; BIERMAN, E. L., and READER, G. G.: The significance of blood lipid alterations in diabetes mellitus. Diabetes, N.Y. *12:* 208–212 (1963).

92 NIKKILÄ, E. A.: Control of plasma and liver triglyceride kinetics by carbohydrate metabolism and insulin. A review. Adv. Lipid Res., vol. 7, pp. 63–134 (Academic Press, New York 1969).

93 Nikkilä, E. A.: Transport of free fatty acids. A review. Progr. biochem. Pharmacol., vol. 6, pp. 102–129 (Karger, Basel 1971).
94 Nikkilä, E. A. and Kekki, M.: Plasma triglyceride transport kinetics in diabetes mellitus. Metabolism *22:* 1–22 (1973).
95 Nikkilä, E. A. and Kekki, M.: Turnover rate of serum triglyceride in diabetes (abstract). Diabetologia *6:* 658 (1970).
96 Nikkilä, E. A.; Miettinen, T. A.; Vesenne, M.-R., and Pelkonen, R.: Plasmainsulin in coronary heart-disease. Response to oral and intravenous glucose and to tolbutamide. Lancet *ii:* 508–511 (1965).
97 Nikkilä, E. A. and Ojala, K.: Unpublished information (1966).
98 Nikkilä, E. A. and Pykälistö, O.: Regulation of adipose tissue lipoprotein lipase synthesis by intracellular free fatty acid. Life Sci. *7:* 1303–1309 (1968).
99 Nikkilä, E. A. and Pykälistö, O.: Influence of lipolytic and antilipolytic agents on synthesis of adipose tissue lipoprotein lipase; in Holmes, Carlson and Paoletti Drugs affecting lipid metabolism, pp. 239–248 (Plenum Press, New York 1969).
100 Nikkilä, E. A. and Taskinen, M.-R.: Hypertriglyceridemia and insulin secretion. A complex causal relationship; in Jones Atherosclerosis, pp. 220–230. Proc. 2nd Int. Symp. (Springer, Berlin 1970).
101 O'Connor, T. P. and Schnatz, J. D.: Lipoprotein lipase activity and hypertriglyceridemia in alloxan diabetic rabbits. Metabolism *17:* 838–844 (1968).
102 Olivecrona, T.: The metabolism of 1-C^{14}-palmitic acid in the rat. Acta physiol. scand. *54:* 295–305 (1962).
103 Ontko, J. A.: Metabolism of free fatty acids in isolated liver cells. Factors affecting the partition between esterification and oxidation. J. biol. Chem. *247:* 1788–1800 (1972).
104 Östman, J.: Effect of nicotinic acid on the fatty acid metabolism of adipose tissue in alloxan diabetic rats. Metabolism *13:* 675–680 (1964).
105 Östman, J.: Studies *in vitro* on fatty acid metabolism of human subcutaneous adipose tissue in diabetes mellitus. Acta med. scand. *177:* 639–655 (1965).
106 Östman, J.: *In vitro* metabolism of omental adipose tissue in lipoatrophic diabetes (abstract); in Proc. 6th Congr. Int. Diab. Fed., p. 162 (Excerpta Medica Foundation, Amsterdam 1967).
107 Parker, F.; Bagdade, J. D.; Odland, G. F., and Bierman, E. L.: Evidence for the chylomicron origin of lipids accumulating in diabetic eruptive xanthomas. A correlative lipid biochemical, histochemical, and electron microscopic study. J. clin. Invest. *49:* 2172–2187 (1970).
108 Patten, R. L.: The reciprocal regulation of lipoprotein lipase activity and hormone-sensitive activity in rat adipocytes. J. biol. Chem. *245:* 5577–5584 (1970).
109 Pelkonen, R.; Nikkilä, E. A., and Kekki, M.: Metabolism of glycerol in diabetes mellitus. Diabetologia *3:* 1–8 (1967).
110 Pelkonen, R.; Aro, A., and Nikkilä, E. A.: Metabolic effects of dietary fructose in insulin dependent diabetes of adults. Acta med. scand., suppl. 542, pp. 187–193 (1972).

111 PERRY, W. F.: Heparin-activated lipase in diabetic, atherosclerotic and healthy subjects. Clin. chim. Acta *16:* 189–194 (1967).

112 PISCATELLI, R. L.; VIEWEG, W. V. R., and HAVEL, R. J.: Partial lipodystrophy. Metabolic studies in three patients. Ann. intern. Med. *73:* 963–970 (1970).

113 PODOLSKY, S.: Lipoatrophic diabetes and miscellaneous conditions related to diabetes mellitus; in MARBLE, WHITE, BRADLEY and KRALL Joslin's diabetes mellitus, p. 722 (Lea & Febiger, Philadelphia 1971).

114 PYKÄLISTÖ, O. and NIKKILÄ, E. A.: Unpublished results (1969).

115 REINHEIMER, W.; BLIFFEN, G.; MCCOY, J.; WALLACE, D., and ALBRINK, M. J.: Weight gain, serum lipids, and vascular disease in diabetics. Amer. J. clin. Nutr. *20:* 986–996 (1967).

116 ROBINSON, D. S.: The function of the plasma triglycerides in fatty acid transport; in FLORKIN and STOTZ Comprehensive biochemistry, vol. 18, pp. 51–116 (Elsevier, Amsterdam 1970).

117 RUBENSTEIN, B. and RUBINSTEIN, D.: The effects of fasting on esterification of palmitate by rat liver *in vitro*. Canad. J. Biochem. *44:* 129–140 (1966).

118 RUDAS, B. und REISSERT, K.: Das Verhalten von Serumlipiden alloxandiabetischer und normaler Ratten nach verschiedenartiger Applikation einer Fettemulsion. Med. Pharmacol. exp. *17:* 11–16 (1967).

119 SAILER, S.; SANDHOFER, F., and BRAUNSTEINER, H.: Overweight and triglyceride level in normal persons and patients with diabetes mellitus. Metabolism *15:* 135–137 (1966).

120 SAILER, S.; SANDHOFER, F. und BRAUNSTEINER, H.: Beziehungen zwischen Blutzuckerspiegel, Umsatzrate der freien Fettsäuren und Fettsäure-Einbau in Plasmatriglyceride bei Diabetikern. Klin. Wschr. *45:* 86–91 (1967).

121 SAILER, S.; SANDHOFER, F.; BOLZANO, K. und BRAUNSTEINER, H.: Ueber den Einfluss der Glucose auf den Umsatz der freien Fettsäuren des Plasmas, die Einbaurate der freien Fettsäuren in Plasmatriglyceride und die Wirkung von Noradrenalin auf diese Stoffwechselgrössen beim Menschen. Klin. Wschr. *45:* 918–924 (1967).

122 SALAMAN, M. R. and ROBINSON, D. S.: Clearing-factor lipase in adipose tissue. A medium in which the enzyme activity of tissue from starved rats increases *in vitro*. Biochem. J. *99:* 640–647 (1966).

123 SANDBERG, H.; SOK MIN, B.; FEINBERG, L., and BELLET, S.: I^{131} triolein tolerance curves in patients with diabetes mellitus. Arch. intern. Med. *105:* 866–872 (1960).

124 SCHERSTÉN, T.; NILSSON, S., and JÖNSSON, J.: Hepatic lipogenesis in two cases with insulin-producing tumor of the pancreas. Acta med. scand. *190:* 353–357 (1971).

125 SCHNATZ, J. D. and WILLIAMS, R. H.: The effect of acute insulin deficiency in the rat on adipose tissue lipolytic activity and plasma lipids. Diabetes, N.Y. *12:* 174–178 (1963).

126 SCHWARTZ, R.; SCHAFFER, I. A., and RENOLD, A. E.: Generalized lipoatrophy, hepatic cirrhosis, disturbed carbohydrate metabolism and accelerated growth (lipoatrophic diabetes). Longitudinal observations and metabolic studies. Amer. J. Med. *28:* 973 (1960).

127 SCOW, R. O. and CHERNICK, S. S.: Mobilization, transport and utilization of

free fatty acids; in FLORKIN and STOTZ Comprehensive biochemistry, vol. 18, pp. 19–49 (Elsevier, Amsterdam 1970).

128 SEGALL, M. and LLOYD, J.: Observation on fat and carbohydrate metabolism in generalized lipodystrophy. Arch. Dis. Child. *44:* 779 (1969).

129 SHIGETA, Y.; NAKAMURA, K.; HOSHI, M.; KIM, M., and ABE, H.: Effect of dextran sulfate on plasma lipoprotein lipase activity in obese subjects and diabetic patients. Diabetes, N.Y. *16:* 238–241 (1967).

130 SÖLING, H. D.; KNEER, P.; DRÄGERT, W. und CREUTZFELDT, W.: Die Wirkung von Insulin auf den Stoffwechsel der isolierten perfundierten Leber Normaler und Alloxandiabetischer Ratten. II. Stoffwechseländerungen unter dem Einfluss intraportaler Insulininfusionen. Diabetologia *2:* 32–44 (1966).

131 SPITZER, J. J. and MCELROY, W. T., jr.: Some hormonal influences on the hepatic uptake of free fatty acids in diabetic dogs. Diabetes, N.Y. *11:* 222–226 (1962).

132 SPITZER, J. J.; NAKAMURA, H.; HORI, S., and GOLD, M.: Hepatic and splanchnic uptake and oxidation of free fatty acids. Proc. Soc. exp. Biol. Med. *132:* 281–286 (1969).

133 STETTEN, D. and BOXER, G. E.: Studies in carbohydrate metabolism. III. Metabolic defects in alloxan diabetes. J. biol. Chem. *156:* 271–278 (1944).

134 STONE, D. B. and CONNOR, W. E.: The prolonged effects of a low cholesterol, high carbohydrate diet upon the serum lipids in diabetic patients. Diabetes, N.Y. *12:* 127–132 (1963).

135 TARRANT, M. E.; MAHLER, R., and ASHMORE, J.: Studies in experimental diabetes. IV. Free fatty acid mobilization. J. biol. Chem. *239:* 1714–1719 (1964).

136 TARRANT, M. E.; THOMPSON, R. H. S., and WRIGHT, P. A.: Some aspects of lipid metabolism in rats treated with anti-insulin serum. Biochem. J. *84:* 6–10 (1962).

137 THANNHAUSER, S. J.: Serum lipids and their value in diagnosis. New Engl. J. Med. *237:* 515 (1947).

138 TOPPING, D. L. and MAYES, P. A.: The immediate effects of insulin and fructose on the metabolism of the perfused liver. Changes in lipoprotein secretion, fatty acid oxidation and esterification, lipogenesis and carbohydrate metabolism. Biochem. J. *126:* 295–311 (1972).

139 TRAGL, K. H. and REAVEN, G. M.: Effect of experimental diabetes mellitus on protein synthesis by liver ribosomes. Diabetes, N.Y. *20:* 27–32 (1971).

140 TUBBS, P. K. and GARLAND, P. B.: Variations in tissue contents of coenzyme A thio esters and possible metabolic implications. Biochem. J. *93:* 550–557 (1964).

141 UNGER, R. H.; AGUILAR-PARADA, E.; MÜLLER, W. A., and EISENTRAUT, A. M.: Studies on pancreatic alpha cell function in normal and diabetic subjects. J. clin. Invest. *49:* 837–848 (1970).

142 VELEMINSKY, J.; BURR, I. M., and STAUFFACHER, W.: Comparative study of early metabolic events resulting from the administration of the two diabetogenic agents alloxan and streptozotocin. Europ. J. clin. Invest. *1:* 104–108 (1970).

143 WADDELL, W. R. and GEYER, R. B.: Effect of insulin on clearance of emulsified fat from the blood in depancreatized dogs. Proc. Soc. exp. Biol. Med. *96:* 251–255 (1957).

144 WENKEOVÁ, J. and PÁV, J.: Release of non-esterified fatty acids from adipose tissue in normal and diabetic rats. Nature, Lond. *184:* 1147 (1959).
145 WIELAND, O.; WEISS, L.; EGER-NEUFELDT, I.; TEINZER, A. und WESTERMANN, B.: Coenzym A-thioester höherer Fettsäuren als mögliche Vermittler enzymatischer Regulationen im Tierkörper. Klin. Wschr. *43:* 645–654 (1965).
146 WILCOX, H. G.; DISHMON, G., and HEIMBERG, M.: Hepatic lipid metabolism in experimental diabetes. IV. Incorporation of amino acid ^{14}C into lipoprotein-protein and triglyceride. J. biol. Chem. *243:* 666–675 (1968).
147 WILSON, D. E.; SCHREIBMAN, P., and ARKY, R. A.: Post-heparin lipolytic activity in diabetic patients with a history of mixed hyperlipemia. Relative rates against artificial substrates and human chylomicrons. Diabetes, N.Y. *18:* 562–566 (1969).
148 WILSON, D. E.; SCHREIBMAN, P. H.; DAY, V. C., and ARKY, R. A.: Hyperlipidemia in an adult diabetic population. J. chron. Dis. *23:* 501–506 (1970).
149 WINAND, J.: Aspects qualitatifs et quantitatifs du métabolisme lipidique de la souris normale et de la souris congénitalement obèse (Ed. Arscia, Bruxelles 1970).
150 WING, D. R. and ROBINSON, D. S.: Clearing-factor lipase in adipose tissue. A possible role of adenosine 3′,5′-(cyclic)-monophosphate in the regulation of its activity. Biochem. J. *109:* 841–850 (1968).
151 WOODSIDE, W. F. and HEIMBERG, M.: Hepatic metabolism of free fatty acids in experimental diabetes. Israel J. med. Sci. *8:* 309–316 (1972).

Author's address: Dr. ESKO A. NIKKILÄ, Third Department of Medicine, University of Helsinki, Central Hospital, *Helsinki* (Finland)

Index